HUMAN FACTORS PSYCHOLOGY

ADVANCES IN PSYCHOLOGY

47

Editors:

G. E. STELMACH

P. A. VROON

NORTH-HOLLAND
AMSTERDAM · NEW YORK · OXFORD · TOKYO

HUMAN FACTORS PSYCHOLOGY

Edited by

Peter A. HANCOCK
University of Southern California
Los Angeles, California
U.S.A.

1987

NORTH-HOLLAND
AMSTERDAM · NEW YORK · OXFORD · TOKYO

ISBN: 0 444 70319 5

Publishers:

ELSEVIER SCIENCE PUBLISHERS B.V.
P.O. Box 1991
1000 BZ Amsterdam
The Netherlands

Sole distributors for the U.S.A. and Canada:

ELSEVIER SCIENCE PUBLISHING COMPANY, INC.
52 Vanderbilt Avenue
New York, N.Y. 10017
U.S.A.

Transferred to Digital Printing 2007

Library of Congress Cataloging-in-Publication Data

Human factors psychology.

(Advances in psychology ; 47)
Includes bibliographies and index.
1. Thought and thinking. 2. Short-term attention.
3. Decision making. 4. Human engineering.
5. Man-machine systems--Psychological aspects.
I. Hancock, Peter A., 1953- . II. Series: Advances in psychology (Amsterdam, Netherlands) ; 47. [DNLM:
1. Attention. 2. Cybernetics. 3. Decision Making.
4. Human Engineering. 5. Memory, Short-Term. 6. Task Performance and Analysis. W1 AD789L v.47 / BF 311 H918]
BF441.H814 1987 620.8'2'019 87-22286
ISBN 0-444-70319-5 (U.S.)

PREFACE

A preface is an appropriate place to present the aspiration, initiation, and production of the text in the readers hands. The present volume is founded upon the belief that the understanding of human behavior is central to progress in Human Factors. It was conceived in association with Dr. George Stelmach, the North American Consulting Editor for the North-Holland Series. The completion of the work was made possible only with the encouragement of Kees Michielsen of North-Holland whose advice was most helpful in all stages of production. The viability of this volume owes much to Barry Kantowitz who not only agreed to contribute a chapter but encouraged a number of colleagues to participate and thus provide the basis for the final work.

The text is designed for the reader concerned with contemporary developments in psychological research as applicable to Human Factors issues. While technological innovation and engineering endeavors have proved a most efficacious avenue of advancement in Human Factors, it is the intricacies of human behavior that pose the more complex challenge and an understanding of these capabilities may, in the long-term, prove to return the greater benefit. Each author, or group of authors, were requested to address their particular speciality and to bring to bear a Human Factors perspective on their contribution. Individual chapters range widely in scope but are associated with sequential processing and energetic components of human action.

In the first chapter, Stuart Klapp deals with the nuances of short-term memory and indicates the dearth of applicational studies concerning this important behavioral characteristic. Christopher Wickens subsequently focuses on attention and elaborates not only on its central role in psychological issues but demonstrates the contemporary impact on many applied problems. In the third chapter, Barry Kantowitz advances a number of provocative observations upon mental workload, an important consideration in both design evaluation and operation of technological systems. Don Kleinmuntz, who parenthetically provided his insightful contribution on particularly short notice, surveys the field of decision-making. This area, somewhat neglected by the contemporary mainstream of Human Factors, possesses considerable practical utility which Kleinmuntz's contribution clearly indicates. Dan Fisk, with Phillip Ackerman and Walter Schneider, explore the practical ramifications of a theory that promises to enervate a number of Human Factors areas, in particular that of training and the understanding of individual differences. Their work, founded upon the initial insights of Schneider and his colleague Richard Shiffrin, are worth careful perusal by Human Factors professionals concerned with the direct application of psychological theory and who often lament the absence of this bridging form of information. Richard Jagacinski, to whom I owe a considerable debt for both his patience and encouragement, has produced with his colleagues Brian Plamondon and Richard Miller, a technically oriented chapter. Their conceptualization of movement as composed of activity modules has considerable value to a discipline whose roots and progress are founded upon an understanding of manual control and the direct manipulation of machine input devices.

In a comprehensive chapter, Ray Eberts summarizes research in the large and rapidly-growing field of human-computer interaction. For bringing some clarity to this burgeoning area alone, the chapter is an important component of the text. But the structuring by a four-part approach is a most helpful framework and gives cohesion to an otherwise largely formless mass of experimentation. The chapter by myself and Mark Chignell deals with the problem of operator stress and task loading and a potential avenue of load problem resolution through dynamic adaptive interface instantiation. As indicated in the chapter by Kantowitz, this avenue of investigation is in its infancy and is only now taking a first series of steps toward actual operational realization. In the final chapter, Hal Hendrick reminds us that what appears as ideas on paper are often sterile conceptualizations until ways are found through the organizational maze to actually enact many of the ideas advanced by Human Factors professionals. In sum, the text is designed to be neither exhaustive in coverage nor formally linked beyond the associations described above, but is constructed to provide information upon a number of important contemporary research topics.

There are many unforeseen problems in producing a uniform camera-ready text as represented in the following pages. I am indebted to the fidelity of the authors who provided prompt and complete copy. Any errors are my responsibility. I was fortunate in the project in having the assistance of a number of individuals who devoted considerable personal effort to the completion of the book. Among these Carolyn Bjerke deserves recognition for her care and application to the production of the text. The book could not have been completed without the computing skills of Nghiem Van Chu and Bob Froehlig, who spent many hours of their personal time and I will always be grateful for their unique efforts. My thanks also go to Cuong Van Chu, Jane Rowcliffe, Laraine Crampton, and Lisa Domme, for their contributions to the production and editing of the manuscript. Also Nancy Knabe made possible the final draft, my appreciation goes to each of these invaluable helpers. At the end of any long term effort it is important to recognize individuals and organizations who made the environment conducive to action. Among these I should thank Professor James O. Pierce and Professor Harry H. Hurt, Jr., for their continued professional and collegial support. I also wish to acknowledge the support of the National Aeronautic and Space Administration (NASA) and my technical monitor Dr. Michael Vidulich. As can be seen in the acknowledgements of a number of chapters, their group on workload headed by Sandra Hart facilitate considerable research in a number of areas represented in this text. I am indebted to Professor John Loy whose companionship and erudition has added immeasurably to my motivation toward scholarly endeavors but principally this work is in part payment to Professor Karl Newell for the many personal hours he invested in both my education and my future, he remains both friend and mentor.

It is often the case that we who produce such work make numerous protestations of the sacrifices made in order to complete our goals. These include many, long hours of work with, as often as not, repeated weekends when the book demands total attention and the office becomes more than a second home. At the risk of sacrilege in respect of this principle, I should admit that I enjoy what I do and as such the real sacrifice is made by my supporting family. So perhaps in part recompense for the many times when my spouse acted as both parents and to my children who sometimes missed out on Saturdays and even Sundays with Daddy, this book is dedicated to my wife and daughters.

DEDICATION

To Frances, Susan, and Gabriella

TABLE OF CONTENTS

HUMAN FACTORS PSYCHOLOGY, Peter A. Hancock (ed.)

1. SHORT-TERM MEMORY LIMITS IN HUMAN PERFORMANCE

Stuart T. Klapp
Department of Psychology
California State University, Hayward
Hayward, CA 94542

1.1. INTRODUCTION

1.1.1. Statement Of The Problem

Imagine that you are a truck dispatcher attempting to schedule trucks which are to leave within the next five minutes. The truck must pick up goods from several sources and make deliveries to several destinations. While you are attempting to produce a schedule, the phone rings and your boss asks you a question which is unrelated to this schedule. In spite of this distraction, you somehow must keep specifics of locations and orders in mind until you have scheduled your drivers and trucks. Then you are free to forget these details, which will never be exactly the same on another day. This is a case of overload of immediate memory. The problem is not a limit in permanent long-term knowledge about your job, but rather the problem is to keep track of a large amount of temporarily-needed information.

Two aspects of this situation can be distinguished: (1) Distraction. Intervening events (e.g., the phone call) can cause loss of material in immediate memory, and (2) Capacity limits. The limited capacity of immediate memory can be exceeded so that the scheduler may fail to incorporate some of the relevant information into the action plan. From the Human Factors perspective, the decision maker can be assisted by appropriate design and scheduling of tasks so that distraction and memory overloads will not be a problem and/or by providing external memory aids (presumably computer-based). Such aids should assist the decision maker without imposing excessive additional distraction and memory load.

This chapter first describes theoretical studies on distraction and capacity limits. This theoretical material forms the background for the subsequent discussion of Human Factors implications of short-term memory. However, before proceeding with this main theme, I discuss the assumption that short-term memory can be considered to be distinct from other forms of memory.

1.1.2. Short-Term Memory As A Separate System

Not everyone would agree with one premise implicit in this analysis. I assume that it is useful to discuss short-term memory as if it were a distinct memory system. However, in recent years it has become fashionable to argue that the distinction between immediate (short-term) memory and other more permanent (long-term) memory is outdated. Perhaps it is the case that, in some sense, the underlying mechanisms of memory are continuous rather than discretely divided into immediate and permanent memory. Nevertheless, the distinction may be valid for an applied emphasis, such as the present discussion. It is not necessary to assume that immediate memory is distinct from permanent memory in order to note that, for people handling situations such as the truck dispatching example, temporary storage of information about the problem of the moment may become distracted or overloaded.

The presumed demise of the distinction between immediate and permanent memory is usually attributed to the influential analysis of Craik and Lockhart (1972). The central theme of this analysis was that the durability of memory depends on how the material was processed when it was received. Rather than characterizing memory as falling into one of two discrete categories, short-term and long-term, duration was viewed as a continuous function of the extent to which the material was made meaningful during its presentation. However, this analysis did not deny the distinction between primary memory (material currently in consciousness) and secondary memory (material which has left current awareness but which can be retrieved), as a close reading will show (Craik & Lockhart, 1972, p. 676). That distinction, originating with William James (James, 1890; Norman, 1976, p. 84), seems valid on intuitive grounds and has not been seriously questioned.

It is not clear how to classify the notion of short-term memory as it applies to keeping track of information in our truck dispatcher example. Perhaps this is primary memory in the sense used by William James, or perhaps it is material which has not been deeply processed in the Craik and Lockhart sense. The present analysis is neutral regarding these theoretical positions, and uses terms like immediate memory, short-term memory (STM), and working memory interchangeably.

1.2. DISTRACTION

Our hypothetical truck dispatcher was faced with two types of problems - loss of STM due to distraction, and the need to retain more information than the capacity of STM allows. This section discusses the distraction problem.

1.2.1. The Brown-Peterson Effect

The role of distraction in causing loss in STM was demonstrated almost simultaneously by Brown (1958) in England and by Peterson and Peterson (1959) in the United States, and the result has become known as the Brown-Peterson effect. In the typical demonstration people are asked to retain three or four letters for 20 seconds while simultaneously counting backwards by threes during the retention interval. The rapid loss of memory for the letters is presumed to be due to the distracting effect of mental arithmetic. This effect is robust enough to be a reliable classroom demonstration, although memory loss is much less pronounced on the first trial compared with later trials (Keppel & Underwood, 1962).

1.2.2. Early vs. Late Rehearsal and Distraction

One simple theory of distraction in STM must be rejected. One might suppose that active attention to one item in memory (e.g., rehearsal) must occur throughout the retention interval in order to maintain the material in STM. This view predicts that distraction should lead to loss in STM regardless of whether the distraction occurs early or late within the retention interval.

Contrary to this view, rehearsal early in the retention interval protects STM from subsequent loss from distraction. For example, Dillon and Reid (1969) used easy versus difficult distraction tasks during the retention interval to study the relative importance of early and late rehearsal. It was assumed that the easy distractor would permit more rehearsal than the difficult distractor. Retention was better with the easy distractor first followed by the difficult distractor than with the reverse order, supporting the view that early rehearsal is more important than late rehearsal. Similarly, the original Peterson and Peterson (1959) report included the finding that recall was better when an initial rehearsal period of three seconds preceded the 18 second distracted delay than when no such initial rehearsal was provided. Furthermore, subjects report more rehearsals early than late in the retention interval (Kroll, Kellicut, & Parks, 1975). Presumably subjects rehearse more early because early rehearsal is more important than late rehearsal. Finally, reaction time (RT) to a secondary task can be used to assess the amount of rehearsal, if one assumes that RT is longer when rehearsals are in progress than with no rehearsal. Consistent with the other studies, secondary task RT was longer early in the retention interval (Stanners, Meunier, & Headley, 1969).

Thus, there is ample reason to suppose that much of the STM retention loss in brief distracted intervals can be prevented by a few rehearsals early in the interval, and that, if given the opportunity, individuals will engage in such early rehearsals.

1.3. CAPACITY LIMIT

In addition to problems attributable to distraction, the second problem facing our hypothetical truck dispatcher had to do with exceeding the information capacity of STM. The idea that such overloads create a critical limit to thinking and decision making is known as the working memory hypothesis. According to this view, STM represents the working space in which thinking and decision making is carried out. This hypothesis can be understood intuitively by considering what seems to happen when we understand a sentence. In order to comprehend the meaning underlying a string of individual words, we must keep the words in immediate memory until all of the sentence has been received. Then, using memory, the words can be combined and integrated into the underlying concept. It is quite obvious that overly long (or overly complex) sentences can overload our capacity to do this.

Working memory has been described as the "site of ongoing cognitive activities - for instance, meaningful elaboration of words, symbol manipulation such as that involved in mental arithmetic, and reasoning" (Klatzky, 1980, p. 87). Thus, working memory is the "space in which information can be stored temporarily while it is being processed" (Klapp, Marshburn, & Lester, 1983, p. 240). It is "the part of the memory system where active information processing takes place" (Chase & Ericsson, 1982, p. 40).

The working-memory hypothesis assumes cognitive roles for working memory in addition to understanding strings of words. For example, people need memory for storage of intermediate results in mental arithmetic (Hitch, 1978). Also, we assume that our truck dispatcher would use working memory to keep track of the details being integrated into an action plan. The complexity and sophistication of this plan would depend on the amount of memory capacity available to keep track of the relevant details which must be incorporated into the plan.

1.3.1. A Particular View of Working Memory

Most current textbooks in Cognitive Psychology (see Klapp, et al., 1983, for a review), and recent texts in Human Factors (Kantowitz & Sorkin, 1983, p. 170; Wickens, 1984, p. 218) identify the concept of working memory with a particular laboratory procedure known as span memory. It is well known that when laboratory subjects attempt to recall random digits in the same order as they were presented, correct performance is only reliably possible up to about seven digits. The presumption is that this laboratory procedure taps the single limited system underlying all of working memory. It should be clear that identification of span memory with working memory is a special case of the working memory viewpoint. That viewpoint could also be retained under the assumption that digit span represents a system which is distinct from working memory or under the assumption that working memory has multiple components including, but not limited to, span memory (Brainerd, 1981; Hitch, 1978; Klapp, et al., 1983).

One prediction of the assumption that working memory corresponds to the digit span procedure is that persons with larger digit spans should generally be more effective in cognitive operations (Hunt, 1978). There is little support for this prediction. The correlation of digit span with full scale intelligence (r=0.56) is not remarkable compared with the correlations for the other subscales, of which all except "object assembly" are higher (Wechsler, 1958, p. 98). Furthermore, patients with severely impaired digit span do not necessarily show the overall cognitive difficulties which would be expected if their entire working memory was also impaired (Shallice & Warrington, 1970). Although traditional span measures fail to correlate with reading comprehension, other memory measures do (Daneman & Carpenter, 1980).

On the other hand, the finding that span memory can be very large for material with which the subject is highly skilled (Chase & Ericsson, 1982) lends some support for the idea of a relation between span memory and cognitive efficiency. I suspect that people develop efficient codes which improve their performance in the skilled task, and that these codes also help in the artificial laboratory span memory situation. In this interpretation, large memory span and cognitive skill are both due to improved coding, but the large memory span is not itself the cause of the level of skill.

Klapp, et al., (1983) presented two additional lines of evidence which cast doubt on the assumption that span memory, as represented by digit span, makes use of the system underlying all of working memory. First, different tasks appear to involve memory systems with distinct properties, suggesting that more than one system of memory is available. Second, some memory tasks show mutual interference when done concurrently, but others do not. Perhaps tasks which use different systems do not interfere with each other. Rather than a single system of working memory, it appears that several systems exist. However, the detailed breakdown and mapping of memory systems onto tasks is not yet understood.

1.3.2. Concurrent Memory Load Method: A Critical Analysis

According to the working memory hypothesis, requiring retention of extraneous material in STM should reduce performance on concurrent tasks which require use of working memory. This prediction has led to experiments in which each of the trials comprise the following sequence of events: (1) receive an extraneous memory "load" such as a string of letters, (2) perform an embedded task which presumably requires working memory, and (3) recall the extraneous memory load. Thus the embedded task is performed concurrently with retention of a memory load. According to the working memory viewpoint, the load and the embedded task should exhibit mutual interference. That is, the presence of the load should reduce performance on the task compared to a no-load control and the presence of the task should reduce recall of the load compared to a no-task control. Results of this form were first reported by Baddeley and his co-workers (Baddeley & Hitch, 1974; Hitch & Baddeley, 1976).

Previously, (section 1.2.1) findings were reviewed showing that rehearsal early in the STM retention interval can eliminate the need for subsequent rehearsal. This conclusion has important implications for interpreting findings from the memory load paradigm. It is possible that effects attributed to interference between retention of the extraneous memory load and performance of the embedded task are really due to interference between initial rehearsal of the load and the embedded task. If the embedded task were postponed until completion of initial rehearsal, then there might be no interference between retention of the load and the embedded task.

Consistent with this interpretation, we have shown for some tasks that mutual interference between memory load and embedded tasks occurs with no initial rehearsal interval. This vanishes if a delay for initial rehearsal is provided between the input of the memory load and the presentation of the embedded task (Klapp, et al., 1983, Experiments 6, 7). Thus, for these situations, the observed interference is between the embedded task and initial rehearsal. Passive retention of the memory load did not interfere with the embedded task. Converging results were reported by Trumbo and Milano (1971). Interference between STM and manual tracking was more pronounced during the encoding (and recall) phases of the STM task than during the retention phase.

On the other hand, mutual interference can occur, even with an initial rehearsal interval, when both the memory load and the embedded task require retention of the order of occurrence of items (Klapp, et al., 1983, Experiment 8; Klapp & Philipoff, 1983). For these situations, retention of the memory load is incompatible with performing the embedded task. This finding shows that initial rehearsal does not cause the load memory items to enter some system (such as long-term memory) where they could not interfere with an embedded task.

This analysis indicates that some experiments using the embedded task paradigm (those without a delay for initial rehearsal) are best interpreted as showing interference between rehearsal of the memory load and performing the embedded task, rather than between retention of the load and performing the task. Although interference between rehearsal and the embedded task is an interesting phenomenon, it does not deal with the issue of working memory. According to the working memory perspective, when memory is filled nearly to capacity there is no "space" left over for other tasks. Therefore, mutual interference will result. Thus, working memory theory makes predictions about interference due to retention, rather than due to initial rehearsal. In our view, only studies for which the interference is not attributable to rehearsal (those with a delay for initial rehearsal) can be interpreted within the working memory perspective. Some such studies show mutual interference and some do not. The next section is a critical review of the memory load paradigm from this perspective.

1.3.3. Review of Memory Load Experiments

This section classifies memory load studies on the basis of whether a delay for initial rehearsal was provided between the memory load input and the embedded task. Those studies which did provide such a delay (and hence are valid according to the present analysis), are further classified as showing or not showing interference between the load and the embedded task. Studies which did not include the rehearsal delay, and hence will not be considered in detail, include: Baddeley and Hitch, (1974); Crowder, (1967); Hitch and Baddeley (1976); Johnston, Greenberg, Fisher, and Martin (1970); Logan (1979); Logan (1980); Reisberg (1983); Shulman and Greenberg (1971); Wanner and Shiner (1976).

Some other studies are difficult to classify. For example, in the experiment by Jonides (1981), subjects initiated the embedded task on their own, and hence could have either provided or not provided the delay. In this study there was mutual interference between memory load and reaction time for an embedded visual attention-shift task. A way to load memory which uses recall uncertainty rather than the embedded task paradigm was developed by Richardson (1984). In this procedure two sets of items were presented simultaneously, and a cue indicated which set was to be recalled. Cues randomly occurred either in advance of or after stimulus presentation, with the latter case assumed to represent concurrent memory loading. This paradigm avoids problems of delay of processing of one task while rehearsing the other, but introduces other problems in interpretation, including the relative roles of selective attention at input versus memory loading as ways to account for the performance decrement for the delayed-cue condition. Another study which is difficult to classify has been reported by Halford, Bain, and Maybery (1984). Although a delay was provided between memory input and the embedded task, subjects were required to rehearse during the embedded task. Thus, the observed interference is probably attributable to rehearsal rather than to retention.

Of greater interest are studies using the embedded task approach which discourage rehearsal of the memory load during the embedded task. Such studies either included a delay for initial rehearsal or used very slow input of the memory load. Some of these studies showed mutual interference between memory load and the embedded task, and others did not. Interference was <u>not</u> observed for the following embedded tasks: (1) manual "Fitts' law" target tapping (Roediger, Knight, & Kantowitz, 1977) and (2) true/false judgements about number relations (Klapp, et al., 1983, Experiment 6). Neither of these embedded tasks appear to require much use of memory, and this may explain the lack of any effect of loading. However, the remaining negative instances clearly do require use of memory: (3) modified Sternberg scanning (Klapp, et al., 1983, Experiment 7). This embedded task involved simultaneous presentation of the embedded task items, and in that sense was different from the other experiments, all of which used sequential presentation of both memory load and embedded task items. Finally, (4) the "missing scan" task in which the subject reports which of the nine non-zero digits had not appeared (Klapp & Philipoff, 1983).

By contrast, mutual interference between memory load does occur for other embedded tasks (even with slow input or an interval for initial rehearsal). The embedded tasks for which interference does occur include: (1) ordered recall (Klapp, et al., 1983, Experiment 8), (2) probed recall in which subjects recall the digit which had occurred just after the tested probe digit (Klapp & Philipoff, 1983), (3) recall of the response term of a paired associate selected at random from among several associated pairs which had appeared (Klapp & Comert, unpublished data) and, (4) identification of consonant-vowel syllables presented for the right ear, but not for the left ear (Hellige & Wong, 1983).

It is quite difficult to see how these facts might be summarized in terms of universal principles. Obviously many questions remain unanswered. However, we are in a position to reject the most simple version of the working memory viewpoint which asserts that there is but one system of STM or working memory. Filling that system should interfere with any other task which also requires use of working memory. The data just reviewed are not consistent with this simple viewpoint. Some cases which should show interference do not, but other cases do. Furthermore, there seem to be no experiments or field studies showing that memory loads interfere with real world tasks such as decision making. From the simple theory of working memory, one would expect that such observations would be commonplace.

These data can be fitted to some form of a multiplex view of STM, which postulates independent systems. If the load and the embedded task use the same system, mutual interference results. If the load and the embedded task enter independent systems, there is no mutual interference. However, the presently available data are not complete enough to justify any detailed theory concerning the number and characteristics of such independent systems.

1.3.4. Types of Memory Which Have Been Proposed

The preceding section concluded that the single-system view of working memory is not supported by data from the memory loading paradigm. A view which would be consistent with these findings assumes more than one system of working memory. The present section follows up on this idea with a review of previous proposals for distinct systems of short-term memory. The following systems and distinctions have been proposed:

1.3.4.1. Auditory

It has long been recognized that at least some of the coding in immediate memory is either auditory or articulatory. Conrad (1964) demonstrated that intrusion errors tend to have a phoneme in common with the correct item, and subsequently Conrad and Hull (1964) showed that lists of items with articulatory features in common are less well-recalled than lists of items with distinct sounds and articulations. It is probably not possible to determine from this confusion technique whether the underlying code is sound or articulation, because items which are articulated in a similar way also sound alike (Wickelgren, 1969).

Another approach provides data consistent with the auditory rather than the articulatory view of immediate memory storage. Requiring irrelevant vocalization during visual presentation of the items to be remembered (e.g., pronouncing "Hi Ya", or "La La") greatly reduces recall (Crowder, 1978; Levy, 1971), an effect initially attributed to articulatory coding (e.g., Baddeley, 1978). However, the following recent findings attribute that effect to auditory rather than articulatory codes: (1) The effect of irrelevant articulation can be eliminated if the articulation is not voiced and the memory items are presented in auditory as well as visual form (Klapp, Griem, & Marshburn, 1981, Experiment 2). (2) Recall is better with auditory input than with unvoiced articulation of the visually presented items, and no better with both auditory and articulatory input than with auditory presentation in the absence of articulation (Klapp, et al., 1983, Experiment 1). (3) Presentation of irrelevant auditory speech disrupts immediate memory (Salame & Baddeley, 1982). (4) The phonological similarity effect (Conrad & Hull, 1964) remains with irrelevant articulation accompanied by relevant auditory input (Baddeley, Lewis, & Vallar, 1984). Apparently the code is auditory rather than articulatory.

1.3.4.2. Spatial (visual)

When material is not linguistic in nature, the coding may often be visual or spatial rather than auditory. The classic evidence for this view is the finding that a task requiring visual imagery is performed better with speech as the output response, compared to spatially-guided pointing as the output. By contrast, a verbal task is performed more easily with pointing than with speech (Brooks, 1968). Apparently performance is best when the output does not interfere with the immediate memory code in use. Although this suggests that coding may depend on the nature of the material, it also appears that even verbal material may be retained visually in the presence of distracting auditory tasks (Saltzberg, Parks, Kroll, & Parkinson, 1971). Baddeley and Lieberman (1980) proposed that the coding which is usually referred to as "visual" should be considered to be spatial. Memory for a spatial-visual situation was disrupted by an auditory-spatial tracking task but not by a visual brightness judgment task which had no spatial component.

1.3.4.3. Motor program

Reaction time (RT) studies suggest that programs for motor-action sequences may be stored in immediate memory, which thereby functions as an output buffer. These studies measure RT prior to the initiation of a motor sequence as a function of the nature of that sequence. Longer sequences are preceded by a longer RT in experiments for which the sequence to be generated is not pre-cued or anticipated, but this effect vanishes for anticipated or pre-cued sequences (Keele, 1981; Klapp, 1977; Klapp, 1981; Klapp & Rodriguez, 1982). This pattern of results is interpretable by assuming that responses must be programmed before they are initiated, and that some responses take longer to program than others. However, when the particular response is anticipated, programming can occur in advance of the RT interval, and the resultant program can be stored in immediate memory. Because the program for an anticipated response is in immediate memory, there is no need to create a new program during the RT interval. Thus, this model uses the concept of immediate memory for motor programs to explain why RT for anticipated responses is independent of the parameters which influence RT for unanticipated responses.

Studies of skilled typists also suggest the existence of a motor code in immediate memory. When given a signal to stop, subjects type an average of about 1.5 characters before actually stopping (Logan, 1982). This can be interpreted as continued response on the basis of a motor program in STM.

1.3.4.4. Actively-formed hypotheses

Most views of working memory, including those reviewed above, suggest that material is retained without transformation or deep analysis. A different perspective is that, rather than passively retaining material which has been heard (or seen), people are "continuously forming hypotheses about the structure of what they are hearing and forming predictions about what they have yet to hear. These are working hypotheses, either confirmed or modified with the arrival of new information" (Wingfield & Butterworth, 1984, p. 352). In this view, working memory stores hypotheses which have been actively formed by the person, so that the representation in working memory may be abstract rather than in the form of auditory, visual, or motor codes.

1.3.4.5. Priming of long-term memory as a code in immediate memory

The process of priming may be illustrated by a typical experiment (Meyer & Schvaneveldt, 1971). Subjects viewed two strings of letters presented simultaneously, responding "yes" if both strings were words, and otherwise responding "no". The mean RT for "yes" responses was shorter for pairs of associated words than for pairs of unassociated words. The first word is assumed to act as a "prime", activating the trace of the second word thereby speeding its retrieval from long-term memory. Thus, the first word is assumed to prime a particular class or category of memory traces (Rosch, 1975). Because some memory items are primed, and others are not, those items which are primed can be thought of as representing a system of immediate memory. Primed permanent memory is sometimes taken to be the form of immediate memory which functions as "working memory". For example, Card, Moran, and Newell (1983, p. 36) assert that "structurally, working memory consists of a subset of the elements in long-term memory that have become activated..."

1.3.4.6. Distinct codes for item and order information

Immediate memory exhibits different properties when the temporal order of the items must be preserved than when the subject need only remember which items were presented (Healy, 1982). For example, whereas order information yields a bow-shaped serial position curve, item information does not (Healy, 1974). See Monsell (1984) for a related distinction.

1.3.4.7. Central executive with slave systems

Baddeley and his colleagues proposed that the function of working memory is carried out by a "central executive" and two "slave systems" called the "articulatory loop" and the "visuo-spatial scratch pad" (Baddeley, 1978; Baddeley & Hitch, 1974; Baddeley & Lieberman, 1980). This position differs from the preceding views in that it proposes a functional relation among several systems of memory. The central executive may correspond to primary memory or current awareness (James, 1890), and is assumed to be very limited in storage. It controls the slave systems which function as auxiliary or supplemental storage. We note, however, that the original "articulatory loop" is probably auditory rather than articulatory (see discussion of auditory code in section 1.3.4.1 above).

This is an interesting and useful viewpoint, and one which can be elaborated and modified while retaining the basic distinction between central executive and slave memory systems. An excellent example of how the framework can accommodate and even encourage new findings is the finding that subjects can learn to use finger response codes as a new slave system (Reisberg, Rappaport, & O'Shaugnessy, 1984).

1.3.5. Evaluation

It is not clear whether this is or is not an exhaustive list of the systems of immediate memory. For example, if there are auditory and visual-spatial systems, perhaps there are also olfactory or kinesthetic systems as well. Or perhaps distinction is between temporally-ordered (e.g., auditory) and spatially-organized (e.g., visual) systems with modality only an incidental feature. It is not clear whether these are separate systems or different codes within the same system (Phillips & Christie, 1977). It is also possible that some of the proposed systems may map onto one another. For example, primed or activated traces in permanent memory may correspond to item memory in the item vs. order distinction, or the motor program buffer may correspond to sensory imagery representing the response goal (Greenwald, 1970) such that the motor program is actually stored as a sensory trace.

Converging evidence for assuming distinct systems of immediate memory could come from memory loading experiments. For example, whereas tasks using systems with different properties (e.g., visual vs. auditory codes) would presumably not exhibit mutual interference, tasks having similar memory properties would. This approach is congruent in spirit with the multiple-resource view of analyzing dual-task situations (Navon & Gopher, 1979; Wickens, Sandry, & Vidulich, 1983). A similar approach has been suggested by Monsell (1984).

1.4. HUMAN FACTORS APPLICATIONS

This chapter began with a description of a hypothetical truck dispatcher suffering from distraction and overload in his STM. It then reviewed the literature from laboratory research and theory related to these sources of memory problems. Now the human factors implications of these research findings are discussed.

1.4.1. Distraction

Clearly material in STM is volatile and subject to loss, especially when distracting events demand that attention be shifted away from STM retention to other activities. Surprisingly, I could find only two reports of an applied perspective on this Brown-Peterson effect, and both of these are laboratory simulations.

An analysis of simulated communication between air traffic controllers and pilots from the STM distraction perspective was reported by Loftus, Dark, and Williams (1979). The authors conclude (p. 179) that, "The fact that forgetting occurs over an interval of 15 [seconds] following the initial reception of a message simply underlines what is intuitively evident to most pilots - that, if at all possible a message should be responded to as soon as possible after it is received." An additional important finding for Human Factors applications is that messages presented in rapid succession were recalled less effectively than were well-spaced messages.

Memory for verbal or visual traffic signs was studied as a function of type of distractor in the 15 second retention interval in a laboratory experiment by Hoffman and MacDonald (1980). As expected, retention was reduced by the presence of a distractor, compared to a no-distractor control. Furthermore, increased interference occurred when the sign and distractor were of similar form (verbal distractor and verbal signs or symbolic distractor and symbolic signs) than when the forms of the signs and distractors were mixed.

Distraction is a more potent cause of forgetting early than late in the retention interval. A few seconds of rehearsal can largely protect STM from the usual loss attributed to distraction. The potential Human Factors implications of this finding appear to have been overlooked. One would suppose that retention of information, such as directives from air traffic control, would be improved by brief rehearsal when that information cannot be used immediately. The extent to which this can lead to successful recommendations which can be implemented in practice needs to be investigated. For example, pilots might be instructed to rehearse directives which can not be implemented immediately, or transmission of additional non-emergency directives might be delayed until rehearsal (or immediate implementation) of the first directive has been accomplished.

1.4.2. Capacity Limit

The literature suggests that the simplistic view that there is a single STM (working memory) is not viable. Rather than a single system which can be overloaded, there appear to be multiple independent systems with different properties, each subject to overload without affecting the other systems. Unfortunately, the notion of breaking working memory into its components on the converging basis of properties and mutual interference is, at present, a future program of research rather than an accomplished fact. However, one can begin to flesh out the human factors implications of this approach even before it is fully completed.

A simple and obvious consequence of this multiplex viewpoint is that material should be distributed among the systems of memory, rather than fitted into just one system. It is not clear just how this might be accomplished in practice, largely because of our ignorance concerning the nature of the memory systems. Furthermore, there are instances in which integrating material into a common representation is better than spreading material among systems (Klapp, Hill, Tyler, Martin, Jagacinski, & Jones, 1985; Wickens & Goettle, 1984).

Another consequence of the multiplex view has been described by Wickens (1984, p. 211-214), who suggested that the possibility of different memory codes has implications concerning the optimum modality of presentation of information. According to his principle of "stimulus/central processing/response compatibility", the display code should match the code used in working memory for the particular task. For example, verbal codes in working memory are best addressed by auditory-verbal display formats, but spatial codes in working memory are best addressed by visual-pictorial codes. This position suggests that air traffic information should be presented in a visual modality, but semantic information such as that pertaining to medical diagnosis should be presented in a verbal format. Tasks such as scheduling in our truck dispatcher example may be spatial, a view consistent with the apparent superiority of spatial over verbal displays in scheduling problems (Gibson & Laios, 1978).

A complementary (reversed) recommendation can be generated for the means of retrieving material from memory aids. The position taken by Wickens refers to modality of information presentation, and holds that this modality should match that of the immediate memory code. By contrast, accessing the retrieval organization of a memory aid system is potentially an interfering event. Thus, perhaps the modality of input required to operate the memory aid should not correspond to the modality of storage in immediate memory. For example, semantic tasks might best be served by spatially-organized memory aids but spatial tasks such as air traffic control might be aided most by semantic-based menu systems for the memory aid. Because most tasks involving computer usage are semantic, it is reasonable that computer manufacturers are moving toward more spatially-organized retrieval systems such as the "mouse" of the Apple Macintosh and the screen pointing system developed by Hewlett Packard.

Certainly an important topic for applied research concerns the conditions under which these complementary recommendations apply. We have proposed a distinction in terms of information content format (matched modalities best) versus retrieval organization (unmatched modalities best).

1.4.3. Workload

A prominent issue in current human factors research is the measurement of "mental workload" in tasks such as flight management, command and control, etc. (Kantowitz & Sorkin, 1983, pp. 187-191; Wickens, 1984, Chapter 8; see Chapter 3 by Kantowitz in this volume). Among the approaches being considered is the use of scales of subjective assessment in which the pilot (or other worker) rates his or her experienced workload (Hart, Childress, & Bortolussi, 1981). The use of such subjective measures assumes that some relevant aspect of mental operations is open to introspection. Thus, this approach raises historical issues of the extent to which humans are consciously aware of the information processing conducted by the nervous system. William

James (1890) defined "primary memory," a form of STM, as equivalent to the content of consciousness. Thus, we might expect that workload created by operations involving STM may be accessible to subjective report even if other aspects of workload are not.

An interesting analysis of a tracking task from this perspective was reported by Vidulich and Wickens (1984). Subjective workload ratings were correlated with performance when control order was manipulated, but not when tracking bandwidth was manipulated. It was assumed that increasing bandwidth required more rapid responding, but did not load primary memory with strategies and planned sequences, as was the case for manipulation of control order.

1.4.4. Conclusion

Although the literature on short-term immediate memory is extensive within Cognitive Psychology, application of these research findings is relatively sparse and often only programatic and suggestive. However, as the field of Human Factors becomes increasingly concerned with cognitive as well as sensory-motor aspects of human performance, this situation is quite likely to improve.

ACKNOWLEDGEMENT

This review was prepared while the author was supported, in part, by Contract F33615-83-K-0039 with the Air Force Human Resources Laboratory. Some of the work was accomplished during the author's tenure with the Air Force Office of Scientific Research Summer Faculty Research Program at Wright Patterson Air Force Base, Ohio. The author is indebted to Dr. Lawrence Reed for many helpful discussions and ideas on this work, and to Drs. Peter Hancock and Nancy Harrison for helpful comments on an earlier version of this chapter.

REFERENCES

Baddeley, A.D., & Hitch, G. (1974). Working memory. In G. Bower (Ed.), *The psychology of learning and motivation: Advances in research and theory*. New York: Academic Press.

Baddeley, A.D. (1978). The trouble with levels: A re-examination of Craik and Lockhart's framework for memory research. *Psychological Review, 85*, 139-152.

Baddeley, A.D., & Lieberman, K. (1980). Spatial working memory. In R.S. Nickerson (Ed.), *Attention and performance VIII*. Hillsdale, New Jersey: Lawrence Erlbaum Associates.

Baddeley, A.D., Lewis, V., & Vallar, G. (1984). Exploring the articulatory loop. *Quarterly Journal of Experimental Psychology, 36A*, 233-252.

Brainerd, C.J. (1981). Working memory and the developmental analysis of probability judgment. *Psychological Review, 88*, 463-502.

Broadbent, D.E. (1983). The functional approach to memory. *Philosophical Transactions of the Royal Society of London, 302*, 239-249.

Brooks, L.R. (1968). Spatial and verbal components in the act of recall. *Canadian Journal of Psychology, 22*, 349-368.

Brown, J. (1958). Some tests of the decay theory of immediate memory. *Quarterly Journal of Experimental Psychology, 10*, 12-21.

Card, S.K., Moran, T.P., & Newell, A. (1983). *The psychology of human-computer interaction*. Hillsdale, New Jersey: Lawrence Erlbaum Associates.

Chase, W.G., & Ericsson, K.A. (1982). Skill and working memory. In G. Bower (Ed.), *The psychology of learning and motivation*. New York: Academic Press.

Conrad, R. (1964). Acoustic confusions in immediate memory. *British Journal of Psychology, 55*, 75-84.

Conrad, R., & Hull, A.J. (1964). Information, acoustic confusion and memory span. *British Journal of Psychology, 55*, 429-432.

Craik, F.I.M., & Lockhart, R.S. (1972). Levels of processing: A framework for memory research. *Journal of Verbal Learning and Verbal Behavior, 11*, 671-684.

Crowder, R.G. (1967). Short-term memory for words with a perceptual-motor interpolated activity. *Journal of Verbal Learning and Verbal Behavior, 6*, 753-761.

Crowder, R.G. (1978). Audition and speech coding in short-term memory: A tutorial review. In J. Requin (Ed.), *Attention and performance VII*. Hillsdale, New Jersey: Lawrence Erlbaum Associates.

Daneman, W., & Carpenter, P. (1980). Individual differences in working memory and reading. *Journal of Verbal Learning and Verbal Behavior, 19*, 450-466.

Dillon, R.F., & Reid, L.S. (1969). Short-term memory as a function of information processing during the retention interval. *Journal of Experimental Psychology, 81*, 261-269.

Gibson, R., & Laios, L. (1978). The presentation of information to the job-shop scheduler. *Human Factors, 20*, 725-732.

Greenwald, A.G. (1970). A choice reaction time test of ideomotor theory. *Journal of Experimental Psychology, 86*, 20-25.

Halford, G.S., Bain, J.D., & Maybery, M. (1984). Working memory and representation processes: Implications for cognitive development. In H. Bouma & D. Bouwhuis (Eds.), *Attention and performance X*. Hillsdale, New Jersey: Lawrence Erlbaum Associates.

Hart, S.G., Childress, M.E., & Bortolussi, M. (1981). Defining the subjective experience of workload. *Proceedings of the Human Factors Society, 25*, 527-531.

Healy, A.F. (1974). Separating item from order information in short-term memory. *Journal of Verbal Learning and Verbal Behavior, 13*, 644-655.

Healy, A.F. (1982). Short-term memory for order information. In G. Bower (Ed.), *The psychology of learning and motivation*. New York: Academic Press.

Hellige, J.B., & Wong, T.M. (1983). Hemisphere specific interference in dichotic listening: Task variables and individual differences. *Journal of Experimental Psychology: General, 112*, 218-239.

Hitch, G.J., & Baddeley, A.D. (1976). Verbal reasoning and working memory. *Quarterly Journal of Experimental Psychology, 28*, 603-621.

Hitch, G.J. (1978). The role of short-term memory in mental arithmetic. *Cognitive Psychology, 10*, 302-323.

Hoffman, E.R., & MacDonald, W.A. (1980). Short-term retention of traffic turn restriction signs. *Human Factors, 22*, 241-251.

Hunt, E. (1978). Mechanics of verbal ability. *Psychological Review, 85*, 109-130.

James, W. (1890). *Principles of psychology*. New York: Holt.

Johnston, W., Greenberg, S., Fisher, R., & Martin, D. (1970). Divided attention: A vehicle for monitoring memory processes. *Journal of Experimental Psychology, 83*, 164-171.

Jonides, J. (1981). Voluntary versus automatic control over the mind's eye's movement. In J. Long & A. Baddeley (Eds.), *Attention and performance IX*. Hillsdale, New Jersey: Lawrence Erlbaum Associates.

Kantowitz, B.H., & Sorkin, R.D. (1983). *Human factors: Understanding people-system relationships*. New

York: Wiley.

Keele, S.W. (1981). Behavioral analysis of movement. In V.B. Brooks (Ed.), *Handbook of physiology: Section 1: The nervous system. Volume II. Motor control. Part 2*. Baltimore: American Physiological Society.

Keppel, G., & Underwood, B.J. (1962). Proactive inhibition in short-term retention of single items. *Journal of Verbal Learning and Verbal Behavior, 1*, 153-161.

Klapp, S.T. (1977). Reaction time analysis of programmed control. In R. Hutton (Ed.), *Exercise and sports sciences reviews*. Santa Barbara, California, Journal Publishing Affiliates.

Klapp, S.T. (1981). Motor programming is not the only process which can influence RT: Some thoughts on the Marteniuk and MacKenzie analysis. *Journal of Motor Behavior, 13*, 320-328.

Klapp, S.T., Greim, D.M., & Marshburn, E.A. (1981). Buffer storage of programmed articulation and articulatory loop: Two names for the same mechanism or two distinct components of short-term memory? In J. Long & A. Baddeley (Eds.), *Attention and performance IX*. Hillsdale, New Jersey: Lawrence Erlbaum Associates.

Klapp, S.T., & Rodriguez, G. (1982). Programming time as a function of response duration: A replication of "dit-dah" without possible guessing artifacts. *Journal of Motor Behavior, 14*, 46-56.

Klapp, S.T., Marshburn, E.A., & Lester, P.T. (1983). Short-term memory does not involve the "working memory" of information processing: The demise of a common assumption. *Journal of Experimental Psychology: General, 112*, 240-264.

Klapp, S.T., & Philipoff, A. (1983). Short-term memory limits in performance. *Proceedings of the Human Factors Society, 27*, 452-454.

Klapp, S.T., Hill, M.D., Tyler, J.G., Martin, Z.E., Jagacinski, R.J., & Jones, M.R. (1985). On marching to two different drummers: Perceptual aspects of the difficulties. *Journal of Experimental Psychology: Human Perception and Performance, 11*, 814-827.

Klatzky, R.L. (1980). *Human memory: Structures and processes*. San Francisco: Freeman.

Kroll, N.E.A., Kellicut, M.H., & Parks, T.E. (1975). Rehearsal of visual and auditory stimuli while shadowing. *Journal of Experimental Psychology: Human Learning and Memory, 1*, 215-222.

Levy, B.A. (1971). Role of articulation in auditory and visual short-term memory. *Journal of Verbal Learning and Verbal Behavior, 10*, 123-132.

Loftus, G.R., Dark, V.J., & Williams, D. (1979). Short-term memory factors in ground controller/pilot communications. *Human Factors, 21*, 169-181.

Logan, G.D. (1979). On the use of a concurrent memory load to measure attention and automaticity. *Journal of Experimental Psychology: Human Perception and Performance, 5*, 189-207.

Logan, G.D. (1980). Short-term memory demands of reaction-time tasks which differ in complexity. *Journal of Experimental Psychology: Human Perception and Performance, 6*, 375-389.

Logan, G.D. (1982). On the ability to inhibit complex movements: A stop-signal study of typewriting. *Journal of Experimental Psychology: Human Perception and Performance, 8*, 778-792.

Meyer, D., & Schvaneveldt, R. (1971). Facilitation in recognizing parts of words: Evidence of dependence between retrieval operations. *Journal of Experimental Psychology, 90*, 227-234.

Monsell, S. (1984). Components of working memory underlying verbal skills: A "distributed capacities" view. In H. Bouma & D. Bouwhuis (Eds.), *Attention and performance X*. Hillsdale, New Jersey: Lawrence Erlbaum Associates.

Navon, D., & Gopher, D. (1979). On the economy of the human processing system. *Psychological Review, 86*, 214-255.

Norman, D.A. (1976). *Memory and attention*. New York: Wiley.

Peterson, L.R., & Peterson, M.J. (1959). Short-term retention of individual verbal items. *Journal of Experimental Psychology, 58*, 193-198.

Phillips, W.A., & Christie, F.M. (1977). Interference with visualization. *Quarterly Journal of Experimental Psychology, 29*, 637-650.

Reisberg, D. (1983). General mental resources and perceptual judgments. *Journal of Experimental Psychology: Human Perception and Performance, 9*, 966-979.

Reisberg, D., Rappaport, I., & O'Shaughnessy, M. (1984). Limits of working memory: The digit digit-span. *Journal of Experimental Psychology: Learning Memory and Cognition, 10*, 203-221.

Richardson, J.T.E. (1984). Developing the theory of working memory. *Memory & Cognition, 12*, 71-83.

Roediger, H.L., III, Knight, J.L., & Kantowitz, B.H. (1977). Inferring decay in short-term memory: The issue of capacity. *Memory & Cognition, 5*, 167-176.

Rosch, E. (1975). (1975). Cognitive representations of semantic categories. *Journal of Experimental Psychology: General, 104*, 192-233.

Salame, P., & Baddeley, A. (1982). Disruption of short-term memory by unattended speech: Implications for the structure of working memory. *Journal of Verbal Learning and Verbal*

Behavior, 21, 150-164.

Saltzberg, P.M., Parks, T.E., Kroll, N.E.A., & Parkinson, S.R. (1971). Retroactive effects of phonemic similarity on short-term recall of visual and auditory stimuli. *Journal of Experimental Psychology, 91*, 43-46.

Shallice, T., & Warrington, E.K. (1970). Independent functioning of verbal memory stores: A neuropsychological study. *Quarterly Journal of Experimental Psychology, 22*, 261-273.

Shulman, H.G., & Greenberg, D. (1971). Perceptual deficit due to division of attention between memory and perception. *Journal of Experimental Psychology, 88*, 171-176.

Stanners, R.F., Meunier, G.F., & Headley, D.B. (1969). Reaction time as an index of rehearsal in short-term memory. *Journal of Experimental Psychology, 82*, 566-570.

Trumbo, D., & Milano, F. (1971). Primary task performance as a function of encoding, retention and recall in a secondary task. *Journal of Experimental Psychology, 91*, 273-279.

Vidulich, M.A., & Wickens, C.D. (1984). Subjective workload assessment and voluntary control of effort in a tracking task. *Proceedings of the Annual Conference on Manual Control, 20*, 57-72.

Wanner, E., & Shiner, S. (1976). Measuring transient memory load. *Journal of Verbal Learning and Verbal Behavior, 15*, 159-167.

Wechsler, D. (1958). *The measurement and appraisal of adult intelligence*. Baltimore: Williams & Wilkins.

Wickelgren, W.A. (1969). Auditory or articulatory coding in verbal short-term memory. *Psychological Review, 76*, 232-235.

Wickens, C.D., Sandry, D.L., & Vidulich, M. (1983). Compatibility and resource competition between modalities of input, control processing, and output. *Human Factors, 25*, 227-248.

Wickens, C.D. (1984). ***Engineering psychology and human performance***. Columbus: Charles E. Merill.

Wickens, C.D., & Goettle, B. (1984). Multiple resources and display formatting: The implications of task integration. ***Proceedings of the Human Factors Society, 28***, 722-726.

Wingfield, A., & Butterworth, B. (1984). Running memory for sentences and parts of sentences: Syntactic parsing as a control function in working memory. In H. Bouma & D. Bouwhuis (Eds.), ***Attention and performance X***. Hillsdale, New Jersey: Lawrence Erlbaum Associates.

HUMAN FACTORS PSYCHOLOGY, Peter A. Hancock (ed.)

2. ATTENTION

Christopher D. Wickens
Aviation Research Laboratory
University of Illinois
Willard Airport
Savoy, Illinois 61874

INTRODUCTION

The concept of attention has a wide variety of meanings, as attested to by a recent volume on the topic entitled Varieties of Attention (Parasuraman & Davies, 1984). We speak of inattention as describing the classic decrement in vigilant performance resulting from long hours on watch (Swets & Kristoffersen, 1970), and lapses of attention as leading to the cause of errors (Reason, 1984). Selective attention is used to describe our behavior in looking and listening selectively to the environment (Moray, 1984). Divided attention is a concept that is employed to describe our successes or failures as we try to perform two tasks at once. Focused attention refers to our concentration on one activity, and its failure is described as "distraction". Attention is also conceived as having the intensive characteristics synonymous with effort (Kahneman, 1973) and this conception is used to describe the notion of mental workload. Finally, attention has sometimes been equated with "consciousness", which produces the dichotomy between conscious and unconscious aspects or behavior.

A theme underlying most of these uses is the notion of some sort of limited capacity within the human operator. It is a capacity that cannot be observed directly, but must be inferred from performance. The existence of these limitations means that to perform effectively, the human must deploy this capacity with skill and understanding. Both the limits themselves, and the manner in which attention is deployed have a tremendous bearing on how we interact with the systems around us. An automobile may crash because attention was momentarily diverted from the road, an aircraft may approach a near collision because the air traffic controller's limited attentional resources were overloaded by excessive demands, or we may forget an important phone number as our attention is distracted by a voice in the next room.

The following chapter outlines the characteristics of attention and how its various manifestations are affected by the nature of the stimuli that must be processed, the kind of tasks or information processing demands that are required, and the characteristics of the human operator. As these characteristics are outlined, they are placed in the context of real world systems and tasks, to demonstrate how knowledge of attention has an important bearing on the field of human engineering.

2.1. METAPHORS OF ATTENTION

Many of the characteristics of attention may be conveniently described by one of two metaphors that have often been employed. Attention in perception may be described by the metaphor of a searchlight (Wachtel, 1967). The momentary direction of our attention is the searchlight beam. The focus of this searchlight falls on that which is in momentary consciousness. Everything within the beam of light is processed whether wanted (successful focusing of attention) or not (failure to focus or distraction). Within the context of this metaphor, two properties of the searchlight are relevant to phenomena of human experience. (1) The "breadth" of the beam (how many objects we can illuminate at once) and the distinction between the desired focus (that which we want to process), and the unwanted objects that are illuminated because they lie close to the wanted objects. These represent the issues of divided and focused attention, respectively. (2) The brain controlling the hand that guides the searchlight. When and why does it shift from one part of the environment to another? This represents the software, or "executive" properties of selective attention. Each of these properties will be considered in some detail below, as we consider examples of how operators search the complex stimulus world for critical pieces of information, and how that information is processed once it has been found.

While the searchlight metaphor emphasizes the unity of attention, the resource metaphor emphasizes its divisibility (Norman & Bobrow, 1975). When performing any task, different mental operations must be carried out (e.g., responding, rehearsing, perceiving, etc.). Performance of each of these activities requires some level of the operator's limited processing resources. Since these resources are limited, the metaphor readily accounts for our failures of time-sharing. Two activities will demand more resources than a single activity, so there will be a greater deficiency between supply and demand in the dual than in the single activity case. The resource metaphor also proposes that some operations may require resources that are different from others (just as some furnaces use oil and others use natural gas or coal). As a consequence, there is less competition between these processes for their enabling resources, and time-sharing between them may be more successful.

2.2. SELECTIVE ATTENTION

When studying selective attention, it is often convenient to adopt a "single channel" model of human information processing, one that makes the convenient assumption that only a single channel of information can be processed at a time (Broadbent, 1958). As we shall see, this assumption is not entirely true, but it does represent a useful heuristic for examining selection.

2.2.1. Optimality Of Selection

In environments such as the aircraft cockpit or the nuclear power plant control room, there are a large number of sources of information that must be sampled periodically. In these situations, engineering psychologists have studied how optimal people can be at selecting the relevant stimuli to attend at the appropriate times. The concept of optimality may be defined in terms of an attention sampling behavior that will maximize an expected value, or minimize an expected cost. For example, an aircraft pilot who continuously fixates (samples) the instrument registering airspeed, but ignores the altimeter when close to the ground is clearly behaving in a non-optimal manner. One who samples both of these, but never checks the visual world outside the aircraft is doing better but is still not optimal, for this pilot is incurring the expected costs of missing important stimuli (e.g., other aircraft) that can only be seen by looking outside the window.

Engineering psychologists who have worked in the area of selective attention divide the environment into channels, along which critical events may periodically occur (e.g., Broadbent, 1971; Moray, 1984; Senders, 1964; Sheridan, 1972). They have assumed that environmental sampling is guided by the expected cost that results when an event is missed. The expected cost of missing an event in turn is directly related to the actual cost of missing the event and the event frequency. Those events that occur often are more likely to be missed if the channels along which they occur are not sampled. Also, in most real world tasks, the probability of missing an event on a channel increases with the amount of time that passes since the channel has last been sampled. This is because our uncertainty about the state of most processes increases with the time since the process was last observed. If a pilot checks out the window to find that the surrounding air space is clear, another glance two seconds later will probably reveal the same picture. But if two minutes pass, the probability that the existence of another aircraft will be missed will have risen substantially.

When optimum sampling is examined in the laboratory, the subject is typically presented with two or more potential channels of stimulus information, along which events may arrive at semi-predictable rates. For example, a channel might be an instrument dial, with an "event" defined as a movement of the needle into a danger zone. The channels may be visual locations in space (e.g., Senders, 1964; Sheridan, 1972), or ears of presentation for auditory stimuli (Moray, Fitter, Ostry, Favreau, & Nagy, 1976). The general conclusions of these studies are well summarized by Moray (1978, 1981, 1984, 1986) and suggest the following:

a. Sampling is guided by a "mental model" of the statistical properties of the environment. The frequency with which different events occur on different channels, and which channels are correlated with each other, determines where and when the operator will sample. Because sampling strategies provide estimates of the operator's internal model of a system under supervision, the patterns of fixations between visual instruments will also be of help to the system designer in locating information displays in an optimal manner.

b. People learn to sample those channels with higher event rates more frequently and lower rates less frequently. However, sampling rate is not adjusted upward or downward with event frequency as much as would be optimal; that is, frequent event channels are not sampled frequently enough, and rare event channels are sampled too frequently (Senders, 1964).

c. Human memory is imperfect and sampling reflects this fact (Sheridan & Rouse, 1971). Hence, people tend to sample an information source more than they would need to if they have perfect memory about the status of that source when it was last sampled. This is, of course, a conservative and "safe" behavior, in that it reflects an accommodation of sampling to one's own cognitive limitations. Also because of limits of memory, people may forget to sample a particular display source if there are a multitude of possible sources.

d. When people are presented with a preview of scheduled events that are likely to occur in the future, then sampling and switching approach an optimal strategy. Now subjects' sampling can be guided in part by an "external model", the display of the previewed events. However, even with preview, there are departures from optimal behavior if the number of channels is large. When the number of channels increases, the "planning horizon"--the use of the future to guide in sampling--is reduced (Tulga & Sheridan, 1980). All in all, human sampling in selective attention may be described as reasonably optimal but not perfect, subject to certain limitations of human memory.

2.2.2. Implications For System Design

The research on selective sampling of the environment is relevant to system design and to human factors whenever an operator is confronted with a task that requires assimilation of information from a wide number of different sources. Two environments come immediately to mind - that of the aircraft pilot, and that of the process control monitor.

Studies of the aircraft cockpit environment, for example, indicate that the pattern of scanning behavior between instruments systematically reflects the well-trained pilot's information needs (Fitts, Jones, & Milton, 1950; Harris & Spady, 1985; McRuer, Jex, Clement, & Graham, 1967; Simmons, 1979). Thus, the general finding that the attitude indicator is the most frequently fixated flight display is consistent with the fact that this instrument presents the most rapidly changing information in the flight control loop. As a consequence, the pilot can gain the greatest amount of information about what the plane is and will be doing from this instrument (McRuer et al., 1967).

Given that scan patterns reflect operator's information needs, the best designed display configurations should be made consistent with these needs by minimizing the total distance traveled between fixations. Frequently fixated displays should be located centrally and close together, while display pairs that are scanned in sequence should also be near each other. These procedures have been employed with some success in display design. For example McRuer et al. (1967) used a model of fixation transitions to derive an optimal layout for a Boeing 707 Instrument Landing System display, and obtained a close correspondence between the model-derived display and that which was actually certified by the FAA.

While the aircraft pilot has six critical instruments that must be monitored in addition to the world outside the cockpit, the supervisor of today's nuclear power plant has literally hundreds of instruments that must be periodically checked. Given the documented failures of memory, it may seem reasonable that with so many instruments, process monitors will scan or look at instruments more often than they need to, having forgotten the value of a particular variable when it was last viewed. However, a potentially greater danger may arise when the operator forgets entirely to check a particular instrument. When the instruments are few in number, this is not much of a problem because the physical presence of each dial in the general field of view serves as a constant reminder that it must be examined. However, when the instruments span three panels, as shown in the control panel of Figure 1, failures of selective attention to guide the inspection of all items becomes more prominent. The potential for such failures, however, may become most severe in the modern computerized control room in which instruments may be viewed on a centralized video display by inputing a keyboard command. In this case, the physical presence of a meter on the wall no longer serves as a reminder to examine it. Instead, the operator must remember when it needs to be examined and, as noted above, human memory does not always serve as a reliable tool to guide selection. Here, is an area of great need for computer assistance to act as a reminder to the operator of the optimal time to sample and inspect numerous informational sources (Moray, 1981).

Figure 2-1: Nuclear Power Control panel, illustrating the heavy attention demands imposed by visual scanning.

The limited characteristics of selective attention have also been applied to environments and tasks that are considerably less exotic and complex than the aircraft or nuclear power plant. Two examples of research in this area have examined the allocation of attention to information at the scene of a crime and to information presented on the home TV-screen. In one study, Wells and Leippe (1981), used the characteristics of selective attention to disprove a common misbelief that jurors often express regarding the testimony of eye witnesses to a crime. The assumption made by such jurists is that the more detail the witness can recall concerning all characteristics at the scene of the crime, the more accurate the identification of the suspect will be. This assumption runs contrary to the idea of a single selective attention beam which, during the instant of the crime, must focus either on the criminal's facial features or on some other part of the environment.

To test the degree of restriction of selective attention in viewing crimes, Wells and Leippe staged violent crimes in front of unsuspecting subjects. Afterwards, the subjects were asked to make reports on irrelevant details at the scene of the crime, and to identify the perpetrators from a line-up. As predicted by a single channel concept of selective attention, Wells and Leippe found that those whose memory for irrelevant detail was good, were in fact less likely to identify the suspect, as attention was directed away from the facial features.

Selective attention can often play an important role in determining the information we extract from television advertisements, as revealed in a study carried out by Warshaw (1978). The typical television commercial usually offers a mix of video and sound track. The former will show people washing clothes, eating cereal, driving cars, and engaging in a variety of other "All-American" activities, while the latter often involves a narration, sometimes carrying important product information. Warshaw tested and confirmed the hypothesis that people are somewhat selective in their attention between the auditory and visual channels. If significant visual information occurs simultaneously with critical auditory information, one or the other will be likely to be missed. Warshaw's finding has some important implications. For example, it suggests that an effective commercial should try to interweave rather than overlap the significant events on both channels. It also brings up a concern that the advertiser who is forced by government regulations to introduce an auditory or visual disclaimer or product warning into the ad ("x" may be hazardous to your health) may choose to do so at times when significant information in the other channel will decrease the chances of the warning being perceived.

2.2.3. Visual Dominance And Sensory Interaction

Warshaw's experiment leads us to ask if there are any significant differences between the visual and auditory modalities in terms of their attention-getting ability. This is a question that can be answered at two levels. First, the auditory modality (along with the sense of touch) is clearly superior to vision as a means of alerting or calling attention to itself when attention has been diverted elsewhere. Hence, auditory warnings are used to alert people of imminent dangers (consider the fire alarm), or a shaking control stick (a tactile channel) may be used to warn the aircraft pilot of dangerous flight conditions.

At a second level, however, there are cases when items of both auditory and visual (or tactile and visual) information arrive with equal frequency; the typical TV commercial is an example. Under these circumstances, humans show a pattern of visual dominance, allowing information in the visual channel to take precedence over information in the auditory or tactile channels. Examples of visual dominance are abundant. For example, Colovita (1971) required subjects to respond as fast as possible to either a light (with one hand) or a tone (with the other hand). On very infrequent occasions, both stimuli were presented simultaneously. When this occurred, subjects invariably made only the response with the hand appropriate to the light. In fact, subjects were often unaware that a tone had been presented at all on these trials! Experiments by Jordan (1972) and Klein and Posner (1974) supported the dominance of vision over proprioception when stimuli on the two channels are presented together. These investigators found that reaction time to a stimulus that was a compound of a light and a proprioceptive displacement of a limb, was slower than reaction time to the proprioceptive stimulus alone. This result suggests that the light "captured" attention and slowed down the more rapid processing of the proprioceptive information. Different examples of visual dominance are observed in experiments in which the sense of vision and proprioception are placed in conflict through prismatically distorted lenses (Rock, 1975). Behavior in these situations suggests that the subject responds appropriately to the visual information and disregards that provided by other modalities.

Some dual task time-sharing situations also appear to manifest a form of visual dominance when an auditory task is performed concurrently with a visual one. Under these circumstances, the auditory task tends to be hurt more by the division of attention, than does the visual task (Massaro & Warner, 1977; Treisman & Davies, 1973).

The phenomenon of visual dominance is viewed by Posner, Nissen, and Klein (1976) as an adaptive and therefore useful adjustment of information processing, made in response to anatomical differences between sensory channels. That is, since vision is naturally less alerting than hearing or touch, visual dominance provides a way of compensating for this handicap. However, there are some circumstances in which visual dominance can lead to nonadaptive behavior. Illusions of movement provide one such example. When the visual system gives ambiguous cues concerning the state of motion, the correct information provided by proprioceptive, vestibular or "seat of the pants" cues is often misinterpreted and distorted. For example, while sitting in a train at a station with an adjacent train just outside the window, the passenger may experience the illusion that his train is moving forward, while in fact, that train is stationary and the adjacent train is moving backwards. The passenger's model of the world has "discounted" the proprioceptive evidence from the "seat of the pants" that informs him that no inertial forces are operating. Hence, the illusion is experienced.

In summary, when an abrupt auditory or tactile stimulus may intrude upon a background of ongoing visual activity, it will probably alert the operator and call attention to itself. However, if visual stimuli are appearing at the same frequency and providing information of the same general type or importance as auditory or tactile stimuli, then biases toward the visual source at the expense of the other two must be expected.

2.3. DIVIDED ATTENTION

Psychologists and system designers have used the single channel approach to attention quite successfully to model human sampling behavior and to develop guidelines for instrument layout, as described in the previous section. It is clear, however, that the single channel model is not a perfectly accurate description of human attention. We all know that we can drive, scan the highway for road signs, and listen to the radio at the same time (although not without some loss in performance of each). Many of us can study while listening to music, and I have often found that my thought processes ("thinking" as a task) are actually improved by listening to music, by walking, or even by running. These examples demonstrate cases of successful divided attention between two tasks. More compelling example of this success occur when our mind is forced to process information channels in parallel, even when it is to our disadvantage to do so. These may be described as failures of focused attention, or distraction and include, for example, the annoying inability to concentrate on a difficult textbook as the television blares in the next room.

Success at dividing and failure at focusing attention thereby represent different aspects of the same general issue - what it is that enables or prevents the human from being a parallel processor of information. The things that improve parallel processing will be a bonus when parallel processing is wanted - time-sharing and dual task performance - but will be a source of irritation and distraction when it is not. The things that prevent parallel processing will make it easier to focus attention without distraction but have a correspondingly disruptive effect when such processing is wanted. With these concepts in mind, four characteristics of the task and the individual shall be discussed that affect the efficiency of parallel processing, noting where appropriate how these variables influence both divided and focused attention tasks, and offering examples of their real world applications. These characteristics are: spatial separation, object integrality, structural task similarity (micro and macro) and, task resource demands.

2.3.1. Spatial Separation

If attention is conceived of as a search light, then using the analogy literally, it should be impossible to process channels of information in parallel that are physically far apart. In the visual world this is of course true, since the eye has a limited field of view within which it can process usable information (about 60°) and a more restricted visual angle, the region of foveal vision, for processing information with high acuity demands (about 2.5°).

For this reason, system designers are quite concerned about locating important instruments relatively close together in the center of the visual field; for example, using a computer terminal to centralize the presentation of critical safety parameter information in nuclear reactor control (Moray, 1981; Wood, Wise, & Hanes, 1981). The ability to parallel process channels that are physically close has also been a major rationale behind the development of the aircraft "head up display" or HUD. This is a means of projecting critical flight instruments directly on the wind screen so that the pilot may monitor these instruments without shifting gaze away from the important visual world outside the aircraft.

Careful laboratory research, however, has indicated that moving material close together in the visual field offers no guarantee of parallel processing, and may even impose costs of its own. For example, in an experiment by Neisser and Becklan (1975) subjects watched a display on which two video games were presented simultaneously, one superimposed over the other. One game showed distant figures tossing a ball, while the other game depicted two pairs of hands slapping each other. One game was designated as "relevant", and critical elements were to be monitored and detected. Neisser and Becklan found that while monitoring one game, subjects failed to see events on the other game, even when these were unusual or novel (e.g., the ball tossers pausing to shake hands). This experiment has direct relevance to the effectiveness of the HUD in aviation. While the intention of the HUD is to guarantee the pilot's parallel processing of the flight instruments and the outside world, Neisser and Becklan's findings suggest that parallel processing of information at the two distances may not necessarily take place. In fact, an experiment by Fischer, Haines, and Price (1980), revealed instances in which the pilot focused on the near instruments while ignoring potentially critical information outside the aircraft simulator. In their experiment, this ignored information was another aircraft sitting on the runway that was to be approached for landing. Obviously, the failure to detect this stimulus in a real-world landing situation could foretell a potential disaster.

Although close proximity will not guarantee parallel processing, sometimes it will present problems of focused attention as well. These problems may occur if for any reason the operator wishes to process only one channel, as others become temporarily irrelevant. If many adjacent channels are too closely spaced, problems related to both perceptual and response conflict arise. Perceptual conflict is a term describing the effects of "display clutter". It describes both the difficulty of locating an item of information that may be closely surrounded by similar items, and the competition for resources in the visual system that may automatically be demanded by the processing of adjacent items (Eriksen & Eriksen, 1974). Response conflict results from the automatic perceptual processing of physically close irrelevant stimuli. This occurs when those adjacent stimuli automatically elicit responses that are incompatible with the responses indicated by the relevant stimulus. For example, consider the response to the word "up" shown in Figure 2. On the left, this task should be easy because of the absence of adjacent stimuli. However, on the right, the presence of two incompatible arrow stimuli will be likely to slow down processing of the relevant word. Also, if the response is to be made under speed stress, this configuration will lead to an erroneous response to the central stimuli present in Figure 2, but consistent with the irrelevant stimulus (i.e., a downward response). The stimuli present in Figure 2 provide an example of the "Stroop Effect", which has been demonstrated in a wide variety of different forms, using both basic laboratory material (i.e., Clark & Brownell, 1975; Eriksen & Eriksen, 1974), as well as stimuli more typical of those encountered in the real world (e.g., Naval officers processing diagrams of relative ship headings; Laxar & Olsen, 1978).

In vision then, the role of close spatial separation appears to be a two-edged sword. Reducing separation can, but will not always,

Figure 2-2: A spatial version of the Stroop task. It is more difficult to read the word on the right than that on the left, because of the response conflict imposed by the flanking arrow stimuli.

encourage parallel processing, yet may lead to distraction as well. In the auditory modality, the unwanted edges of this sword become sharpened further. Here, because the ears are not limited by a restricted spatial area like the fovea, parallel processing of the raw perceptual information is possible no matter from where the two channels originate. On the other hand, if auditory stimulus channels are brought close together, the ease of parallel processing is not substantially increased, but the difficulties of focal attention are made considerably greater. Two auditory stimulus channels may be brought together either by reducing their spatial separation (e.g., monaural versus stereo listening), or by making some other defining characteristic more similar (for example, presenting two voices of the same, rather than different sex, a change which will increase the similarity of the fundamental pitch (Treisman, 1964)). Research has also found that the differences in semantic content or meaning between two messages play a role in one's ability to focus attention on one of them, as if "semantic space" is similar to physical space. It is more difficult to focus attention on one of two messages dealing with the same general topic, than on one message when the other deals with a different topic.

Collectively these results have an important implication for the design of systems in which an operator must deal with a number of different auditory messages, each emanating from different channels. This situation might describe for example, the pilot of an aircraft who has three distinct audio channels available: one for messages from the co-pilot, one for messages from air traffic control, and a third reserved for messages from other aircraft or for computer-synthesized voice messages. The pilot could not process the three channels in parallel since all would call for common semantic analysis, but could at least focus attention on one, without the unwanted intrusion of information from the others. The possibility of intrusions would be reduced then if the channels were made as physically different from each other as possible. For example, one message to the left ear might be in a high pitch, one to the right ear in a low pitch, and one addressed to the "middle of the head" (by equal intensity binaural presentation) could be spoken in a middle pitch.

2.3.2. Object Integrality

The discussion in the previous section indicated that moving stimuli close together in visual space was a necessary but not sufficient condition to foster parallel processing. There is some evidence, however, that when two or more stimulus channels are represented as attributes of a single object, this quality is a sufficient condition to guarantee that some parallel processing can take place (Treisman, 1969).

In arguing this position, Kahneman and Henik (1981) have introduced the concept of the "object file", a set of data pertaining to all aspects of an object (size, shape, color, orientation, etc.) that becomes available in parallel as the object is perceived. Attention focused on one attribute of the object will cooperatively facilitate the processing of its other attributes. An example of the success of the object in fostering parallel processing is an experiment carried out by Lappin (1967) who had subjects try to report three stimulus attributes presented in a very brief tachistoscopic exposure. When the three attributes were the size, shape, and color of a single geometric form, the report was far more accurate than when the size of one object the shape of a second and the color of a third were to be reported.

The concept of parallel processing of object dimensions has a number of potential implications for display design, although many of these have not yet been systematically explored. It has implications for static symbolic coding, in which two or more pieces of data are displayed concerning a single element, (for example, the size, political preference and mean age of city residents on a political map). The object concept suggests the value of using two or three dimensions of a single symbol to represent the critical dimensions of information.

There has been some recent interest in using the object display concept as means of providing integrated information to the monitors and supervisors of nuclear reactors, who currently face great banks of separate instruments, each providing its own channel of information as shown in Figure 1. One version of the object display is the iconic display of safety parameter information, examined by Woods, Wise, and Hanes (1981), see Figure 3. Here the value of each parameter is represented by the length of a spoke, forming the axes of a polygon.

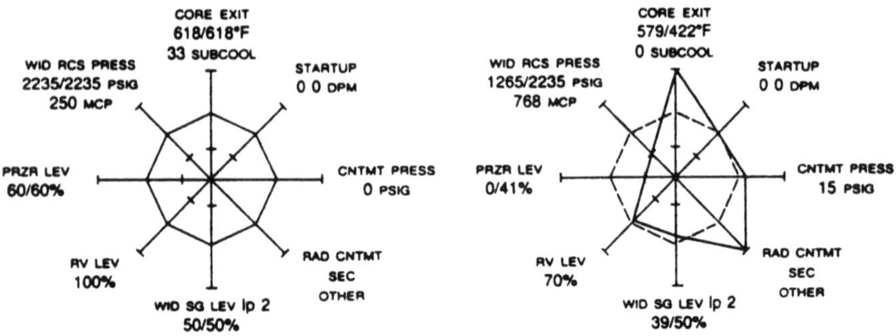

Figure 2-3: Object display of safety parameter information in nuclear power plants (after Woods, Wise, & Hanes, 1980).

While this configuration would clearly seem superior to the original multiple dial version, a controlled evaluation of the difference between the conventional separate instrument display and an integrated object display of safety parameter information carried out by Petersen, Banks, and Gertman (1981) indicated that the object display is not always best. In one experiment they found that the object display was superior when subjects (trained nuclear reactor operators) had to determine if any parameter was out of tolerance. However, in a second experiment, when subjects were required to designate how many and which of nine indicators was out of alignment, the original separate meter display proved to be superior. The differences between these two conditions is important. The first condition, requiring detection, allows the subjects to integrate all of the data into a pattern of normal or abnormal appearance, or to compare one indicator with all of the others. The second condition, asking for identification, requires that each out-of-tolerance indicator be treated separately, so that its identity can be labeled. Thus, it would seem that when integration of variables is possible, an object display will work best. When separation must be maintained, a separate display will work best (Casey & Wickens, 1986; Petersen, Banks, & Gertman, 1981; Wickens, Carswell, Fracker, Barnett, & Kramer, 1986).

Two further investigations also suggest that the benefits of an object integration may be better realized when the task requires a comparison between and across the various dimensions (i.e., integration). In the first of these, Carswell and Wickens (1987) asked subjects to diagnose failures in a simulated process control task. Simulated dynamic systems (in which two time-varying inputs drove a time-varying output) were monitored in order for the subject to detect failures or changes in the dynamic input-output equations. The displays of the three variables were either separated, in conventional bargraph form, or integrated in a configural object-like form (Figure 4). In the latter condition, the value of each variable was represented by the distance of the apex of a triangle from the triangle's center. Data were collected from systems using a variety of different dynamic equations and detection performance was found to be significantly better with the more integrated "closer" display configuration, than with the separated bargraphs. However, when the task was changed so that integration was no longer required and attention needed to be focused on each variable in isolation from the others, the bargraph displays were superior, thus replicating the results of Petersen, Banks, and Gertman (1981).

The second study involved a more comprehensive evaluation of four different multi-element displays. In this study, Jacob, Egeth, and Bevon

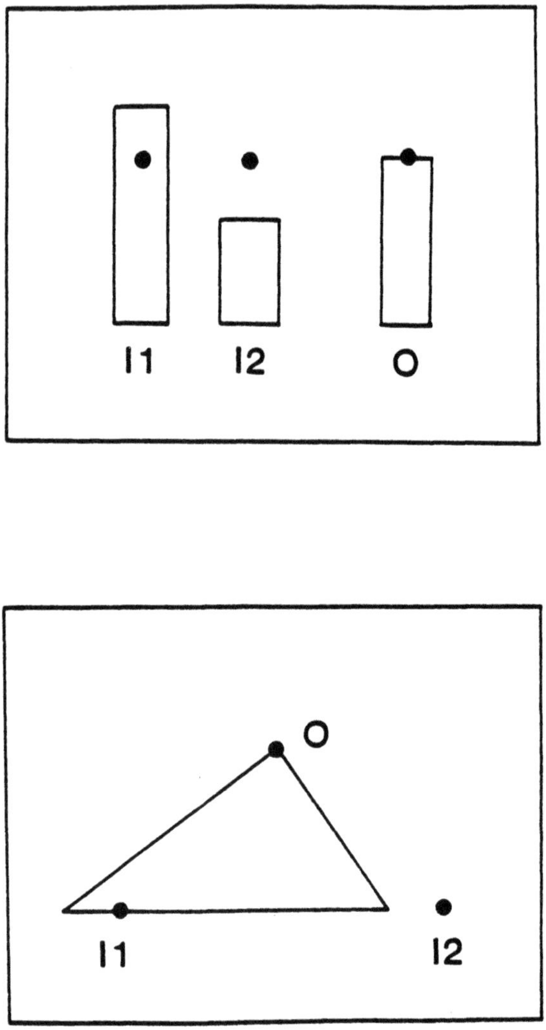

Figure 2-4: Separate bargraph display (top) and integrated object display (bottom), used by Carswell and Wickens (1987). In the triangle display at the bottom, the output (O) is equal to the height of the triangle, while the two inputs are equal to the distance of the two legs at the base, from the point below "O".

(1976) had subjects view a series of stimuli that varied along nine correlated dimensions. They were required to sort these stimuli rapidly into one of five categories, each defined by correlated levels along the nine dimensions. This categorization might correspond to five different failure "syndromes" in a complex process or five different diseases that are each characterized by a particular pattern of symptoms.

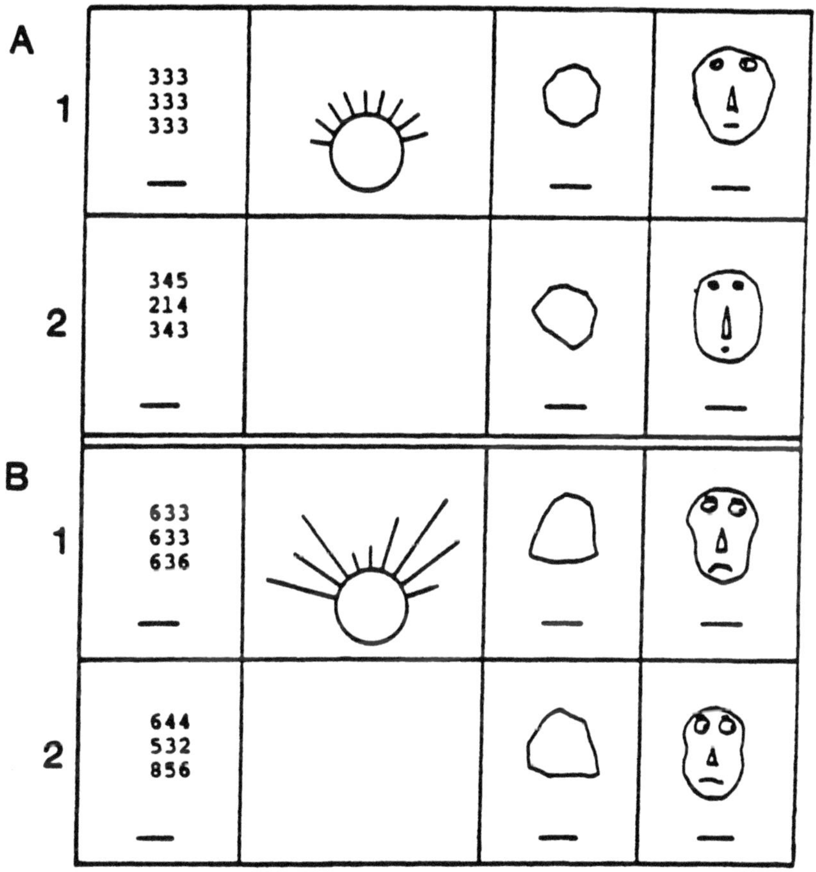

Figure 2-5: Four categories of stimuli used by Jacob, Egeth and Bevon (1976). Within each column (except column 2) are two examples each of two assigned categories (A & B).

The critical independent variable in Jacob et al.'s investigation was the manner in which the nine dimensions were displayed. These are shown across the columns in Figure 5 in an order that essentially defines their degree of "objectness". Number displays in the left column show each dimension separately in a verbal numerical format. The "glyph" display, in the next column, portrays each dimension as an analog quantity, but the quantities are not integrated into any holistic configuration. The polygon display in the next column, like the "iconic" display shown in Figure 3, and the triangle display in Figure 4, capitalizes on both an analog format and integrated contours. Finally, in the right column, the face display adds the dimension of familiarity. Each dimension is created by a feature of the face (e.g., the mouth) and levels on that dimension are coded by the shape of the feature. A smile and a frown, for example, would represent two levels of the "mouth feature." Within each column of Figure 5 are shown two representative categories A and B of the five to be sorted. Within each category (except the glyphs) are two examples of member 1 and 2. The similarity of values between the two members of the category reflects the correlation of the dimensions. The results of Jacob et al.'s study clearly indicated that more object-like representations to the right of Figure 5 assisted classification. This advantage was reflected in both speed and accuracy.

The advantage of the face display over numerical values has also been supported in experiments by Hahn, Morgan, and Lorensen (1983), and Moriarity (1979). Moriarity presented subjects (both business students and professional accountants) with information about a series of corporations on 13 different dimensions of financial interest, from the Dunn and Bradstreet business ratios. As in the experiment by Jacob et al., these dimensions are typically correlated. Subjects were asked to process this information to decide if a company was in a position to file for bankruptcy. Moriarity found that performance of both groups was faster and more accurate when the 13 dimensions described variations of features in a face display than when they were presented as numerical values. However, Casey and Wickens (1986) found that the face display was inferior to bargraph displays in a continuous process monitoring task.

2.3.3. Task Similarity

The previous section described the influence of similarity or proximity between input channels on the ability to process these channels in parallel. The data generally suggested that increased similarity of visual stimuli made parallel processing more likely, although in the case of auditory information, this was not necessarily the case. In the present section, we consider the broader effect of the similarity between tasks, rather than between input channels, and will see that the effects of similarity are reversed. With a few exceptions, two tasks that are more similar will be time-shared less effectively with less parallel processing possible. This conclusion seems to be consistent with our everyday experience. It is certainly harder to listen to two conversations at once, (two identical tasks) than it is to listen to one conversation while driving an automobile (two quite different tasks). To be of use in system design, however, the engineer would need to have available a more precise theory of exactly what it means for two tasks to be "more similar," so that more precise levels of time-sharing capability between two tasks can be predicted.

The efforts to develop such theories have suggested that the concept of "similarity" can be defined at both a micro and macro level; in terms of the specific processing mechanisms employed, and the more general human information processing resources demanded by the tasks.

2.3.4. Micro Level Similarity

At a micro level, the similarity between two tasks may be described by the nature of the material employed on the tasks, or the nature of the mental subroutines that are used. Here the principle of similarity holds up to some extent. For example, in memory research it is well documented that a series of items that sound alike, look alike, or have similar meanings will often be confused at recall (Conrad, 1964; Shulman, 1972). Research on interference in long-term memory has found that memory for one sort of material will be adversely affected if other similar material is experienced before or just after the first material is learned.

However, there are some circumstances in which aspects of similarity between tasks will increase, rather than decrease their time-sharing efficiency, and these should be noted as well. Increase in time-sharing efficiency usually occurs in situations in which a common or shared attribute of both tasks can be used cooperatively (Kramer, Wickens, & Donchin, 1983). For example, one such shareable attribute is timing. It is quite difficult to tap two rhythms that are not harmonically related but very easy to do so when they are related (Peters, 1977). Making the rhythm more similar will foster better time sharing. It is also hard to make two simultaneous movements that have different timing parameters, but easy if the timing is similar (Kelso, Southard, & Goodman, 1979). A factor closely related to timing is the similarity of the control dynamics between two axes of tracking. For example, dual axis tracking is more difficult when the operator must track a position control system with one hand and an acceleration system with the other, than when both hands control a position or both control an acceleration system (Chernikoff, Duey, & Taylor, 1960).

The reason why time-sharing is improved with greater similarity in these examples is because similarity allows certain aspects of the task to be cooperatively shared. This reduces what was a dual task requirement down to a situation in which less than two (but more than one) tasks are actually required.

2.3.5. Macro Level Similarity: Multiple Resource Theory

While the previous section discussed task competition, or sharing in terms of specific structures and processes, the similarity between tasks may also be defined in terms of the more global resources which they demand. Figure 6 provides representation of attention as a resource that can be distributed to different stages of processing depending upon task demands (Kahneman, 1973). Tasks that are more difficult demand more resources, allowing fewer to be available for other concurrently performed tasks. Several investigations, summarized by Wickens (1980; 1984) and Navon and Gopher (1979), have suggested that there are multiple processing resources. All tasks do not compete for the same "undifferentiated pool" of resources, just as all heating systems do not rely on the same source of fuel: some use coal, some gas, and some oil.

The implications of the multiple resources concept is that two tasks relying upon separate resources will be time-shared effectively, while two tasks with common resource demands will interfere with each other. This representation is shown in Figure 7. Here two resources are identified. Tasks A and B both compete for common resources and so will interfere with each other. Tasks A and C do not, and hence will be time-shared quite efficiently.

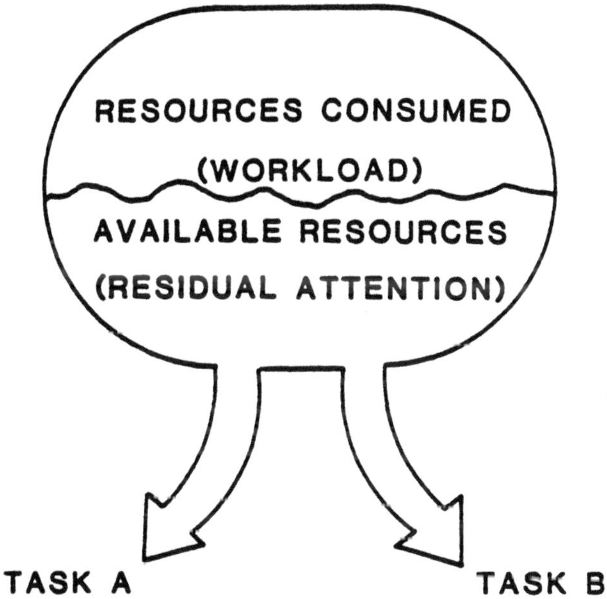

Figure 2-6: Representation of relation between resources demanded by a task (workload) and residual attention.

In fact, however, there appear to be more than two resources within the processing system. One proposal identifying the structural dimensions of the human processing system that defines these resources was presented by Wickens (1980, 1984), and is shown in Figure 8. According to this view, separate resources may be defined in terms of three dichotomous dimensions. These are: (1) Stages of information processing: perception and central processing (i.e., working memory)

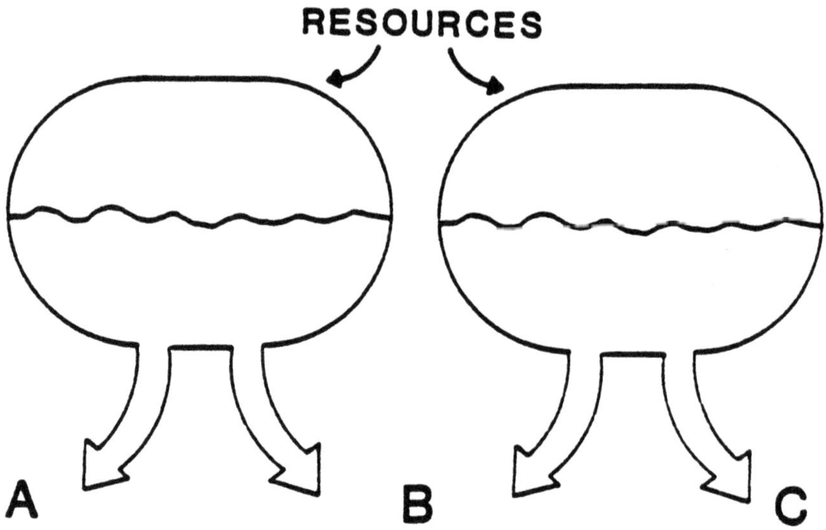

Figure 2-7: A simplified representation of multiple resources. Tasks A and C use separate resources and will not interfere. Difficulty increases in either one will have little effect on performance of the other: Task B, however; shares resource demands with both A and C and will compete with each.

operations use different resources from those underlying response processes. Notice here that, while perception and central processing are defined as separate stages, potentially occurring at different points in time as information is processed for a particular task, they nevertheless demand common resources to function. (2) Modalities of input: visual processing employs different resources from auditory processing. (3) Codes of information processing: handling spatial and analog information uses different resources from those involved with handling verbal and linguistic information. This third dichotomy is applicable to perception (speech and print vs. graphics and pictures), central processing (spatial working memory versus memory for linguistic information), and response processes (speech output versus spatially guided manual responses). The multiple resources model presented in Figure 8 does not predict that tasks demanding separate resources will be perfectly time-shared. It does predict that time-sharing efficiency will be improved (the decrement from single task to dual task performance will be reduced) to the extent that a pair of time-shared tasks use different levels along the three dichotomous dimensions shown in Figure 8. The following presents some examples of tests of the theory along the three dimensions, carried out in applied contexts.

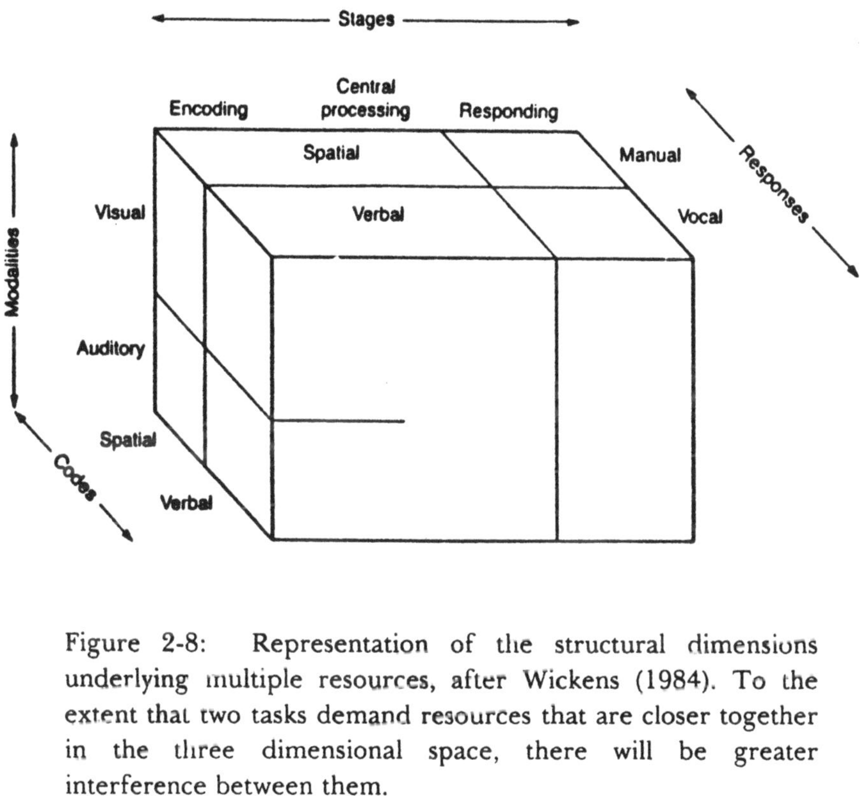

Figure 2-8: Representation of the structural dimensions underlying multiple resources, after Wickens (1984). To the extent that two tasks demand resources that are closer together in the three dimensional space, there will be greater interference between them.

Wickens, Sandry, and Vidulich (1983) examined differences both between the two input modalities (auditory versus visual) and the two output codes (speech versus manual control) as means for interfacing a verbal task in a flight simulator environment. Subjects had to listen to or see a series of navigational commands, and enter these using either voice or keyboard while flying the simulator. Both the auditory input and voice response were found to produce better time-sharing performance than their visual-manual counterparts. This finding is consistent with the multiple resource model, because the voice input-output system will use different resources from the visual-manual flight control task. General findings from this and other studies also suggest that the benefits to speech control in dual task settings are better realized in the concurrent task than in the speech-controlled task itself (Aretz, 1983; Wickens, Vidulich, & Sandry-Garza, 1984). When adhering to multiple resources theory as a guideline for implementing voice recognition and synthesis (VRAS) technology, however, it is important that the tasks that are interfaced are compatible with those modalities. We have argued elsewhere for example, (Wickens et al., 1983; 1984) that VRAS technology will be most beneficial in both single and dual task settings when verbal, rather than spatial tasks are interfaced. Use of VRAS technology with incompatible spatial tasks will often lead to more task interference.

The role of central processing codes (spatial and verbal) in influencing task interference is such that tasks demanding spatial central processing activity will interfere with concurrent tasks that are spatial and visual and will be time-shared fairly efficiently with tasks that are verbal, linguistic, and phonetic, while the converse relation will hold for verbal tasks. For example, Wickens and Weingartner (1985) found that the spatial task of process control monitoring could be carried out quite efficiently with a concurrent auditory linguistic task, but not with a concurrent task that required the use of spatial working memory. As a design guideline for predicting when tasks will and will not interfere, it can be assumed that almost any task that requires moving, positioning, or orienting objects in space, or performing other analog transformations, is predominantly spatial. Tasks for which words, language, or logical operations are natural mediators are verbal.

The dichotomy of processing resources by the stage of processing dimensions is used to account for the fact that tasks demanding perception and memory often compete with each other, but not with tasks whose primary demands are related to responding. As a consequence, for example, requiring the monitor of a complex visual display to respond to events on the display will not disrupt performance of the monitoring itself (particularly if speech responses are used), because this involves time-sharing of "early" (perceptual/cognitive) with "late" (response) processing demands. However, adding an additional perceptual channel to monitor, or imposing working memory demands, will be likely to increase the degrees of interference.

Multiple resource theory has sometimes been faulted for its unparsimonious nature, see the chapter by Kantowitz in this volume. The concern is well founded because, as Kantowitz notes, a good theory must be testable, and to be testable, it must be possible to be disconfirmed. The deficiency with multiple resource theory in this regard is that it is always possible to define another resource, if one's current version of the theory cannot account for the data at hand. An inexhaustible supply of "things" that can be labeled resources, is like an unlimited number of degrees of freedom in fitting a polynomial function to a set of data points. With this unlimited freedom, the multiple resources theory can never be disconfirmed. In response to this vital criticism, two answers can be given.

First, a testable theory of multiple resources must define only a limited number of resources, and possess independent means (i.e., other than using the basis of dual task performance data), of specifying the identity of the resources. In this way, one avoids, the circularity of saying "that which makes good time-sharing is separate resources, and the resources must be separate because the time-sharing was good". In the case of the theory proposed in Figure 8, all of the dichotomous dimensions possess anatomical substrates or independent single task behavioral data which can be used to define the resources demanded by a task a-priori. It makes a certain amount of logical sense that portions of the brain that are "farther apart" will be more likely to be able to function with greater independence. Thus, the eyes and ears are clearly distinct, as are the voice and the hands. With regard to processing stages, a clear discrimination can be made between motor and non-motor related regions of the brain. The dichotomy between verbal and spatial processes correlates (although not perfectly) with the function of the two cerebral hemispheres, and is also defined independently by individual differences in cognitive ability.

The second answer that can be given is simply that the alternative, more parsimonious concept - single resources or undifferentiated capacity theory - simply cannot account for the data on dual task performance. For example, within a single resource theory it is difficult to explain how an increase in the difficulty level of one task can leave performance of a concurrent task unaffected, a phenomenon described as difficulty insensitivity (Wickens, 1980; 1984). Yet, for example, there is ample documentation of instances in which a tracking task is time-shared with a verbal task involving discrete voice responses, the difficulty (resource demand) of either or these may be increased, and yet performance of the other does not deteriorate (e.g., Wickens et al., 1983). If all demanding tasks do compete for the same resources, then a tradeoff between performance and difficulty must be expected to occur. But if the visual, spatial, manual, and response-related resources required of tracking are separate from the auditory, verbal, vocal, and cognitive resources of the discrete task, then the absence of a tradeoff can be readily expected.

Thus, a single resource theory, like the single channel view of attention described earlier in the chapter, may be viewed as a useful simplifying heuristic for understanding certain phenomena such as automaticity and workload, see Kantowitz this volume. In fact, this conception will be discussed below. But as a useful tool for the system designer, to aid in predicting when tasks will and will not interfere, a model of resources must take into account their multiplicity. At the same time, the modeler must bear in mind the need to keep the number of these resources modest, and their identity specifiable in a way that resource demands of a particular task can be labeled by the human factors engineer without the need for calling in the trained experimental psychologist as a consultant.

2.4. RESOURCES, PRACTICE, AND DIFFICULTY

The previous section emphasized certain structural characteristics of tasks that will influence how easy it is to divide attention between them; that is, whether they are auditory or visual, cognitive or motor, verbal or spatial, and so forth. The concept of resources has been invoked to describe the entity within each structure that enables task performance. It should be apparent that time-sharing efficiency will be influenced not only by which resources are demanded by a task, but also by the extent of resource demands. Time-sharing will be more successful between two easy tasks than between two difficult ones. It is also a relatively straightforward matter to identify a series of task variables that will increase the demands of performing a task. Kantowitz in this volume has distinguished between two types of task demands: task difficulty; the resources required to achieve a criterion level of performance, and task complexity; the mental architecture necessary to perform the task. Table 1 lists a large number of task variables that can be expected to increase the difficulty or complexity of a task, followed by a letter indicating which factor will be increased.

Table 2-1: Variables Influencing Resource Demand

Stimulus Quality	D
Stimulus Discriminability	D
Stimulus Information Content	C
Stimulus-Response Compatibility	C
Working Memory Load	D
Irrelevant Perceptual Noise	C
Information Condensation	C
Visual Scanning	C
Absence of Preview	C
Control Stability (Tracking)	C, D
Generating Lead [a] (Tracking)	C, D
Reducing Interstimulus Interval	D
Increasing Response Precision	D
Practice	C, D

[a]Results from adding acceleration dynamics to a tracking task.

(C=Complexity, D=Difficulty).

There are probably more fundamental dimensions that can be added to this list, but the important point is that each of them will influence the resource demands of a task, and hence will be expected to affect time-sharing efficiency.

While it is easy to list variables, it is far more challenging to predict exactly how much of an influence on dual task performance a given manipulation will have. In an effort to do so, the concept of the performance-resource function has been invoked. This function shown in Figure 9 is a hypothetical construct that plots the quality of performance as a function of the processing resources invested in a task (Norman & Bobrow, 1975). The two functions in Figure 9 might represent the performance of two versions of a single task that differs in its level of difficulty (or the level of practice of an operator performing the task).

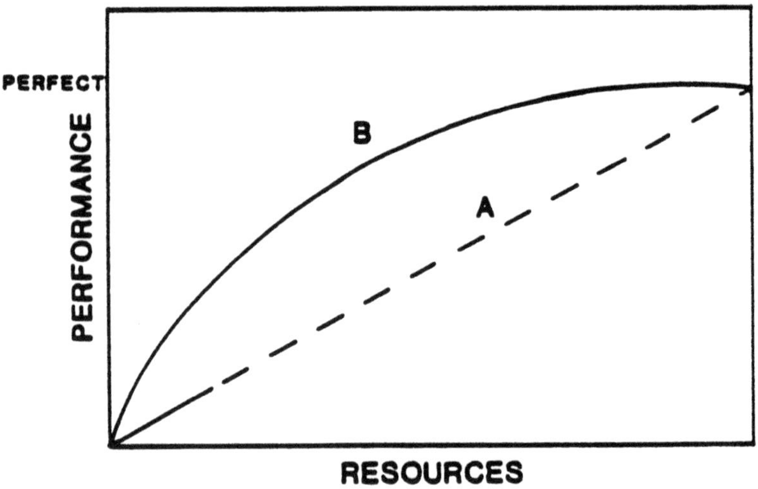

Figure 2-9: Two performance resource functions. Curves A and B both give rise to the same level of single task performance, when full resources are invested. But task B is more data limited, and will yield a smaller decrement when resources are withdrawn.

The performance resource function (PRF) shown by curve B may be divided into two regions. On the left, the quality of performance may be improved as more resources are invested. Here the task is said to be "resource limited". On the right, where the function is basically flat, performance cannot improve with more resource investment. The level of performance is thereby limited by the perceptual or memory data available, and no amount of effort can improve things further. Hence, the function is said to be data limited. An example of a data limited task is that of trying to understand a whispered conversation across the room. No amount of "straining one's ears" (resource investment) can improve one's level of comprehension. Correspondingly, trying to recall a phone number after it has been forgotten in working memory is also a data-limited task. Further effort (resource investment) cannot bring back the forgotten digits. Function A in Figure 9 is resource limited across its entire range.

Two tasks, such as A and B in Figure 9 may be performed perfectly under single task conditions, but show marked differences when some resources must be diverted from their single task level under dual task conditions. The resource concept allows this important distinction between primary task performance level, and resource investment to be made. It is a distinction that is quite critical to the concept of mental workload discussed by Kantowitz in this volume.

Unfortunately, the use of the PRF to predict precise levels of time-sharing efficiency is limited by three circumstances. (1) The amount of resources, the variable on the x-axis, cannot be measured directly, but can only be inferred by what a subject is instructed to do (e.g., "try to devote 50% of your effort to this task", or "divide resources 70% to one task and 30% to the other" (Navon & Gopher, 1979; Vidulich & Wickens, 1984)). (2) The level of performance will be defined on different scales, with different units for different tasks. (3) As noted in the previous section, the knowledge of how many resources a task may demand represents only half of the answer. Which common resources are demanded by a pair of tasks will also effect time-sharing, and as yet there are not good algorithms for combining these qualitative and quantitative factors in an overall measure of the absolute level of predicted performance. Despite these shortcomings, the resource concept is a viable one, because it allows one to go beneath the surface of primary task performance, and consider the level of resources demanded by that performance.

At this point, the reader may wonder how useful the concept of resources is to the system designer, given (1) the hypothetical state of resources (they are not a commodity that can be observed sloshing around in the head), and (2) the unparsimonious characteristic of their multiplicity. Clearly the multiple resource model has not yet achieved the satisfactory status of a quantitative scientific model.

In this regard, the statement made by Kantowitz in his chapter on workload is true. The best tool for application is a well validated and precise theory. Yet, the present state of imprecision of the multiple resource theory is not in itself sufficient reason to render it useless as a tool for design applications. In its current state, the theory remains qualitative, with its performance predictions generally allowed to be made only on an ordinal scale (i.e., task configuration A will be performed better than configuration B, or will demand fewer resources than configuration B). Yet, even enabling these imprecise statements to be made provides an advantage over a blind, or intuitive choice between systems, displays, or configurations. Furthermore, resource theory tells the system designer the importance of using workload measures to supplement primary task performance measurement when evaluating systems, a point that Kantowitz elaborates in his chapter.

2.5. ATTENTION AND HUMAN ERROR

The previous discussion painted the general picture that task performance will benefit to the extent that resource demands are light. While this statement holds particularly true in dual task situations, there is at least one interesting exception to this characteristic that occurs under single task conditions. This exception describes the sources of human error that occur in highly automated, highly skilled behavior. Normally, of course, error is viewed as resulting from low, rather than high levels of skill. The beginning flight student will be more likely than the skilled pilot to misperceive the altimeter reading, just as the medical student will be more likely to arrive at a faulty diagnosis, than the trained physician. In fact, much of the reason that performance of the novice is slower than that of the expert, is because the novice needs to maintain a continuous attention-demanding monitoring of the quality of performance, so that errors can be corrected if they occur.

Norman (1981a) has labeled these errors of misperception that are often made at low levels of skill as mistakes. In contrast with mistakes, however, a second class of errors called mode errors occur more frequently when performance is more automated. These errors result when a particular action that is highly appropriate in one mode of operation is performed in a different, inappropriate mode because the operator has not correctly perceived the appropriate context. A potentially disastrous example of a mode error would be a pilot who attempts to raise the landing gear while still on the runway, having wrongly perceived that the plane is airborne. Mode errors are of increasing concern in human-computer interaction when the operator occasionally must deal with keys that serve different functions depending upon the setting of another part of the system. Mode errors may occur in computer text editing, in which a command that is intended to delete a line of text may instead delete an entire data file because the command was executed while the system was in the wrong mode. Certain computer text-editing systems are particularly unforgiving in this respect because mode errors are quite likely to occur, and their consequences are drastic (Norman, 1981b).

Mode errors are a joint consequence of relatively automated performance--when the operator fails to be aware of which mode is in operation--and of improperly conceived system design in which such mode confusions can have major consequences. The reason, of course, that mode errors can occur is that a single action may be made in both appropriate and inappropriate circumstances.

Capture errors, or slips, like mode errors are also prominent when the operator is functioning at a highly automated level of behavior (Norman, 1981a; Reason, 1984), and often occur when three conditions are present in a sequence of action: (1) An inappropriate action is selected and performed because the tendencies for this action may already be high. (2) The stimuli that trigger the wrong action are quite similar to the stimuli that would trigger the appropriate action. (3) The operator is functioning in a fairly automatic mode of behavior in which selection of the action is not closely monitored. For example, we may reach out to pour the pitcher of orange juice on our pancakes; the two pitchers being similar, the pouring action of each identical, and our attention running elsewhere as we carry out this automated action.

Examples of slips caused by faulty instrument design are notorious when two controls are placed close together. When an operator who is functioning in an automated mode needs to adjust the less frequently used control, the fact that the two stimuli (the controls) are close together may cause the more frequent response to be made. We sometimes confuse windshield wipers and cars lights in this fashion in poorly designed dashboards. A more serious example is provided by the close proximity of the landing gear control and the flaps in some propeller-driven aircraft.

2.6. CHANGES IN ATTENTIONAL FUNCTION

The previous sections described a number of characteristics of attention, as measured in the average unstressed human. All of these characteristics clearly provide useful guidelines to the human factors engineer as to how to design displays and format tasks in a way that will ensure the best parallel processing and the fewest distractions. It is equally important to understand how the characteristics of attention may change or differ between people and within an individual (Wickens, Braune, Stokes, & Strayer, 1985). Otherwise, a system designed to be optimal for one occasion will not be for another. In this final section we will consider three sources of change or differences in attentional functioning--change due to stress, due to practice, and between individuals.

Sound knowledge of any of these changes can help the human factors engineer to accommodate them. For example, knowing that attention may "narrow" under stress will enable the designer of displays for a high performance aircraft to anticipate this narrowing and, as a result, position critical instruments that are needed under high stress conditions close together. Knowing that attention can be trained as a skill will allow the human factors specialist to develop generic "time-sharing trainers" from which performance may transfer to large numbers of different complex task environments. Finally, knowing that a divided attention or time-sharing ability may be an enduring and stable "trait", will allow the selector of personnel to improve the prediction of complex task performance by designing generic tests of time-sharing ability and using performances on these as a basis for selection (Wickens, Mountford, & Schriener, 1981). Each of these three changes will now be examined.

2.6.1. Stress And Attentional Narrowing

Over the past few decades a reasonable body of experimental evidence has been generated indicating that the "breadth" of the selective attention beam, and the number of sources of information it processes both decline under those conditions of stress that increase the operator's level of arousal (Easterbrook, 1959; Kahneman, 1973; Williams, 1982, 1986). Furthermore, the qualitative nature of this perceptual narrowing is such that cues of greater importance and greater centrality are processed, at the expense of peripheral cues of lesser importance (Bacon, 1974).

Throughout these studies, the concept of "level of arousal" is employed as a hypothetical construct or intervening variable that is associated with a number of experimental manipulations and measurable characteristics. These include stressors such as noise (which increases arousal) and sleep loss (which decreases it), incentives (which increase arousal), emotional states (which generally increase arousal), drugs such as caffeine and amphetamines (increase) or atropine (decrease), personality variables, and physiological responses that indicate arousal level such as the galvanic skin response or the electroencephalogram.

For example, Hockey (1970) compared the changes in performance on central tracking and a peripheral monitoring task produced by noise stress. He found that increasing noise facilitated tracking but disrupted peripheral monitoring performance. Calloway (1959; Calloway & Stone, 1960) concluded that amphetamines facilitate central monitoring performance more than peripheral monitoring, while atropine, in reducing arousal, exerted the opposite effect. Anderson and Revelle (1982) investigated the effects of arousal (as manipulated by personality type and caffeine intake) on proofreading errors. The authors reasoned that the detection of errors in the relations between words, such as grammatical or semantic mistakes, requires a broader beam of attention than the detection of errors within a word such as misspellings. Consistent with the narrowing hypothesis, they found that the manipulations that increased arousal and narrowed the beam, produced a greater disruption of the detection of grammatical errors than of the detection of spelling errors. Although the arousal increases lowered the accuracy of detecting both kinds of errors, the speed of performance in detecting spelling error was increased, while the speed for grammatical errors requiring a broader attentional beam, like the accuracy was decreased.

The phenomenon of perceptual narrowing with arousal increase has received only few experimental demonstrations in more applied multi-cue situations such as the aircraft cockpit, or the industrial monitoring station, although anecdotal reports indicate that it is present there as well (Sheridan, 1981). One more applied context is the stress imposed by underwater diving. In a simulation of this hazardous environment, Weltman, Smith, and Egstrom (1971) found that a diver's ability to detect peripheral stimuli was impaired. It is important to note, however, that the phenomenon represents a mixture of optimal and non-optimal behavior. Arousal produces a non-optimal response by limiting the breadth of attention. But subject to this limit, the human appears to respond optimally by focusing the restricted searchlight of attention on those environmental sources that are judged to be most important.

2.6.2. Attention Changes With Skill Development

Continued practice in a dual task environment will soon lead to improved dual task performance. Naturally, some component of this gain could result simply from an improvement in the single task component skills. As the two skills demand fewer resources and become more data-limited through practice (shifting from curve A to curve B in Figure 9), their combined resource demand will be diminished, and their dual task efficiency will improve accordingly. It should also be recalled that the attention demands of single task performance resulting from practice does not necessarily entail an increased level of single task performance, but only an increase in the data-limited region of a PRF that asymptotes at the same level (curve B in Figure 9).

This form of improvement does not actually result from acquisition of an attentional or time-sharing "skill", but from a reduced resource demand of single task performance. The improvement may be acquired, in short, from extensive single task performance. Such a mechanism explains the "automatic" processing of familiar perceptual stimuli such as letters (LaBerge, 1973), of consistently assigned targets (Schneider & Shiffrin, 1977; Schneider, 1985; see the chapter by Fisk, Ackerman, and Schneider in this book), of repeated sequences of stimuli (Bahrick, Noble, & Fitts, 1954; Bahrick & Shelly, 1958), or the automatic performance of habitual motor acts such as signing one's own name. When describing the response end of processing, this automaticity of performance is referred as the motor program (Summers, 1981), a concept that is discussed in more detail in the chapter by Jagacinski, Plamondon, and Miller in this volume.

Distinct from automation is the true skill in time-sharing that results explicitly and exclusively from multiple task practice. Three investigations here seem to have gone farthest in establishing the existence of time-sharing skills acquired through practice. Damos and Wickens (1980) required subjects to perform a pair of discrete digit processing tasks (one involving memory, the other classification) singly and in combination. Over practice, dual task performance continued to improve even as single task performance remained constant. A microscopic analysis of the dual task performance data suggested that the former improvement was not simply the result of single task automation. Further evidence that a time-sharing skill developed, and that this was a somewhat generic skill, not tied to a particular pair of tasks, was provided by observing that subjects who had undergone this dual task training with two discrete tasks showed positive transfer to a dual axis tracking task. It was as if the skill acquired in a dual task processing of discrete stimuli also assisted performance of the time-shared continuous tasks.

Reick, Ogden, and Anderson (1980) also found evidence for the acquisition of time-sharing skills. They administered to subjects differing amounts of single and dual task practice on a digit classification task paired with a discrete tracking task. The final level of dual task performance was related to the amount of dual task but not single task training, suggesting that only dual task practice was useful for developing the time-sharing skill. Equally relevant to the issue of the generality of the skill was their finding that subsequent dual task performance on a different dual task combination (continuous tracking and choice reaction time) benefited from the amount of dual task practice on the first pair.

Finally, two studies by Neisser and his collaborators (Neisser, Hirst, Spelke, Reaves, & Caharack, 1980; Spelke, Hirst, & Neisser, 1976) have also demonstrated the acquisition of a time-sharing skill. With extensive training their subjects were able to perfectly time-share the tasks of reading one message for comprehension and writing a separate message from oral dictation. In subsequent experiments, the investigators established that the learning was not simply achieved through the automation of single-task skills. In fact, their subjects showed a reasonably good level of understanding both of the meaning of the passage read, and of the relation between the orally dictated words. This level of comprehension would not have been obtained if either or both tasks had been processed at an "automated" level.

Exactly what is learned under periods of dual task practice is less clear. Experimental evidence however, indicates that at least one major component that is acquired is the optimal managing and allocating of attentional resources between relevant tasks at appropriate times. The concept of an internal model that guides attention switching and sampling was discussed earlier in the chapter. There is no doubt that the development of an internal model plays an important role in time-sharing skills whenever a variety of visual channels must be sampled, as in the aircraft cockpit. There is also evidence that the optimal sampling and allocation of resources plays a role in skill development even when stimulus channels are not spatially separated. The results of two more experiments lend support to this conclusion.

Gopher (1980) observed that subjects who were trained in a series of dual task trials which placed different degrees of emphasis on (and therefore changed resource allocation between) the two tasks, became more efficient time-sharers in general, than did a group trained only with equal priorities. The former group was also better able to adjust performance in response to changes in dual task difficulty.

Schneider and Fisk (1982) found that subjects could time-share an automated and a resource demanding letter detection task with perfect efficiency if they received training to allocate their attention away from the automatic task. In the absence of this training, subjects allocated resources in a non-optimal fashion by providing more resources to the automated task than it needed, at the expense of the resource-limited task.

It appears safe to conclude that the very efficient time-sharing performance of the expert results not only from the more automated performance of component tasks, but also from a true skill in time-sharing; knowing when to sample what off of the display, when to make which response, and how to better integrate the flow of information in the two tasks. To what extent the time-sharing skill acquired in one environment is generalizable to others is not well established. The data obtained by Damos and Wickens (1980) and by Reick, Ogden, and Anderson (1980), as noted, did indicate some transfer. In both experiments, however, the amount of transfer was not large, relative to the amount of skill learning demonstrated on the new tasks. It seems then that most time-sharing skills that are learned, are probably fairly specific to a give task combination, and are not of the generic kind.

2.6.3. Attention Differences As An Ability

Differences between individuals in time-sharing efficiency may, to a large extent, be related either to differences in automation of single task skills or to the practice-related acquisition of time-sharing skills described in the previous section. Thus, the observation made by Damos (1978) that flight instructors have greater reserve capacity than novices, as measured by performance of a secondary task, was undoubtedly related to the greater degree of automation of the flight task for the instructors.

On the other hand, there is some evidence for more stable differences between people in attentional characteristics that are not related simply to practice. For example, Gopher and his colleagues report that measures of flexibility of attention switching in a dichotic listening task predict complex task performance. The dichotic listening task employed by Gopher, in which two independent sources of auditory information are presented to the two ears, was separated into two phases. During the first phase, subjects heard a series of dichotically presented Hebrew verbs, interspersed by occasional digits. Their task was to report the digits when they occurred in one ear that was designated as relevant. Gopher and Kahneman found errors of both intrusion (reporting a target on the relevant ear when one had occurred on the other ear) and of omission.

When the task was administered to flight cadets (N=100 in Gopher & Kahneman, 1971; N=2000 in Gopher, 1982), performance on both of these measures was found to predict the level of proficiency obtained by the cadets in later flight training. During the second phase of the dichotic listening task, the relevant ear is suddenly switched and performance assesses how rapidly attention can switch accordingly. A further investigation by Kahneman, Ben-Ishai, and Lotan (1973) found the switching performance could predict the accident rate of bus drivers.

To date, the evidence for or against a general time-sharing ability has been mixed. On the negative side, Wickens, Mountford, and Schreiner (1981), examined differences in time-sharing ability of 40 subjects performing four different information processing tasks in nine different pairwise combinations. Although they observed substantial individual differences in the efficiency of time-sharing a given task pair, these differences did not correlate highly across the different task combinations. These findings, along with similar results obtained by Jennings and Chiles (1977), and Sverko (1977) suggest that accountability for differences between individuals in time-sharing ability is related either to differences in automation of component tasks, or to differences in the ability to time-share a specific task pair, or to differences in a component process (e.g., switching) that is relevant to some, but not all multiple task combinations (Braune & Wickens, 1986; Keele & Hawkins, 1983).

As an example of the second kind of ability, Damos, Smist, and

Bittner (1983) have identified a fairly stable difference in the ability with which people can process information in parallel between two discrete tasks. Some can easily process stimuli and responses of the two at the same time (although not perfectly), while others must deal with epochs of one task followed by epochs of the other. Once again, however, this dichotomy is probably somewhat task specific. It is not likely that it also accounts for differences in the ability of people to perform two tracking tasks at once or to the ability to read one message while shadowing another.

On the positive side, Ackerman, Schneider, and Wickens (1984) examined the analytical procedures employed in the studies discussed above and found a number of flaws which they felt might have prevented the investigators from finding a general factor of time-sharing. Re-analyzing the data from Wickens et al. (1981), using a different factor analytic technique that makes some prior assumptions about what the data should look like if a time-sharing factor does exist, they found that the data were consistent with the existence of a general time-sharing ability underlying the various dual-task combinations.

Fogarty and Stankov (1982) administered a series of 16 task configurations to 91 subjects. These included ten single tasks, and six dual task combinations. All tasks were auditory, involving different operations performed on either letter series or tone series, presented dichotically in the dual task configurations. In each of the dual task trials, one task was emphasized and the other deemphasized. Factor analysis revealed that variance in primary task performance was accounted for by single task variance, but that secondary task variance was not, thereby presumably reflecting a time-sharing factor. How general this factor is cannot be established from this data because all dual task configurations had in common the sharing of auditorily presented letter material with auditorily presented tonal material. It is possible that the dual task ability reflected in these data was similar to the attention-switching factor (again involving two auditory sources) measured by the dichotic listening task.

Finally, Braune and Wickens (1986) administered a series of dual axis tracking tasks to 40 subjects. The tasks were presented in different configurations designed systematically to assess different components of attention (e.g., switching, parallel processing, use of an internal model). Factor analysis revealed that these components did indeed account for individual differences in performance; furthermore they were able to predict, with some success, performance on a reaction time and working memory task combination, sharing no common single task elements. Hence, these data also supported the existence of a general time-sharing ability.

In summary, there appears to be little doubt that some stable individual differences in time-sharing exist. To what extent, however, these are task-specific abilities, rather than general abilities relating to resource allocation, resource management, and attention switching awaits further research.

2.7. CONCLUSION

The topic of human attention has been of interest to psychologists for over a century (Paulen, 1887). While human knowledge has clearly expanded since that time, the basic human constraints and limitations in processing information have remained unchanged. At the same time, the amount of information that humans are being asked to process, integrate, and understand as they interact with today's complex systems is increasing exponentially. Ironically, this fact remains true even as computer automation takes over many of the functions conventionally assigned to humans. This is because a human must now monitor and understand the automating computers and gracefully assume control if and when the automated system fails, as it often does (Rasmussen & Rouse, 1981). As system complexity grows, the number of things that must be monitored grows with it. The better understanding of human attention will not provide all of the answers necessary for coping with system complexity, but it will certainly offer a good start.

REFERENCES

Ackerman, P.L., Schneider, W., & Wickens, C.D. (1984). Deciding the existence of time-sharing ability: A combined methodological and theoretical approach. *Human Factors, 26*, 71-82.

Anderson, K.S., & Revelle, W. (1982). Impulsivity, caffeine, and proofreading: A test of the Easterbrook hypothesis. *Journal of Experimental Psychology: Human Perception and Performance, 8*, 614-624.

Aretz, A.J. (1983). A comparison of manual and vocal response modes for the control of aircraft systems. *Proceedings of the Human Factors Society, 27*, 97-101.

Bacon, S.J. (1974). Arousal and range of cue utilization. *Journal of Experimental Psychology, 102*, 81-87.

Bahrick, H.P., Noble, M., & Fitts, P.M. (1954). Extra task performance as a measure of learning a primary task. *Journal of Experimental Psychology, 48*, 298-302.

Bahrick, H.P., & Shelly, C. (1958). Time-sharing as an index of automation. *Journal of Experimental Psychology, 56*, 288-293.

Braune, R., & Wickens, C.D. (1986). Time-sharing revisited: Test of a componential model for the assessment of individual differences. *Ergonomics, 29*, 1399-1414.

Broadbent, D.E. (1958). *Perception and communication*. New York: Pergamon Press.

Broadbent, D.E. (1971). *Decision and stress*. New York: Academic Press.

Calloway, E. (1959). The influence of amobarbital and methamphetamine on the focus of attention. *Journal of Medical Science, 105*, 382-392.

Calloway, E., & Stone, G. (1960). Re-evaluating the focus of attention. In L. Uhr & S. Miller (Eds.), *Drugs and behavior*. New

York: Wiley.

Carswell, C.M., & Wickens, C.D. (1987). Information integration and the object display: An interaction of task demands and display integrality. *Ergonomics, 30*, 53-78.

Casey, E., & Wickens, C.D. (1986). *Visual display representation of multidimensional systems: The effect of information correlation and display integrality* (Tech. Rep. CPL-86-2). , Prepared for the U.S. Army Research Institute.

Chernikoff, R., Duey, J.W., & Taylor, F.V. (1960). Effects of various display-control configurations on tracking with identical and different coordinate dynamics. *Journal of Experimental Psychology, 60*, 318-322.

Clark, H.H., & Brownell, H.H. (1975). Judging up and down. *Journal of Experimental Psychology: Human Perception and Performance, 1*, 339-352.

Colovita, F.B. (1971). Human sensory dominance. *Perception & Psychophysics, 16*, 409-412.

Conrad, R. (1964). Acoustic comparison in immediate memory. *British Journal of Psychology, 55*, 75-84.

Damos, D. (1978). Residual attention as a predictor of pilot performance. *Human Factors, 20*, 435-440.

Damos, D., & Wickens, C.D. (1980). The acquisition and transfer of time-sharing skills. *Acta Psychologica, 6*, 569-577.

Damos, D., Smist, T., & Bittner, A.C. (1983). Individual differences in multiple task performance as a function of response strategies. *Human Factors, 25*, 215-226.

Easterbrook, J.A. (1959). The effect of emotion on cue utilization and the organization of behavior. *Psychological Review, 66*, 183-207.

Eriksen, B.A., & Eriksen, C.W. (1974). Effects of noise letters upon the identification of a target letter in a non-search task. *Perception & Psychophysics, 16*, 143-149.

Fischer, E., Haines, R., & Price, T. (1980). *Cognitive issues in head-up displays (NASA Technical Paper 1711)*. Washington, DC: NASA.

Fitts, P. M., Jones, R.E., & Milton, E. (1950). Eye movements of aircraft pilots during instrument landing approaches. *Aeronautical Engineering Review, 9*, 24-29.

Fogarty, G., & Stankov, L. (1982). Competing tasks as an index of intelligence. *Personality and Individual Differences, 3*, 407-422.

Gopher, D., & Kahneman, D. (1971). Individual differences in attention and the prediction of flight criteria. *Perceptual and Motor Skills, 33*, 1335-1342.

Gopher, D. (1980). On the training of time-sharing skills: An attention viewpoint. *Proceedings of the Human Factors Society, 24*, 259-263.

Gopher, D. (1982). A selective attention test as a prediction of success in flight training. *Human Factors, 24*, 173-184.

Hahn, G., Morgan, C., & Lorensen, W.E. (1983). . *Datamation,* , 23-29.

Harris, R.L., & Spady, A.A. (1985). Visual scanning behavior. *Proceedings of NAECON,* , 1032-1039. New York: IEEE.

Hockey, G.R. (1970). Signal probability and spatial location as possible bases for increased selectivity in noise. *Quarterly Journal of Experimental Psychology, 22*, 37-42.

Jacob, R.J.K., Egeth, H.E., & Bevon, W. (1976). The face as a data display. *Human Factors, 18*, 189-200.

Jennings, A.E., & Chiles, W.D. (1977). An investigation of time-sharing ability as a factor in complex performance. *Human Factors, 19*, 535-547.

Jordan, T.C. (1972). Characteristics of visual and proprioceptive response times in the learning of a motor skill. *Quarterly Journal of Experimental Psychology, 24*, 536-543.

Kahneman, D. (1973). *Attention and effort*. Englewood Cliffs, New Jersey: Prentice-Hall.

Kahneman, D., Ben-Ishai, R., & Lotan, M. (1973). Relation of a test of attention to road accidents. *Journal of Applied Psychology, 58*, 113-115.

Kahneman, D., & Henik, A. (1981). Perceptual organization and attention. In M. Kubovy & J.R. Pomerantz (Eds.), *Perceptual organization*. Hillsdale, New Jersey: Lawrence Erlbaum Associates. (pp. 181-209).

Keele, S.W., & Hawkins, H.L. (1983). Explorations of individual differences relevant to high level skill. *Journal of Motor Behavior, 14*, 3-32.

Kelso, J.A.S., Southard, D.L., & Goodman, D. (1979). On the coordination of two-handed movements. *Journal of Experimental Psychology: Human Perception and Performance, 2*, 229-259.

Klein, R.M., & Posner, M.I. (1974). Attention to visual and kinesthetic components of skills. *Brain Research, 71*, 401-411.

Kramer, A., Wickens, C.D., & Donchin, E. (1983). An analysis of the processing demands of a complex perceptual-motor task. *Human Factors, 25*, 597-621.

Lappin, J.S. (1967). Attention in the identification of stimuli in complex visual displays. *Journal of Experimental Psychology, 75*, 321-328.

Laxar, K., & Olsen, G.M. (1978). Human information processing in navigational displays. *Journal of Applied Psychology, 63*, 734-740.

LeBerge, D. (1973). Attention and the measurement of perceptual learning. *Memory & Cognition, 1*, 268-276.

Massaro, D.W., & Warner, D.S. (1977). Dividing attention between auditory and visual perception. *Perception & Psychophysics, 21*, 569-574.

McRuer, D.T., Jex, H.E., Clement, W.F., & Graham, W.F. (1967). *Development of a Systems Analysis Theory of Manual Control Displays*. Systems Technology, Inc., TR-163-1.

Moray, N., Fitter, M., Ostry, D., Favreau, D., & Nagy, V. (1976). Attention to pure tones. *Quarterly Journal of Experimental Psychology, 28*, 271-285.

Moray, N. (1978). Strategic control of information processing. In G. Underwood (Ed.), *Human information processing*. New York: Academic Press.

Moray, N. (1981). The role of attention in the detection of errors and the diagnosis of errors in man-machine systems. In J. Rasmussen & W. Rouse (Eds.), *Human detection and diagnosis of system failures*. New York: Plenum Press.

Moray, N. (1984). Attention to dynamic visual displays in man-machine systems. In R. Parasuraman & R. Davies (Eds.), *Varieties of attention*. San Diego: Academic Press.

Moray, N. (1986). Monitoring behavior and supervisory control. In K. Boff, L. Kaufmann, & J. Thomas (Eds.), *Handbook of human perception and performance*. New York: Wiley.

Moriarity, S. (1979). Communicating financial information through multidimensional graphics. *Journal of Accounting Research, 17*, 205-224.

Navon, D., & Gopher, D. (1979). On the economy of the human processing system. *Psychological Review, 86*, 254-284.

Neisser, U., & Becklan, R. (1975). Selective looking: Attention to visually specified events. *Cognitive Psychology, 7*,

480-494.

Neisser, U., Hirst, W., Spelke, E.S., Reaves, C.C., & Caharack, G. (1980). Dividing attention without alternation or automaticity. *Journal of Experimental Psychology: General, 109*, 98-117.

Norman, D.A., & Bobrow, D. (1975). On data-limited and resource-limited processing. *Journal of Cognitive Psychology, 7*, 44-60.

Norman, D.A. (1981). Categorization of action slips. *Psychological Review, 88*, 1-15. (a).

Norman, D.A. (1981). *Steps towards a cognitive engineering: Systems images, system friendliness, mental models*. Presented at Symposium. Models of Human Performance ONR Contractors Meeting. La Jolla, California. (b).

Parasuraman, R., & Davies, D.R. (1984). *Varieties of attention*. San Diego: Academic Press.

Paulen, M. (1887). La simultanéité des actes psychiques. *Revue Scientifique, 39*, 684-689.

Peters, M. (1977). Simultaneous performance of two motor activities: The factor of timing. *Neuropsychologica, 15*, 461-465.

Petersen, P.J., Banks, W.W., & Gertman, D.I. (1981). *Performance-based evaluation of graphic displays from Nuclear Power Plant Control Room*. Proceedings of the Conference on Human Factors in Computer Systems ACM.

Posner, M.I., Nissen, J.M., & Klein, R. (1976). Visual dominance: An information processing account of its origins and significance. *Psychological Review, 83*, 157-171.

Rasmussen, J., & Rouse, W.B. (1981). *Human detection and diagnosis of system failures*. New York: Plenum

Psychological Bulletin, 68, 417-419.

Warshaw, P.R. (1978). Application of selective attention theory to television advertising displays. *Journal of Applied Psychology, 63*, 336-372.

Wells, G.L., & Leippe, M.R. (1981). How do triers of fact infer the accuracy of eyewitness identification? *Journal of Applied Psychology, 67*, 682-687.

Weltman, G., Smith, J.E., & Egstrom, G.H. (1971). Perceptual narrowing during simulated pressure-chamber exposure. *Human Factors, 13*, 99-107.

Wickens, C.D. (1980). The structure of attentional resources. In R. Nickerson (Ed.), *Attention and performance VIII*. Hillsdale, New Jersey: Lawrence Erlbaum Associates. (pp. 239-257).

Wickens, C.D., Mountford, S.J., & Schreiner, W. (1981). Multiple resources, task hemispheric integrity, and individual differences in time-sharing. *Human Factors, 23*, 221-229.

Wickens, C.D., Sandry, D., & Vidulich, M. (1983). Compatibility and resource competition between modalities of input, central processing, and output: Testing a model of complex performance. *Human Factors, 25*, 227-248.

Wickens, C.D. (1984). *Engineering psychology and human performance*. Merrill Publishing Company: Columbus, Ohio.

Wickens, C.D., Vidulich, M., & Sandry-Garza, D. (1984). Principles of S-C-R compatibility with spatial and verbal tasks: The role of display-control location and voice-interactive display-control interfacing. *Human Factors, 26*, 533-543.

Wickens, C.D., & Weingartner, A. (1984). Process control monitoring: The effects of spatial and verbal ability and current task demand. In R. Eberts & C. Eberts (Eds.), *Trends in Ergonomics/Human Factors II*. Amsterdam: North-Holland.

Wickens, C.D., Braune, R., Stokes, A., & Strayer, D. (1985). *Individual differences and age-related changes: Refinement and elaboration of an information processing performance battery with aviation relevant task structures*. Engineering-Psychology Research Laboratory. University of Illinois. EPL-85-1.

Wickens, C.D., Carswell, M., Fracker, L., Barnett, B., & Kramer, A. (1986). Display/cognitive interface: The role of display proximity in information integration tasks relevant to command, control, and communications (C3) displays. In R. Ginnett (Ed.), *Proceedings, 10th Symposium on Psychology in the Department of Defense*. U.S.A.F. Academy.

Williams, L. (1982). Cognitive load and the functional visual field. *Human Factors, 24*, 683-692.

Williams, L. (1985). Tunnel vision induced by foveal load manipulation. *Human Factors, 27*, 221-227.

Woods, D.D., Wise, J.A., & Hanes, L.F. (1981). An evaluation of nuclear power plant safety parameter display systems. *Proceedings of Human Factors Society, 25*, 110-114.

HUMAN FACTORS PSYCHOLOGY, Peter A. Hancock (ed.)
©Elsevier Science Publishers B.V. (North-Holland), 1987

3. MENTAL WORKLOAD

Barry H. Kantowitz
Purdue University and BITS, Inc.
Department of Psychological Sciences
Purdue University
West Lafayette, Indiana 47907

INTRODUCTION

For the past few years, one particular question has occupied most of my waking hours. I have been plagued by this question. Indeed, no sooner do I emerge from an airplane on my way to attend a conference or to present a colloquium, than I know for certain that within the next hour I will again be asked the eternal question: "What is mental workload?" Experience has taught me that a blunt forthright answer, "I don't know!," is often misinterpreted as either a desire to hoard knowledge on my part, or even worse, as a lack of interest in the question. Elaboration, "Nobody else knows either," seldom satisfies the questioner. So, like P.B. Medawar's scientist who has been asked what he conceives the scientific method to be, I am compelled to adopt an expression that is at once solemn and shifty-eyed; solemn, because I know I should declare an opinion about so grave a question, yet shifty-eyed because I am not confident that a scientifically accurate and complete answer can be formulated at this time. This chapter represents my attempt at catharsis, wherein I purge myself of many of the good and bad thoughts I have entertained concerning mental workload as both a pragmatic and scientific concept. Furthermore, the next time I am asked "What is mental workload?", I can simply thrust a copy of this chapter into the beseeching hands of the questioner and beat a hasty retreat.

3.1. WHY PEOPLE ASK "WHAT IS MENTAL WORKLOAD"

A substantial part of my job as a professor requires me to answer odd, albeit well-intentioned, questions from people who vary considerably in their experimental psychology sophistication and background. When faced with a question I cannot answer, e.g., is the limited-capacity channel innate or learned?, it often helps to understand why the question was posed in the first place. Thus, if the questioner really wants to know if practice improves the ability of the limited-channel to transmit information through the human, I can still be helpful, even though I don't know whether the limited-capacity is innate or learned. Sometimes, the conceptual framework of the questioner is so far removed from that of the professional experimental psychologist that I cannot initially bridge the gap. Then I ask them a question to make the gap apparent to both of us, for example, "Where does the light go when it goes out?" Most of my students are sufficiently sophisticated about electricity and light bulbs so that the problem in this question is immediately apparent. I can then point out that from my point of view their question is much like asking where the light went.

"What is mental workload?" is a well-intentioned question. But anyone who expects a succinct answer has about the same odds of receiving it as finding out succinctly where the light went. People expect a succinct answer because of an implied analogy with physical workload. We can measure physical workload simply and accurately. Any traditional human factors textbook (e.g., McCormick & Sanders, 1982) will explain how energy costs for physical work can be found by measuring oxygen consumption and kilocalories. Physical work can be scaled according to these values. So we can define (and measure) very light work as that requiring an energy expenditure of less than 2.5 kcal/min and an approximate oxygen consumption under 0.5 liters/min. Unduly heavy work demands an energy expenditure over 12.5 kcal/min and consumes over 2.5 liters/min of oxygen. Workload falling between these two extremes can be scaled accordingly. This is a simple scale and accords well with our notions of work learned in studying elementary physics, where college freshmen roll balls up and down inclined planes.

Thus, when someone asks me how much mental workload is associated with some particular job demand, I understand what kind of an answer they would like. I am supposed to be able to say that I will attach electrodes to an airplane pilot coming in for a landing with one engine out and the nose wheel locked up, and that the meter now reads 12 more megafrizzes of mental workload compared to flying under autopilot control. Alas, this is not the state of the art today and I am not confident that this kind of simple solution is just around the corner. Where does the light go when it goes out?

The technical difficulties associated with measuring mental workload have not diminished researchers' efforts in this area. An entire NATO conference on mental workload (Moray, 1979) was devoted to defining and proposing measurement techniques. While no clear solution to the problem was developed, the proceedings accurately represent the enthusiasm of researchers about this topic.

At the risk of only moderate exaggeration, efforts in science can be divided into two approaches. These are illustrated by the tale of the drunkard looking for his car keys under a streetlamp. When asked where his keys were lost, he pointed to a dark alley. When asked why he was not looking for his keys in the alley, the drunkard replied that the light was better under the streetlamp. One view of science, held especially dear by basic researchers, is that science operates like the drunkard. It works best in the light even if the light is not yet strong enough to illuminate the practical problem at hand. Another view, held by applied researchers, is that important problems must be tackled now, even if the scientific tools are not all that one would like. Problems will not wait for basic researchers to develop more powerful lights, and even partial solutions are better than intuitive guesses. It is important to look in the dark alley right now, even if the only light available is from a weak candle.

This conflict between basic and applied approaches to human factors has been discussed by me at length (Kantowitz, 1982), so I will limit my remarks here to the specific issue of mental workload where both approaches have been used. My interpretation of the NATO conference places the work reported more toward the basic research end of the continuum, with emphasis upon the development of appropriate scientific tools and models. However, most human factors journal articles that attempt to measure workload associated with specific tasks are closer to the applied end, as are most technical reports when an answer must be found. My own resolution of this conflict is perhaps overly simple. Practitioners who need immediate design solutions should use the best available "quick and dirty" estimates of mental workload, probably subjective estimates, even thought these techniques have several flaws (to be discussed later in this chapter). Basic researchers who have the luxury of longer time frames should develop better tools as rapidly as possible. This chapter emphasizes the problems faced by the basic researcher. Readers who are more interested in estimating mental workload right now would be better served (in the short run) by consulting a chapter aimed at practitioners, especially those who need estimates of mental workload in the aviation industry (Kantowitz & Casper, 1986). A brief primer on mental workload for those with minimal training in human factors can be found in Kantowitz (1985b). The present chapter is aimed at professionals and students with knowledge of human factors and experimental psychology.

3.2. HEURISTICS FOR STUDYING MENTAL WORKLOAD

Mental workload is essentially a pragmatic topic. As technological improvements increase, people spend more and more of their working day handling information. Mental work is gradually replacing physical work as the dominant mode of American industry. The justification for current research efforts in mental workload, including substantial doses of public funds administered by several federal agencies, is the expectation that this research will prove useful in setting standards for appropriate levels of mental workload in industry. Employers and designers urgently need to know how to create safe and reasonable levels of mental workload on the job.

In order to accomplish this pragmatic goal, we need a pragmatic tool. <u>The best practical tool is a good theory of mental workload</u>. Theory fills in the gaps between empirical results. Human factors will never have enough data to cover all the problems that must be solved. There will never be a handbook that removes the need for thought. So long as practitioners feel that theory is irrelevant, they will continue to function based upon experience and intuition when data are not available. This is indeed unfortunate, since it implies that human factors will never develop scientifically. Any two researchers may have different theories. Science provides a systematic way of reconciling differences among theories. As C.S. Peirce noted in 1877, well before the advent of human factors as a discipline, the scientific method is the only method of fixing belief that is self-correcting. Practitioners who eschew theory must rely upon the methods of authority or the a priori method (e.g., the earth is flat because that seems reasonable) to fix their beliefs. These methods are inadequate because they offer no procedure for establishing the superiority of one belief over another. Moreover, as applied to human factors, they necessitate the continual reinvention of the wheel, as each new problem is solved without regard to its relationship to broader theoretical issues. Thus, I believe that theory is necessary before we can use the concept of mental workload to achieve practical goals.

As will be discussed later, mental workload is a multidimensional concept. Many theories of behavior could be relevant. Indeed, one might even argue that before we can truly explain mental workload we need a fairly complete theory of all human performance. Unfortunately, this will not be developed in time to help practitioners now extant. In order to develop a useful theoretical framework as rapidly as possible, I believe it prudent to limit our theorizing to one major topic, that of attention. Although mental workload involves more than attention, I feel that the body of knowledge regarding attention offers the best opportunity for helping the practitioner. Since another chapter in this book by Wickens reviews attention, I will proceed directly to the relationship between attention and mental workload, without summarizing research efforts in attention.

3.3. ATTENTION AND MENTAL WORKLOAD

Experimental psychology offers several models of attention, each with its unique set of advantages and disadvantages. Many of these models have been reviewed and compared (e.g., Kantowitz, 1974; 1985a; Lane, 1982; Schweickert & Boggs, 1984) yet it is not clear which model, if any, is best. Of course, the practitioner who merely wants a tool to help determine mental workload can satisfice, rather than optimize. For this purpose, any of several models of attention offer a sound starting place. Even the hoary limited-capacity channel model proposed by Broadbent (1958), although no longer state of the art, can help the practitioner if considered a first approximation. It is more important to start with some credible model of attention than to worry overly about which particular model to select.

But rather than selecting a particular model of attention that I find friendly (e.g., Kantowitz & Knight, 1976 hybrid model), I will first sketch some characteristics of classes of models of attention. Theory, to be useful, must be used. Otherwise, it is like having a vaccine without a hypodermic to administer it. Indeed, for me the potential for practical use of a theory is an important criterion that distinguishes among competing classes of theory. All other things being equal, I believe that researchers should try to formulate their models in ways that make their practical implications easier for the practitioner to discover. This does not mean that basic researchers must necessarily concern themselves with applied problems, although this might not hurt them (Kantowitz, 1982), only that some minimal sensitivity to the needs of users of the theories would be nice. Theoreticians who deny that there are users of theory are engaged in narcissistic activity that is not worthy of public support.

The typical model of attention represents the internal human information-processing system as a series of boxes (or stages) joined by an array of arrows indicating both information flow and control processes. A detailed analysis of these models (Kantowitz, 1985a) concluded that two general concepts are required in any viable model of attention: capacity and structure. Structure refers to the arrangement of internal processing stages postulated by the model. Typical structures are: serial processing, where stages are cascaded so that the output of one becomes the input of a succeeding stage; parallel processing, where an output can address several succeeding inputs; and hybrid processing, where elements can be processed either serially or in parallel. Capacity refers to the "price" a stage charges to perform its transformation of information. Models differ according to the number of possible pools or sources of capacity and how capacity expands or remains fixed within a given pool. Without a sufficient source of capacity, a stage will take longer to complete its transformation and may not be able to complete it at all.

A complete model of attention will specify (1) the number or configuration (structure) of postulated internal stages, (2) the capacity requirements of individual stages, and (3) the total availability of capacity (number of pools and location on the limited/partially-limited/unlimited continuum) and rules that govern allocation of capacity to individual stages. Few models are this explicit and complete. Merely listing the various possibilities for models would take a rather large piece of paper. At one extreme, we would have relatively uncomplicated models with few stages arranged serially, with each stage having its own independent and sufficient source of capacity. Changing only one axiom, by removing the independence of stages, creates another class of models. Removing the sufficiency restriction creates another class. At the other extreme are models with many pools of capacity, several parallel (or hybrid) processing paths, and complicated dependencies for allocation of capacity to individual stages or elements. Even without adding such technical refinements as whether transmission of information from one stage to the next is discrete (all-or-none) or incremental (continuous), how hazard functions can be used to represent the internal operation of stages, etc., there is room for years and years of frolicsome debate among theoreticians. All of this complexity can lead the practitioner, who "merely" wants to measure mental workload in some specific setting, to throw up his or her hands in utter frustration, abandoning all hope of theoretical assistance.

But this relinquishment of theory would be premature. This chapter will contain an example of theory improving our ability to measure the mental workload of airplane pilots. So, with this carrot left dangling, we continue our foray into the turgid depths of attention theory in search of topics that are relevant to mental workload.

3.3.1. Spare Capacity

The astute reader will have noticed that I have neglected to offer a technical definition of "capacity". This omission is intentional (but see Kantowitz, 1985a for a detailed discussion). For present purposes it is sufficient to regard capacity as an analogy: let it be the information transmission rate (in bits/sec) of the entire processing system. If this rate exceeds that demanded by some particular task, the system is said to have "spare capacity". So long as our concern is with the entire system, this definition presents no problem. However, when we descend into the crucible of stages and sub-systems there is no simple way to meter internal transmission rates, so this definition of capacity loses some of its appeal (see Kantowitz, 1985a). Nevertheless, we shall slog on, bravely ignoring this difficulty, since our goal is not to formulate a model of attention but only to apply existing concepts.

Most human factors experts would agree that spare capacity and mental workload are inversely related. Zero spare capacity implies high mental workload and negative spare capacity implies even greater mental workload. However, while this is true at the level of the entire system (e.g., the human operator), some care is required when this relationship between spare capacity and mental workload is invoked at the sub-system level. In particular, one must remember that capacity and structure are not independent concepts, but must be considered jointly. Certain classes of attention models create the functional equivalent of spare capacity by modifications of structure rather than by assumptions about the flow of capacity. These models may be less suitable for driving predictions about mental workload. For example, consider a model that has an unlimited number of parallel processing paths. As external load is increased, the system switches on more parallel processors as required. Even though each parallel path can process but a finite amount of transmitted information per unit of time, the entire system has no such limit. (To make this a realistic model of the human, some finite limit on the number of parallel processing paths need be invoked, but that does not change the nature of the current argument).

We could try to equate mental workload with the number of parallel processors that are operating, but this neglects the capacity requirements of individual parallel processors. Thus, having five parallel processors in operation, each carrying a load of four bits/sec, would yield greater mental workload than only two processors each carrying eleven bits/sec. I would be reluctant to advocate this type of model until simpler classes of models (e.g., one pool of capacity with a fixed limit) proved to be inadequate. While structural and capacity assumptions can be manipulated to mimic various classes of attention models (Townsend & Ashby, 1984), this flexibility may not aid our goal of determining mental workload. It is more parsimonious to identify mental workload with a single construct, that of capacity, than with the more general concept of attention which refers to some combination of capacity, structure, and allocation policy. This simpler goal may not be fully realizable, but it seems better to me to start small. This implies that our search for relevant models of attention should at first be limited to the classes of models that postulate a single pool of capacity rather than multiple-resource models. Indeed, the nonparsimonious nature of multiple-resource models creates difficulties for the study of mental workload. So this chapter will focus only upon single-pool attention models.

The grandfather single-pool model of attention is, of course, the limited-channel model popularized by Broadbent (1958). It is generally agreed that although its historical importance cannot be overestimated, this model is no longer in the forefront (Kantowitz, 1985a; Kinsbourne, 1981; Sanders, 1979). The data generated by researchers inspired by the model have surpassed the model's explanatory powers. This is not an unhappy fate for a model, especially one that has been so influential and the basis for so much research. Nevertheless, our search must extend beyond this important model.

Sanders (1979) discussed three types of internal processor (called modestly A, B, and C processors) relevant to mental workload. Processor A was a single-pool processor where capacity was allocated as needed. Sanders rejected this processor as a correct model of human attention, primarily because it required unconstrained allocation of capacity. I agree with his analysis. He then proposed processor B, which operated in the time domain rather than the capacity domain.

Processor B has no spare capacity because any extra capacity would be applied immediately to reduce processing time. While Sanders found some data to be consistent with Processor B, he reserved his greatest approbation for Processor C, a multiple-pool model of attention. I eliminate Processor C on the grounds of parsimony: we should not entertain multiple-pool model until single-pool models have been clearly rejected. Processor B has some good features but eliminates the rate aspect of capacity that I consider to be crucial. Capacity is best considered a limitation upon rate of information processing (e.g., bits/sec) rather than upon amount of information transmitted. To cite an example used by Broadbent (1965), who makes this key point quite forcefully, it is more useful to characterize a roadway by the number of cars that can pass a given point in an hour, than by the number of cars that the highway can hold at any given time. Highway capacity expressed as a rate (cars/hour) is a far more useful indicant than the static measure of the highway's capacity as a parking lot. Processor B, derived from single-channel (rather than limited-channel) theory, only cares how much time is required for some internal operation. Furthermore, substituting this static measure for a rate measure also eliminates spare capacity as a viable construct. I cannot accept a model that eradicates so useful a concept as spare capacity.

Yet Sanders is not alone in his rejection of Processor A. Wickens (1979) discussed two phenomena that are quite awkward for single-pool models with unconstrained capacity allocation. These are "difficulty insensitivity", where manipulation of primary task difficulty alters performance on some concurrent secondary task but not on others, and "structural change", where dual-task interference is dramatically altered by changing the modality (input or output) of one of the simultaneous tasks. A brief quotation from Wickens (1979, p. 80) illustrates his point:

> "These two phenomena, difficulty insensitivity and structural change, seemingly require that the concept of processing resources be elaborated to account for what will be referred to as structure specific resources-a concept closely associated to the ideas of Kantowitz and Knight (1976). Under this conception, resources that are associated with a particular processing structure or mode of processing (e.g., perceptual,

response, memory, analog, semantic), are available in restricted quantity to tasks sharing those particular structures, but not (or to a lesser extent) to tasks requiring exclusively different structures."

Wickens later elaborated this idea to formulate a multiple-resource model of attention with several distinct pools of capacity (e.g., Wickens, 1984).

It would seem that once having rejected processors of class A, there is little choice but to postulate multiple pools of capacity, as has been done by Sanders, by Wickens, and by other researchers (e.g., Navon & Gopher, 1979). However, as I have noted elsewhere (Kantowitz, 1985a), I am quite reluctant to take this step. Multiple-resource models are too powerful and too difficult to reject. I do not trust a model that cannot be falsified. If data cannot be fitted with three or perhaps four pools, all that need to be done is to add yet another pool. This reminds me of MacDougall's hormic psychology with its endless proliferation of instincts. How many distinct pools of capacity will fit on the head of a pin? Where does the light go when it goes out?

3.3.2. Kantowitz-Knight Hybrid Model

Is it possible to have a single pool model that avoids the terrors of unconstrained capacity allocation? Briefly, the answer is yes. Such a model was proposed by me and James Knight (Figure 1) to account for the difficulties with unconstrained capacity allocation quoted above. At the time I thought that going beyond the traditional serial model to a hybrid structure was a daring step that greatly increased the complexity of the model. Indeed, Knight and I never published a series of computer simulations of the model, because we were embarrassed that the model was so hard to reject. With the clear view of hindsight, I now realize that this was a very modest step compared to inventing as many sources of capacity as one would like.

The Kantowitz-Knight (1976) hybrid model depended upon a Static Capacity Allocator (SCA) to divide its single pool of capacity between perceptual and response stages of information processing. Once the

SCA setting was established, front-end processing stages could not dynamically exchange capacity. In effect, the single pool of capacity was partitioned into two sub-pools that acted separately. However, this is not quite the same as having two separate and independent pools of capacity. The total capacity of both sub-pools was limited to that of the single pool. Any sub-pool could gain capacity only at the expense of other sub-pools. This mechanism allowed the model some of the flexibility of multiple-resource models without the serious disadvantage of allowing capacity to increase unbounded as the number of pools increased. Therefore, it is meaningful within the context of this hybrid model to speak of spare capacity both in a molar sense relating to the total pool (i.e., the sum of the sub-pools), and in a smaller partitioning relating to the spare capacity of any given sub-pool.

Perhaps a fanciful analogy will make this important point more salient. In the multiple-resource class of models, some processing stages are fueled by gasoline, some by coal, some by oil, some by nuclear power, etc. Interference can only occur between stages that use the same type of fuel (or resource).

In the Kantowitz-Knight hybrid model, all stages are fueled by solar power. Interference between stages is governed by allocation policies that already have been negotiated before the start of a trial. Due to allocation policy some sets of stages may have spare capacity while others are impoverished and lack sufficient capacity. This feature mimics some of the results of multiple-resource models where resources are not substitutable (or only substitutable at great marginal cost). So long as the total capacity required by all the sub-systems is less than that provided by the single pool of capacity, it is possible to develop an optimal allocation policy so that all stages are satisfied. However, once total required capacity exceeds the single pool, some or all sub-systems will be starved and operate at less than maximum efficiency. This allows the model to handle the various patterns of interaction and additivity obtained in the experiments reported by Kantowitz and Knight (1976).

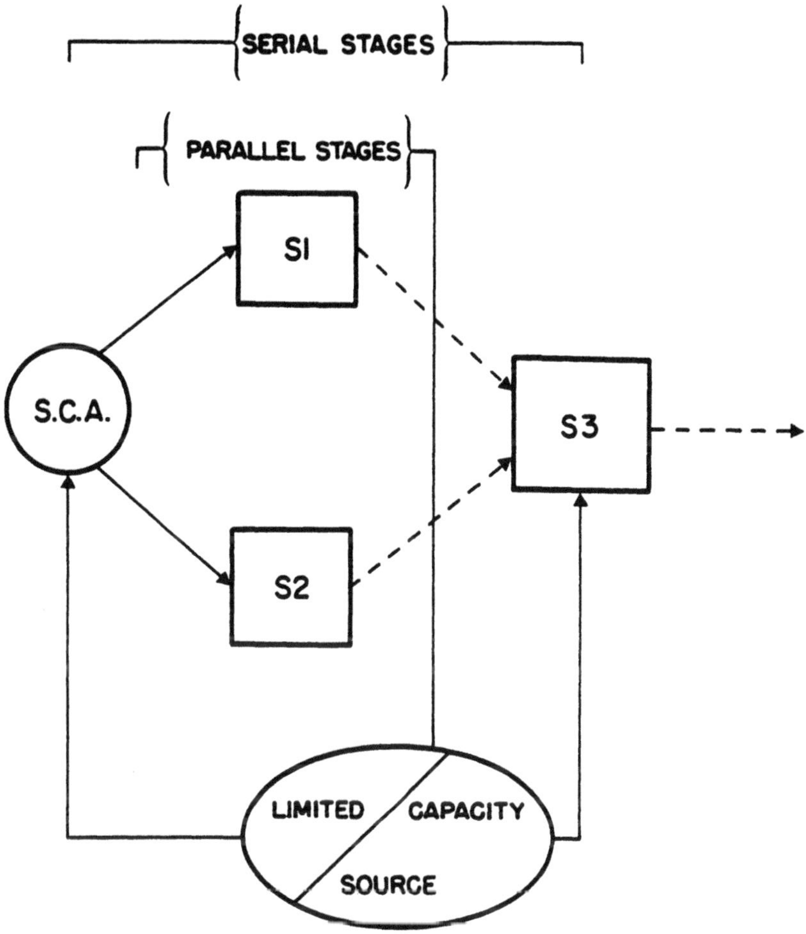

Figure 3-1: The Kantowitz-Knight hybrid model of attention and time sharing. The parallel stages represent early (perceptual) processing. They are controlled by a static capacity between them (from Kantowitz & Knight, 1976; Copyright North Holland. Reprinted with permission).

3.3.3. Summary

This section has reviewed the concept of spare capacity. Although I believe that spare capacity is an important component of mental workload, spare capacity can only be evaluated within the context of some particular model of attention. It is meaningless to talk of spare capacity in isolation as a free-form concept. Some models of attention, such as Sanders' type B processor, are inconsistent with a concept of spare capacity. Other models, such as multiple-resource models with several pools of capacity, weaken the concept of spare capacity: if additional pools can be added as needed it seems less important to distinguish between spare capacity and "ordinary" capacity.

Models are created to solve problems. (See Kantowitz & Roediger, 1984, Chapter 1 for an explanation of how models compare to other kinds of theories in experimental psychology). It is futile to state that some particular model is best without stating the goals to be achieved by the model. The practitioner interested in measuring mental workload must select a model of attention that will promote this goal. I have argued that those models of attention that have only a single pool of capacity are most likely to be useful because they make the most of the construct of spare capacity. The Kantowitz-Knight hybrid model is used as an exemplar of this class of model, not as a panacea for the attentional ills of mental workload. I have not claimed here that my hybrid model is the best model of attention (but see Kantowitz, 1985a), because my goal has been to find models that offer congenial constructs for the study of mental workload. I do believe that someone interested in mental workload will find the Kantowitz-Knight hybrid model a better starting place than the remaining attention models discussed above.

3.4. TASK DIFFICULTY VERSUS TASK COMPLEXITY

It is common for researchers to use the term "task difficulty" to describe any modification of an experimental task that either increases the time required to perform the task, or decreases the accuracy or efficiency of task performance. This imprecision has led to debates about the proper definition of task difficulty, and its relationship to models of attention and the theoretical construct of capacity (e.g., Lane 1977; Kantowitz & Knight, 1978a, b). Arguments about definitions cannot be resolved empirically. Instead, one definition will prevail over another because it fits better within a more useful theoretical framework. Since such a framework takes years to be developed and evaluated, hindsight is often the best way to resolve conflicts between definitions.

I have argued, and still maintain, that it is useful to distinguish between task complexity and task difficulty (Kantowitz & Knight, 1976; 1978a, b). Task difficulty must be inferred from performance; it is not an intrinsic property of a task existing in isolation. So a necessary condition for evaluating task difficulty requires setting performance criteria. The Potency Principle (Kantowitz & Knight, 1978b) relates performance and capacity: performance increases monotonically with capacity investment. Task difficulty, termed task performance difficulty by Kantowitz and Knight (1978b), combines the Potency Principle and a performance criterion. Task (performance) difficulty increases monotonically with the capacity investment needed to obtain a specified level of performance. This implies that two tasks are equal in difficulty if they require equal amounts of processing capacity. Note that this definition is most reasonable for a single-pool model of attention. Multiple-resource models cannot use such a definition without equating different pools of capacity, which is awkward. Therefore, multiple-resource theorists tend to use a definition of task difficulty based upon marginal efficiencies, but this creates new problems that are at least as awkward as equating different pools (see Navon, 1984; Kantowitz, 1985a).

Task complexity refers to the (hypothetical) mental system architecture used to perform a task. As additional internal stages or processes are required, task complexity increases. Some examples will help to distinguish between task difficulty and complexity. Increasing the distance between two targets in a Fitts' law tapping task is a manipulation of task difficulty for two reasons. First, there is no a priori basis for suspecting that this change alters system architecture (provided, of course, that movement amplitudes fall within the bounds of Fitts' law). Second, there is empirical evidence that such a change increases the attentional requirements of the task (e.g., Kantowitz & Knight, 1978a), and so calls for more capacity. However, requiring a person to subtract 9 from a visually-presented two-digit number represents a manipulation of task complexity compared to naming the number. It is likely that extra processes or stages are required to perform the mental subtraction. (However, proving this empirically is difficult, because our methods for discovering the number of internal processing stages necessarily make strong assumptions about capacity. For example, the additive factors method popularized by Sternberg assumed that each stage has its own sufficient source of capacity. See Kantowitz, 1985a, for more discussion of the relationship between capacity and system architecture). While one might be tempted to informally state that subtracting 9 from a number is more difficult than merely naming the number, I eschew this imprecise use of language. Instead, I would state that subtracting 9 is more complex than only naming.

Is mental workload affected more by task difficulty or by task complexity? The short answer is, "I don't know". The long answer is that this depends upon how much spare capacity remains after the task manipulation. Increasing task complexity by adding an internal stage or process that requires no spare capacity (e.g., is "automatic" if you'll pardon another expression that increases our vocabulary rather than our understanding (Kantowitz, 1985a)), should not increase mental workload. Increasing task difficulty always increases mental workload since by definition, an increase of task difficulty demands additional capacity. Increasing task complexity may or may not increase mental workload. The logical possibility that manipulations of task difficulty versus task complexity may have different effects upon mental workload, is yet another justification for distinguishing between task difficulty and task complexity.

3.4.1. Summary

I have argued that pursuit of mental workload is best accomplished from a theoretical perspective. In particular, I have suggested that the theoretical concept of spare capacity is a keystone for mental workload. Spare capacity is not equally represented in all models of attention, so single-pool models appear to be better suited as frameworks for the study of mental workload. Another reason for preferring single-pool models is that they are most consonant with the distinction between task difficulty and task complexity. Of course, mental workload implies more than attention. The following section discusses other aspects of mental workload.

3.5. MULTI-DIMENSIONAL ASPECTS OF MENTAL WORKLOAD

In any operational setting, mental workload is but one component of a broader construct called "workload". There have been many definitions of workload which include not only mental workload, but physical effort, emotional reactions to task and environmental demands, etc. This chapter is concerned with the mental operations used to perform a task, and not with the broader issues associated with workload in general. Nevertheless, some of the aspects of workload may be relevant to mental workload as well. Therefore, I have included suggestions in this section that have implications for mental workload, even though these suggestions may have been originally formulated in the more general context of workload.

3.5.1. Feelings

Many researchers believe, either explicitly or implicitly, that mental workload is a private mental experience or feeling. It is difficult to find published studies that do not ask for subjective ratings of mental workload or workload in general. The following section, on measurement of mental workload, will discuss ratings methods in detail. For now, merely note that the prevalence of this kind of data in workload experiments indicates general agreement that workload either is a subjective experience or at least is well-measured by subjective ratings. For example, Hart (1982) defined workload as follows (p. 456):

> "Workload is a subjective experience caused by external and internal factors such as motivation, ability, expectations, training, timing, stress, fatigue, and circumstances in addition to the number, type, and difficulty of tasks performed, effort expended, and success in meeting requirements."

As was noted before, definitions cannot be tested empirically. Certainly, the above definition covers most, if not all, of the relevant aspects of workload. Each of its components bears a reasonable a priori link to the general concept of workload. I doubt that many researchers, especially those concerned with practical solutions, would take issue with this definition; certainly not I.

However, this conception of workload as a subjective experience implies that some kinds of measurement techniques will more directly, and therefore probably more effectively, tap workload. In particular, subjective ratings seem especially useful. Moray (1982) has argued that even more research is needed on the subjective aspects of mental workload. Yet, a brass instruments type of experimental psychologist feels a certain discomfort when subjective experience is pursued so relentlessly. Psychology does not have an outstanding track record in the effective study of subjective experience. The method of introspection has been soundly rejected, despite recent attempts to bring it back (in the guise of protocol analysis) in artificial intelligence studies aimed at simulating human problem solving. While the pendulum has returned from the extreme behaviorism of Watson that rejected all subjective experience as fit for psychological study, I must confess that I am far more comfortable with the study of objective behavior than that of subjective experience. While I do not deny that people are capable of describing their subjective experiences of what they call mental workload, I do not share Moray's enthusiasm that this is the most useful and effective path for researchers to take. Instead, I believe that existing techniques for mapping internal information flow (a topic that Watson would probably have rejected along with introspection) are far superior to existing techniques for measuring subjective experience. At Purdue University, we offer several required graduate psychology courses pertaining to mapping internal information pathways, but only educate students about introspection as part of a history of psychology course. While I do not reject the study of subjective experience as did Watson, I believe that our present tools are better suited to studying objective behavior. Thus, I favor a definition of mental workload as an intervening variable, similar to attention, that modulates or indexes the tuning between the demands of the environment and the capabilities of the organism. This definition is related to models of attention and does not focus upon subjective experience.

3.5.2. Effort

Effort may be a concept that bridges the gap between subjective experience and objective behavior. As a subjective experience, people can readily report how much effort they are exerting. Also, experimenters have devised many manipulations that control objective indices of effort. For example, early work in social psychology on social loafing by a German psychologist named Ringelmann (reported in Moede, 1927) had people pull on ropes. The pressure they exerted was a clear, objective measure of effort. Ringelmann discovered that when groups of people pulled ropes, the total pressure was less than the sum of the individual pressures measured when people pulled by themselves. More recent research (Latane, Harkins, & Williams, 1980) found that measuring individual performance in a group situation eliminated social loafing.

In addition to empirical results, there are also several theories that at least point to a starting direction for explaining how manipulations of effort are mediated. Kahneman's (1973) book, entitled Attention and effort, is probably the best known theoretical effort. Effort was related to both capacity and arousal. This leads naturally to measuring effort by physiological concomitants of behavior. Kahneman was able to relate pupil diameter to effort. However, some years later Kahneman (1984, p.137) summarized his efforts as follows:

> "I have had my bits of experience with the study of pupillary dilation and heart rate, and there have been other techniques, such as eye-movement recordings. The net input into cognitive psychology, to an excellent first approximation, has been zero. It appears that some kinds of evidence are simply not acceptable, because they use tools and techniques that do not fit elegantly into particular paradigms with which cognitive psychologists are concerned."

I feel that this is a legitimate criticism of cognitive psychology. Fortunately, human factors specialists are more pragmatic than are cognitive psychologists. Researchers interested in mental workload have been far more willing to try tools and techniques, such as heart rate, etc., that do not fit elegantly into cognitive paradigms than have been most cognitive psychologists studying attention. Therefore, I expect that Kahneman's efforts at studying effort are more likely to bear fruit in applied settings than in the sometimes faddish arena of cognitive psychology. Practical problems like mental workload remain until they are solved, whereas particular paradigms in cognitive psychology tend to replicate themselves into exhaustion. I find it more useful to peruse last decades' articles in human factors rather than last decades' articles in experimental psychology, even though the latter are often of a higher scientific quality.

3.5.3. Biocybernetics

Another approach to mental workload involves such psychophysiological measures as heart rate, evoked potentials, and more recently, magnetic-evoked potentials. Indeed, the Air Force Office of Scientific Research has a successful research program titled "Biocybernetics and Workload". For some researchers (including me) these various biocybernetic measures are best regarded as functional signals, in principle no different from a behavioral measure such as reaction time, that are indicants of mental workload. Other researchers, with more of a psychobiological or neuropsychological approach, find the brain or neural mechanisms that produce biocybernetic signals to be of intrinsic interest. The two approaches are complimentary because greater understanding of the production of biocybernetic signals will also aid in interpreting them as indicants of mental workload.

However, I regard these indices of workload as correlational, and not causational. Furthermore, the assortment of brain wave components that may be related to mental workload (e.g., P300, N2, and so on) are but esoteric averages reflecting some unknown weighting of three-dimensional brain activity. They are even plotted inverted (e.g., positive is down!). I doubt that single-cell stimulation and recording with human subjects will soon replace these averaging methods. Hence, the physiological researcher is in much the same position as an electrical engineer who is given a voltmeter to study the operation of an intact computer running a FORTRAN program. It would be very difficult to reconstruct the software from the voltage measurements. My bias is that the optimal way to study behavior is through behavior.

3.5.4. Stress and Strain

The concepts of stress, strain, biocybernetics measures, and arousal are often linked. The Yerkes-Dodson law is well-known as a functional link between performance and arousal. However, there are many ambiguities associated with arousal, even with the Yerkes-Dodson law. These have been summarized by Hockey (1984), and I concur with his analysis. I do not see great value in biological indicants of stress or strain. For example, increased heart rate is associated with both digging ditches and making love. Yet, these are hardly equivalent behavioral activities. I suspect a poll of readers of this chapter would reveal an overwhelming preference for one of these activities over the other.

Strain is defined as the effects of stress, as when the deformation of a steel structural member is measured as a function of some external force or stress. This concept is far less useful in studying human behavior and mental workload. While structural engineers can predict the breaking strain for a steel beam, human factors researchers cannot predict, on the basis of physiological measures, how much strain is unsafe for a human operator. I agree with Rasmussen (1979) that stress and strain are ambiguous concepts, and that "operator resources must be evaluated within a much more refined framework" [p.31].

3.5.5. Other Aspects

In the interest of brevity, I conclude this section by mentioning only a few more aspects of mental workload. The issue of individual differences often arises in operational settings. For example, I was asked if a worker's IQ was related to his ability to handle levels of mental workload when I served as an expert witness in a wrongful-death lawsuit. Human factors is only beginning to relate mental workload and individual differences. Damos (1987) has reported that Type A and Type B personalities differ in their reaction to workload and timesharing. It is not at all clear to what extent individual differences in handling mental workload can be attributed to differences in strategy. Sanders (1967) obtained different results for simultaneous versus sequential handling in a psychological refractory period experiment. This manipulation of strategy was effective, even though the stimuli and responses were the same for both group. Similar results of strategy can be found in virtually any experiment where a pay-off matrix is used to control decision criteria in studies of speed-accuracy trade-off, signal detection, decision making, etc. It seems reasonable that some strategies may reduce mental workload more than others. This notion is particularly appealing within a theoretical context, such as advocated in this chapter, where spare capacity is related to internal processing architecture. Perhaps network models of mental processing (Schweickert & Boggs, 1984) can be used to relate strategy to mental workload.

Finally, it seems clear that practice is also related to mental workload. A task that seems formidable at first often becomes far more manageable after some practice. Researchers not concerned with training often avoid this issue by using highly skilled research subjects such as professional airplane pilots. But our understanding of mental workload will not be complete until we can also explain and predict the effects of practice.

3.6. MEASURING MENTAL WORKLOAD

There are three principle methods available to the researcher or practitioner who wishes to measure mental workload empirically. These are (1) subjective ratings, (2) objective secondary-task techniques, and (3) biocybernetic indices. Assorted minor methods will be grouped into a fourth subsection. Discussion is intended to be illustrative, rather than an exhaustive review of studies in each category.

3.6.1. Subjective Ratings

This has been the dominant method of measuring mental workload . Its main advantages are ease of obtaining data (especially in operational settings), minimal disruption of primary-task performance (especially when ratings are obtained after task performance rather than contemporaneously), and minimal equipment requirements.

Its disadvantages include the lack of a theoretical basis for explaining rating results, psychometric difficulties in generating interval (rather than ordinal) data scales, inability to compare ratings results obtained by different experimenters who used different ratings scales, and the problem of ratings yielding relative (rather than absolute) results, even when the same rating scale has been used in different experimental or operational settings.

3.6.1.1. Cooper-Harper Scales.

The original rating scale proposed by Cooper and Harper (1969) was designed to measure aircraft handling characteristics. It employed a decision tree with multidimensional descriptors of handling difficulty. The final result was a rating on a scale of 1 to 10. Note that the unidimensional final rating is an untransformed datum; the pilot actually gives a rating response that is a number from 1 to 10. Thus, any unidimensional characteristics of this datum arise by fiat, rather that being the result of psychometric scaling operations. Similarly, there is no evidence that more than an ordinal scale has been achieved, even if one is willing to grant that the final scale values (for which this scale are also the raw data reported by pilots) avoid problems of transitivity, etc., associated with collapsing a multidimensional world onto a unidimensional scale.

These potential difficulties notwithstanding, the Cooper-Harper scale has been widely modified and adjusted to measure workload in both aviation and other settings. A representative modification is that tested by Wierwille and Casali (1983), presented in Figure 2. They obtained favorable results in three experiments and concluded that their modified Cooper-Harper scale ratings "are a valid, statistically-reliable indicator of overall mental workload (p. 113)".

3.6.1.2. SWAT Ratings.

A rare attempt to apply psychometric analysis to the subjective rating of mental workload has been essayed by a group of researchers at the Air Force Aerospace Medical Research Laboratory. Their Subjective Workload Assessment Technique (SWAT) is based upon an esoteric branch of psychometric theory called "conjoint measurement"

(Reid, Eggemeier, Shingledecker, 1981). SWAT ratings are derived from three major factors: time load, mental effort, and psychological stress. Each of these is rated on a 3-point scale. This yields 27 (3x3x3) basic scale points. Conjoint measurement techniques (Krantz & Tversky, 1971) are then applied to create a unitary measurement scale with interval properties. Implicit in this procedure are several mathematical axioms that must be satisfied before conjoint measurement can be applied. While the research group is well aware of these axioms, I have seen no published data that allow the reader to evaluate how well these axioms apply. However, informal statements from members of the research group indicate that they plan to present such data eventually. In the meantime, their experiments with SWAT have given at least as good grounds for encouragement as, for example, results with Cooper-Harper scales. A summary of recent successes can be found in Shingledecker (1983).

3.6.1.3. Weighted Bi-Polar Ratings.

After many studies requiring pilots to rate subjective workload on several (unweighted) rating scales (e.g., Hart, Childress, & Bortolussi, 1981), a group of researchers found even better results when these individual scales were combined into a single, weighted measure of workload. Their better, but also more expensive, technique involved using several sub-scales that refer to different mental components of the task at hand. These sub-scales can be combined statistically to create a single number representing overall calculated mental workload. Table 1 summarizes this technique as used by researchers in the Human Performance Group, NASA-Ames Research Center. In order to calculate an overall workload score, each sub-scale is multiplied by an appropriate weight. In general, the weights are different for different people. For any one person, the weights are obtained by asking people to select which of a pair of sub-scales is more important as a contributor to overall mental workload. All possible pairs of sub-scales are compared in this manner. The number of times a particular sub-scale was preferred to other sub-scales is its weight. The overall workload rating is calculated as the weighted sum of each sub-scale rating, that is, each person's rating on a sub-scale is multiplied by the weight of that sub-scale and all these products are added together.

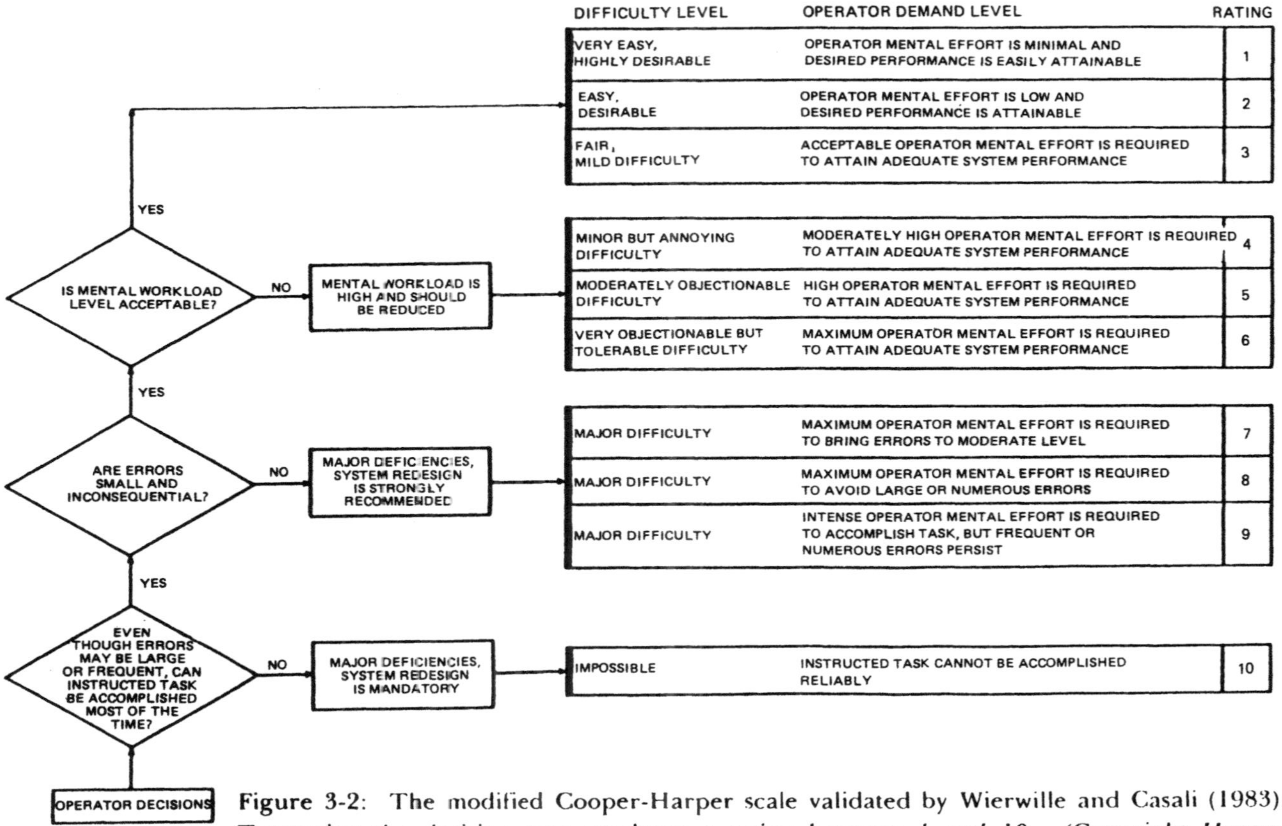

Figure 3-2: The modified Cooper-Harper scale validated by Wierwille and Casali (1983). Traversing the decision tree produces a rating between 1 and 10. (Copyright Human Factors Society. Reprinted with permission).

Recent results using this weighted rating appear to indicate that the subjective workload of particular flight sub-tasks (e.g., fly at constant altitude, fly at constant heading, etc.) can be combined linearly to predict subjective workload for combinations of sub-tasks (Kantowitz, Hart, Bortolussi, Shively, & Kantowitz, 1984). The most impressive finding is that the constant of linearity used to combine sub-tasks is independent of the particular sub-tasks being combined. Thus, for example, the subjective workload reported for <u>any</u> combination of two subtasks is quite close to 0.70 of the total subjective workload of the individual sub-tasks performed in isolation. This outcome would be expected, if an interval workload scale was produced by the weighting process. However, since there is no psychometric justification for claiming an interval scale, this result must be interpreted with caution. Retrospective application of this technique to a data bank containing three years worth of subjective ratings found similar results across many experiments. This technique offers great promise for operational settings. I would feel more confident if I knew why it works so well.

Table 3-1: Rating Scale Description

Title	Endpoints	Descriptions
OVERALL WORKLOAD	Low, High	The total workload associated with task, considering all sources and components.
TASK DIFFICULTY	Low, High	Whether the task was easy or demanding, simple or complex, exacting or forgiving.
TIME PRESSURE	None, Rushed	The amount of pressure you felt due to the rate at which the task elements occurred. Was the task slow and leisurely or rapid and

		frantic?
PERFORMANCE	Failure, Perfect	How successful you think you were in doing what we asked you to do and how satisfied you were with what you accomplished.
MENTAL/ SENSORY	None Impossible	The amount of mental and/or perceptual activity that was required (e.g., thinking, deciding, calculating, remembering, looking, searching, etc.).
PHYSICAL EFFORT	None Impossible	The amount of physical activity that was required (e.g., pushing, pulling, turning, controlling, activating, etc.).
FRUSTRATION	Fulfilled, Exasperated	How insecure, discouraged, irritated, and annoyed versus secure, gratified, content, and complacent you felt.
STRESS LEVEL	Relaxed, Tense	How anxious, worried, uptight, and harassed or calm, tranquil, placid and relaxed you felt.
FATIGUE	Exhausted, Alert	How tired, weary, worn out, and exhausted or fresh, vigorous, and energetic you felt.
ACTIVITY TYPE	Skill-Based, Ruled-Based,	The degree to which the task required mindless reaction to

Knowledge-Based	well-learned routines or required the application of known rules or required problem solving and decision making

3.6.2. Secondary Tasks

The secondary-task paradigm attempts to obtain direct estimates of spare capacity, and hence mental workload, by requiring an additional task to be performed in concert with the primary task of interest (Rolfe, 1971). Decrements in secondary-task performance are interpreted as indicating increased mental workload associated with increases in primary task demands. The secondary task has been occasionally termed a side task or a loading task in human factors literature.

The major advantages of the secondary-task paradigm are its objective nature and the availability of a large corpus of theoretical models of attention (that aid in interpreting data and designing experiments). Its disadvantages include substantial equipment requirements, the obtrusive nature of a secondary task in operational settings, and the impossibility of obtaining a universal secondary task suitable as a benchmark in all environments (Pew, 1979). Secondary-task measures of mental workload tend to be used in laboratory and simulator studies more than in operational settings.

A host of assumptions requiring several control and experimental conditions are needed before one can safely interpret data obtained from secondary-task paradigms. These have been reviewed and analyzed in tendentious detail (Kantowitz, 1985a; Ogden et al., 1979). It is difficult to find published studies, especially in the human factors literature, that contain all the requisite conditions. The most obvious assumption in secondary-task paradigms is that adding a secondary task does not alter either performance or the internal information flow within the operator. The minimum requirement for testing this assumption about performance is a single-stimulation control condition where the primary task (e.g., the task so labeled by the experimenter and hopefully treated as such by the subject) is performed alone. Similarly, secondary-task decrement cannot be established without a single-stimulation condition where the secondary-task is performed alone. Finally, it is helpful to include at least two levels of secondary-task difficulty or complexity in order to make inferences about the internal system architecture; this point has been elaborated by me previously (see Kantowitz, 1985a; Kantowitz & Knight, 1976).

The most serious practical difficulty in using secondary-task techniques to measure mental workload arises from the multiplicity of secondary tasks that have been studied (Ogden, Levine, & Eisner, 1979). This creates the same problem mentioned in the context of subjective rating scales: It is hard to compare results based upon different secondary tasks. However, this problem can be minimized by using theory to select secondary tasks. If we understand how some particular secondary task relates to some particular class of attention model, we can compare different secondary tasks by invoking the model. This point was made salient in a popular human factors textbook (Kantowitz & Sorkin, 1983, p. 188):

> "While a few researchers have tried to use theoretical models to guide selection of secondary tasks (Wickens, 1979; 1980), most have used arbitrary combinations of primary and secondary tasks that prevent generalization of results beyond the particular combination of tasks selected. Since it is not practical to perform experiments with all possible combinations of the wide variety of tasks used to assess spare mental capacity, it is clear that progress in this area will depend more on new theoretical developments than on acquisition of more data based upon the incorrect limited- or single-channel model."

At the risk of appearing immodest, I shall use a series of studies performed at Ames Research Center to illustrate the secondary-task technique (Kantowitz, Hart, & Bortolussi, 1983; Kantowitz, Hart, Bortolussi, Shively, & Kantowitz, 1984). In all these studies the primary task was flying a moving-base (GAT) flight simulator. A choice-reaction secondary task with 2, 3, or 4 alternatives was used to index mental workload. At first blush, this selection of secondary task seems unwise. Most researchers have used probe (i.e., simple reaction time) tasks to minimize possible interference with the primary task. However, I believed, based upon a hybrid model of attention (Kantowitz & Knight, 1976), that simple probe tasks would be insensitive, and furthermore suffer from an assortment of methodological problems (see Kantowitz, 1985a). Thus, I was willing to take the risk that the primary task might be disrupted, and included appropriate control conditions to test for this unhappy possibility. But I expected this risk to be much less than for an experiment conducted on college sophomores. We used highly-skilled pilots and I was banking (pardon the pun) on their training and experience. Pilots are taught first and foremost to fly the airplane, no matter what else is happening. The first responsibility of a pilot is always to keep the airplane safely in the air. Therefore, pilots seemed an ideal population for taking the risks associated with a choice-reaction secondary task.

Results to date have been felicitous. Not only did the dread interference with the primary task fail to materialize, but the secondary-task was able to discriminate among levels of mental workload in many situations. Since virtually every model of attention has something to say about choice-reaction time, I believe that the choice reaction-time task should be high on every experimenter's list of preferred secondary tasks.

Of course, many other behavioral secondary tasks have been, and should be, utilized. But instead of grabbing a market basket of the top twenty tasks that have been used historically, I would prefer an approach that selects a far smaller set of secondary tasks based upon their theoretical relevance. Theory is the best practical tool. I would rather put all my eggs in one theoretical basket than to scatter hosts of secondary tasks in hopes that (a) some of them might yield significant effects, and (b) that I could then explain why tasks 2 and 17 worked while the rest did not.

3.6.3. Biocybernetic Measures

Biocybernetic indices of mental workload include such psychophysiological dependent variables as heart rate, pulse wave velocity, skin temperature, electromyogram, evoked potentials, magnetic evoked potentials, and pupil diameter. Their major advantage is unobtrusiveness, although it is hard to argue that any technique that requires electrodes to be attached to assorted parts of the human body is completely unobtrusive. While there are some theoretical explanations associated with biocybernetic measures, there is no unified theory that attempts to explain all of these psychophysiological correlates simultaneously. Most laboratories finesse this difficulty by only measuring one or two biocybernetic indices at a time. In recent years, perhaps as a result of federal research funding, there has been considerable interest in applying biocybernetic measures to cognitive psychology in general (Donchin, 1984) and to mental workload in particular. Results appear promising, but is is too early to state that these measures are superior to other indices of mental workload (Wierwille, 1979).

When compared to subjective ratings, biocybernetic measures offer the virtue of objectivity. An airplane pilot may walk out of a simulator, soaked in sweat, and calmly tell the investigator that the simulated flight was a piece of cake. But having the "right stuff" will not protect the pilot from manifesting effects of the flight via biocybernetic measures. When compared to secondary-task techniques, biocybernetic measures offer less opportunity to interfere with performance of the primary task. Furthermore, they hold the hope of detecting mental overload earlier and more "automatically" than would a behavioral measure. For example, a jet fighter pilot might not be aware (or willing) to allocate some functions to an onboard computer as his mental workload increased, but a device that monitored his biocybernetic signals might be able to alter allocation of function without manual operator-intervention, see the chapter by Hancock and Chignell in this volume. I hasten to add that such a device has yet to be demonstrated in an operational setting.

Disadvantages of biocybernetic indices are related to methods of signal analysis and their theoretical interpretations. Biocybernetic signals are usually embedded in noise so that assorted averaging and more complex statistical techniques are required to increased the visibility of the signal. This difficulty is further complicated by the large number of analog data points that must be sampled, stored, and analyzed. Some of the more sophisticated analysis techniques require considerable computational power and so cannot be accomplished on-line or even rapidly, off-line. While these problems can be handled in laboratory environments, they pose significant problems for operational settings. On the other hand, improvements in hardware are making it far easier to monitor biocybernetic signals in noisy non-laboratory environments. Finally, biocybernetic models of behavior are not nearly as well-formulated as, for example, behavioral models of attention. Skeptics might argue that taking two phenomena we do not yet understand--mental workload and biocybernetic signals--merely doubles our difficulties. My own view is that biocybernetic indices can be quite useful when combined with other measures of mental workload.

The large number of potential biocybernetic measure precludes finding any single representative published study. Therefore, I shall fall back upon describing one of my own studies-in-progress that measures heart rate and evoked-potential in a psychological refractory period paradigm (Kantowitz, 1985c). In the psychological refractory period, two stimuli are presented in close temporal succession. We used inter-stimulus intervals of 120 and 480 msec. It is well known that reaction time is a decreasing function of inter-stimulus interval (Kantowitz, 1974). Indeed, this paradigm was selected precisely because it is well understood behaviorally. It is intended to serve as a kind of benchmark allowing us to calibrate the utility of the biocybernetic measures. The strategy underlying this experimental approach is simple. Once biocybernetic measures have been calibrated in the (behavioral) context of an attention paradigm that has been well studied and explained, we can go on to use these biocybernetic tools in more complex cognitive paradigms, where explanations have been more elusive. Mental workload was varied by manipulating inter-stimulus interval (shorter interval produces greater workload) and number of stimulus-response alternatives (two versus four). We do not expect both cybernetic measures to be sensitive to both manipulations of mental workload.

Since evoked potentials, especially P300, seem to be indicants of perceptual operations, we expect the inter-stimulus interval to alter P300. Since number of alternatives is primarily a response variable, we expect P300 not to reflect this manipulation, whereas heart rate should be an indicant of response load. Should results agree with our predictions, we will have validated a tool for distinguishing between front- and back-end stages of human information-processing. This can also be done by purely behavioral methods, but only at the cost of several experiments designed to produce converging operations. How much nicer it is to answer this question by recording biocybernetic signals in a single experiment.

Our preliminary results with evoked potentials were surprising. No P300 effects were found for either number of choices or inter-stimulus interval. However, the amplitude of the N200 component of the evoked potential was sensitive to number of choices. While N200 amplitude was also influenced by inter-stimulus interval, the form of the relationship varied according to the location of the electrode. Heart rate variance, another biocybernetic signal recorded in the experiment, was inversely related to the number of choices as predicted. While these results indicate that biocybernetic signals are indeed useful in studies of mental workload, they also indicate that the present state of biocybernetic theory has yet to reach the point where confident predictions about specific components of the evoked potential can be generated and verified. We are presently engaged in more sophisticated analyses of our biocybernetic data in the hope that this situation can be improved.

3.6.4. Other Measures

While the three indices reviewed above are the most important techniques for establishing mental workload, they are hardly the only techniques that have been used or suggested. Another possibility is measuring primary-task performance by itself, with decrements indicating increased mental workload. At least one researcher has taken this technique to its ultimate end by suggesting that there is no overload whenever a mission has been successfully completed (Albanese, 1977). I regard this as an extremist viewpoint that is unwarranted. I have never parachuted from an airplane but I admit the possibility that I might do so (once) without incurring grave bodily harm. But I would not interpret this fortunate outcome as evidence for no overload. When a primary task is difficult, variations in performance may reflect levels of mental workload. But analysis of a primary task that is associated with low-to-medium levels of mental workload is often too insensitive to reveal the effects of additional tasks or changes in procedures.

A novel approach suggested by Billings (personal communication) involves administering a drug, such as alcohol, known to produce performance impairment. There are several ways in which this could help to scale mental workload. The simplest is to observe primary-task performance and note how much of the drug is required to produce measurable decrements under different levels of primary-task requirements. The more drug required, the less the mental workload. The drug approach could also be readily combined with the major techniques discussed previously. This approach could be quite valuable for drugs that have good theoretical models relating them to human information processing.

3.7. CONCLUSIONS

Mental workload is a complex construct. It is most unlikely that a single metric will be able to index mental workload in a wide variety of situations. The best hope for developing practical of measure mental workload calls for extensive use of theory.

My own strategy is to build upon existing theoretical models of attention. Furthermore, I believe that the class of attention models that postulates a single pool of capacity is better suited for developing indices for mental workload than are multiple-pool models. While multiple-pool models are so powerful that it has been easy to achieve some initial successes with them, in the long run, their lack of parsimony makes them less useful in achieving a practical goal. Measuring mental workload is a practical goal. The construct of spare capacity, derived from models of attention, is the most important component of mental workload. Spare capacity is a more useful concept when embedded in single-pool, rather than multiple-pool, models of attention.

However, mental workload is more than just spare capacity. Additional aspects of mental workload include subjective feelings, effort, individual differences, strategy and practice. Current measurement techniques tap different aspects of mental workload. Subjective rating are best suited for "quick and dirty" estimates, which may be quite adequate for operational settings. Objective secondary tasks are more refined, but are also more cumbersome. Nevertheless, they offer the best long-run possibilities for effective prediction of mental workload. Biocybernetic measures can be extremely useful as ancillary means, especially when used in concert with secondary tasks. But it is too early for them to stand alone. Perhaps, in the near future, after sufficient behavioral cross-validation, biocybernetic measures may emerge as the best choice in operational settings.

In my own research I am most comfortable with secondary tasks as indicants of mental workload. Even so, I doubt that secondary tasks alone will fill the bill. Thus, I have started to use biocybernetic measures. I also use subjective ratings, primarily to compare my findings with those of other researchers, although I freely admit being uncomfortable with them because of psychometric problems and the lack of a strong theoretical framework.

Solving a practical problem often requires compromises relative to laboratory standards. But the lack of complete experimental control in an operational setting, or even in a simulator, should not be taken as a license justifying sloppy experimental procedures and rejection of theory. Indeed, the need for theory increases as our experimental control decreases. Judging only by error variance, my own research conducted in the Ames Research Center flight simulator is an order of magnitude worse than my research in the Human Information Processing Laboratory at Purdue University. Even so, we have obtained important results. Judging by the time in years needed to apply my research to achieve a practical end, my Purdue research will take an order of magnitude longer than my Ames research. Even so, our models and results in the area of attention may eventually provide better ways to measure mental workload than our simulator research. Furthermore, each approach aids the other.

The benefit to theoreticians (e.g., people who study attention) of an applied perspective (e.g., what is mental workload?) has been fluently stated by Broadbent (1971, p. 4):

> "In situations arising from technology our attention is compelled to the major variables in human behavior, and we cannot ignore them in favor of some artificial distinction as can a theorist working purely from the ivory tower. The researcher remote from immediate practical pressures may indeed be free to study major variables in which at this instant society does not seem to be interested; but he should not use this freedom in order to study minor variables, until there are no major ones within reach of our techniques. The necessity for some relevance to real life is a worthwhile intellectual discipline."

I would only add that the converse is equally true. Practitioners will be aided by theory. Theory keeps them from reinventing the wheel every time a seemingly new problem must be solved. Although theory may not solve a problem immediately, it can reduce the number of alternatives that need to be investigated. The best practical tool is a good theory.

ACKNOWLEDGEMENT

This research was supported by Cooperative Agreement NCC 2-228 from the National Aeronautics and Space Administration (NASA); S.G. Hart was the Technical Monitor. I am grateful for their willingness to support both applied and theoretical goals within the same package.

REFERENCES

Albanese, R.S. (1977). Mathematical analysis and computer simulation in military mission workload assessment. *Proceedings of the AGARD Conference on methods to access workload, AGARD-CPP*, 2166.

Broadbent, D.E. (1958). *Perception and communication*. London: Pergamon Press.

Broadbent, D.E. (1965). Application of information theory and decision theory to human perception and reaction. In N. Wiener & J.P. Schade (Eds.), *Cybernetics of the nervous system*. Amsterdam: Elsevier.

Broadbent, D.E. (1971). *Decision and stress*. New York: Academic Press.

Cooper, G.E., & Harper, R.P. Jr. (1969). *The use of pilot rating in the evaluation of aircraft handling qualities* (Tech. Rep.). National Aeronautics and Space Administration, Ames Research Center, TN-D-5153.

Damos, D.L. (1987). Individual differences in subjective estimates of workload. In P.A. Hancock & N. Meshkati (Eds.), *Human mental workload*. Amsterdam: North Holland.

Donchin, E. (1984). *Cognitive psychophysiology*. Hillsdale, New Jersey: Lawrence Erlbaum Associates.

Hart, S.G., Childress, M.E., & Bortolussi, M. (1981). Defining the subjective experience of workload. *Proceedings of the Human Factors Society, 25*, 527-531.

Hart, S.G. (1982). Theoretical basis for workload assessment research at NASA-Ames research center. In M.L. Frazier & R.B. Crombie (Eds.), *Proceedings of the workshop on flight testing to identify pilot workload and pilot dynamics*. Edward Air Force Base, CA, AFFTC-TR-82-5.

Hockey, R. (1984). Varieties of attentional state: The effects of environment. In R. Parasuraman & D. R. Davies (Eds.), *Varieties of attention*. New York: Academic Press.

Kahneman, D. (1973). *Attention and effort*. Englewood Cliffs, New Jersey: Prentice Hall.

Kahneman, D. (1984). Discussion. In E. Donchin (Ed.), *Cognitive psychophysiology*. Hillsdale, New Jersey: Lawrence Erlbaum and Associates.

Kantowitz, B.H. (1974). Double stimulation. In B.H. Kantowitz (Ed.), *Human information processing*. Hillsdale, New Jersey: Lawrence Erlbaum Associates.

Kantowitz, B.H., & Knight, J.L. (1976). Testing tapping time-sharing, II: Auditory secondary task. *Acta Psychologica, 40*, 340-362.

Kantowitz, B.H., & Knight, J.L. (1978). Testing tapping time-sharing: Attention demands of movement amplitude and target width. In G. Stelmach (Ed.), *Information processing and motor control*. New York: Academic Press. (a).

Kantowitz, B.H., & Knight, J.L. (1978). When is an easy task difficult and vice versa? *Acta Psychologica, 42*, 162-170. (b).

Kantowitz, B.H. (1982). Interfacing human information processing and engineering psychology. In W.C. Howell & E.A. Fleishman (Eds.), *Human performance and productivity (Vol. 2)*. Hillsdale, New Jersey: Lawrence Erlbaum Associates.

Kantowitz, B.H., Hart, S.G., & Bortolussi, M.R. (1983). Measuring pilot workload in a moving base simulator: I. Asynchronous secondary choice-reaction task. *Proceedings of the Human Factors Society, 27*, 319-322.

Kantowitz, B.H., & Sorkin, R.D. (1983). *Human factors: Understanding people-system relationships*. New York: Wiley.

Kanowitz, B.H., Hart, S.G., Bortolussi, M.R., Shively, R.J., & Kantowitz, S.C. (1984). Measuring pilot workload in a moving base-simulator: II. Building levels of workload. *Proceedings of the Annual Conference on Manual Control, 20*, 359-371. Ames Research Center, Moffet Field, CA: NASA CP23-1.

Kantowitz, B.H., & Roediger, H.L. (1984). *Experimental psychology 2nd ed*. St. Paul, Minnesota: West Publishing.

Kantowitz, B.H. (1985). Stages and channels in human information processing: A limited analysis of theory and methodology. *Journal of Mathematical Psychology, 29*, 135-174. (a).

Kantowitz, B.H. (1985). Mental work. In B.M. Pulat & D.C. Alexander (Eds.), *Industrial ergonomics*. Institute of Industrial Engineering. (b).

Kantowitz, B.H. (1985). *Biocybernetic analysis of a hybrid workload model* (Tech. Rep.). Air Force Office of Scientific Research, (c).

Kantowitz, B.H., & Casper, P. (1986). Mental workload in aviation. In E.L. Wiener & D.C. Nagel (Eds.), *Human factors in modern aviation*. . In preparation.

Kinsbourne, M. (1981). Single-channel theory. In D. Holding (Ed.), *Human skills*. New York: Wiley.

Krantz, D.H., & Tversky, A. (1971). Conjoint measurement analysis of composition rules in psychology. *Psychological Review, 78*, 151-169.

Lane, D. (1977). Attention allocation and the relationship between primary and secondary task difficulty. *Acta Psychologica, 41*, 493-95.

Lane, D. (1982). Attention. In W.C. Howell & E.A. Fleishman (Eds.), *Human performance and productivity (Vol. 2)*. Hillsdale, New Jersey: Lawrence Erlbaum Associates.

Latane, B., Harkin, S.G., & Williams, K. (1980). Many hands make light the work: Social loafing as a social disease. Manuscript awarded the 1980 Socio-Psychological prize by the American Association for the Advancement of Science.

McCormick, E.J., & Sanders, M. (1982). *Human factors in engineering and design (6th Ed.)*. New York: McGraw Hill.

Moede, W. (1927). Die richtlinien der leistungs-Psychologie.

Industrielle Psychotechnik, 4, 193-207.

Moray, N. (Ed.). (1979). *Mental workload*. New York: Plenum Press.

Moray, N. (1982). Subjective mental workload. *Human Factors, 24*, 25-40.

Navon, D., & Gopher, D. (1979). On the economy of the human processing system. *Psychological Review, 86*, 214-255.

Navon, D. (1984). Resources: A theoretic soupstone. *Psychological Review, 91*, 214-224.

Ogden, G.D., Levine, G.M., & Eisner, E.J. (1979). Measurement of workload by secondary task. *Human Factors, 21*, 529-548.

Pew, R.W. (1979). Secondary task and workload measurements. In N. Moray (Ed.), *Mental workload*. New York: Plenum Press.

Rasmussen, J. (1979). Reflections on the concept of operator workload. In N. Moray (Ed.), *Mental workload*. New York: Plenum Press.

Reid, G.B., Eggemeier, T., & Shingledecker, C.A. (1982). Subjective workload assessment techniques. In M.L. Frazier & R.B. Crombie (Eds.), *Proceedings of the workshop on flight testing to identify pilot workload and pilot dynamics*. Edward Air Force Base, CA: AFFTC-TR-82-5.

Rolfe, J.M. (1971). The secondary task as a measure of workload. In W.T. Singleton, J.C. Fox, & D.W. Whitfield (Eds.), *Measurement of man at work*. London: Taylor and Francis.

Sanders, A.F. (1967). The effect of compatibility on grouping successively presented signals. *Acta Psychologica, 26*, 373-382.

Sanders, A.F. (1979). Some remarks on mental load. In N. Moray (Ed.), *Mental workload*. New York: Plenum Press.

Schweickert, R., & Boggs, G.J. (1984). Models of central capacity and concurrency. *Journal of Mathematical Psychology, 28*, 223-282.

Shingledecker, C.A. (1983). Behavioral and subjective workload metrics for operational environments. *Sustained intensive air operations: Physiological and performance aspects, AGARD*, Conference Proceedings No. 338. Neuilly sur Seine, France.

Townsend, J.T., & Ashby, F.G. (1983). *Stochastic modeling of elementary psychological processes*. New York: Cambridge University Press.

Wickens, C.D. (1979). Measures of workload. In N. Moray (Ed.), *Mental workload*. New York: Plenum Press.

Wickens, C.D. (1980). The structure of attentional resources. In R. Nickerson (Ed.), *Attention and performance VIII*. Hillsdale, New Jersey: Lawrence Erlbaum Associates.

Wickens, C.D. (1984). Processing resources in attention. In R. Parasuraman & D.R. Davis (Eds.), *Varieties of attention*. New York: Academic Press.

Wierwille, W.W. (1979). Physiological measures of aircrew mental workload. *Human Factors, 21*, 575-594.

Wierwille, W.W., & Casali, J.G. (1983). A validated rating scale for global mental workload measurement applications. *Proceedings of the Human Factors Society, 27*, 129-133.

HUMAN FACTORS PSYCHOLOGY, Peter A. Hancock (ed.)
©Elsevier Science Publishers B.V. (North-Holland), 1987

4. HUMAN DECISION PROCESSES: HEURISTICS AND TASK STRUCTURE

Don N. Kleinmuntz
Sloan School of Management
Massachusetts Institute of Technology
Cambridge, MA 02139

4.1. INTRODUCTION

Decision making is a pervasive and central component of human behavior. Individuals must cope with increasingly complex, changing environments that make the effectiveness of decision making critical. At the same time, an increasing amount of new technology is available that is intended to aid and improve the quality of decision making. These human-machine systems vary in sophistication and complexity, from relatively simple information retrieval systems to state-of-the-art techniques drawn from disciplines like artificial intelligence, operations research and decision theory.

Against this background of increasing complexity and proliferating decision aids, the importance of understanding human decision processes is self-evident. In particular, researchers in the area of behavioral decision theory have focused on the cognitive processes of individual decision makers. By studying the nature and limitations of the human information processing system, it is possible to predict when and why decision making is effective (or not), and consequently, when and how decision aids may be most helpful.

The purpose of this chapter is to provide an overview of some key findings and methods of behavioral decision theory. The discussion will be organized as follows: First, research on the nature of cognitive judgment and choice strategies will be outlined. Of particular interest is the influence that the structure of the decision task has on the effectiveness of decision strategies. In addition, some research on a particularly important category of tasks, dynamic decisions, will be discussed. Following this, alternate methodologies for investigating cognitive strategies will be compared. Finally, some implications for the design and implementation of human-machine decision support systems will be discussed. Note that this chapter does not attempt a comprehensive review of the literature. More complete coverage can be found elsewhere (Einhorn & Hogarth, 1981; Hogarth, 1980; Kahneman, Slovic, & Tversky, 1982; Pitz & Sachs, 1984; Slovic, Fischhoff, & Lichtenstein, 1977; Slovic, Lichtenstein, & Fischhoff, in press).

4.2. JUDGMENT UNDER UNCERTAINTY

One unique feature of decision research is its reliance on normative theories of decision making. A common paradigm is to define a judgment or choice task, determine an optimal or rational response using some variant of utility theory or Bayesian decision theory, and observe the extent to which actual behavior deviates from the prescribed response. Early research was primarily concerned with establishing the existence of deviations, with particular emphasis on deviations caused by the inconsistency of human decision makers (Meehl, 1954; Slovic & Lichtenstein, 1971). However, as research accumulated to show that decision behavior deviated systematically from normative guidelines (Edwards, 1968; Slovic, et al., 1977), the central research question shifted from documenting the existence of suboptimality to describing the underlying information processing strategies that cause that suboptimality.

4.2.1. Heuristics

The work of Tversky and Kahneman (1974) was particularly influential in bringing about this change in focus. They initially identified three cognitive heuristics used in a variety of probabilistic judgment tasks. These simple strategies are easy to use and work well in a variety of settings, but have the potential to lead to "severe and systematic errors" (Tversky & Kahneman, 1974, p. 1124). For instance, people often judge the likelihood of an uncertain event by the degree to which it is representative of the process generating it. An example would be judging the probability that an individual was in a particular occupation (e.g., engineer or social worker) based on a personality sketch of that person. A convenient procedure would be to assess the similarity of the sketch to one's stereotypes of engineers and social workers. Unfortunately, this procedure can also lead to serious errors since it neglects important statistical considerations, including the prior probability distributions of job categories, and the poor predictive validity of the sketch. In the same vein, judgments utilizing the availability of similar instances or a process of anchoring on initial values and adjusting can be shown to lead to systematic violations of basic normative principles.

Tversky and Kahneman's work stimulated a large body of research on heuristics in judgment under uncertainty (for an overview, see Kahneman, et al., 1982). The initial demonstrations of biased judgments have been replicated in a variety of settings and the catalog of potential information processing biases extended. For instance, Hogarth (1980, Chapter 9) lists approximately thirty different biases and relates each to different stages of information processing. The limitations of judgmental heuristics have been considered in other areas of psychology, including social attributions (Nisbett & Ross, 1980) and consumer behavior (Bettman, 1979). They have also been communicated to decision makers in a number of other professions, including medical practitioners (Eddy, 1982; Politser, 1981), public policy and risk analysts (Slovic, Fischhoff, & Lichtenstein, 1976; Slovic, Kunreuther, & White, 1974), and statisticians (Hogarth, 1975; Kahneman & Tversky, 1979a).

4.2.2. The Prevalence of Biased Judgments

One conclusion sometimes drawn from this body of research is that people are generally inept decision makers, whose feeble cognitive abilities fall far short of what they believe them to be (e.g., Dawes, 1976). Some caution must be exercised before accepting this inference. For instance, some researchers have questioned the external validity of laboratory demonstrations of biased decisions. While Slovic et al. (1977, pp. 14-17) reviewed a number of studies that showed similar findings in laboratory and field settings, Ebbesen and Konecni (1980) described a number of instances where laboratory research failed to generalize. Einhorn and Hogarth (1981) suggest:

> "The issue of external validity is not liable to be resolved without recourse to theory that attempts to answer how tasks vary between the laboratory and natural environment and what kinds of effects can be expected from such differences." (p. 81)

This theoretical framework would permit an assessment of the general conditions under which heuristics work well or badly as well as the distribution of those conditions in the natural environment.

This framework may also guide attempts to "debias" judgment. Fischhoff (1982) suggests a classification scheme for efforts to improve or correct judgment processes. One category of efforts aims to correct problems caused by the task, by either removing unfair task demands or clarifying misunderstood elements of the task. Another category finds fault with the judge's limited cognitive capacity, with the recourse being either to engage in training exercises or to replace the judge with a mechanical aid. A final category involves mismatches between the task and the judge; the judge has the requisite cognitive abilities to perform the task, but for some reason those abilities are not used. Possible corrective strategies include restructuring the task through decomposition or alternate formulations.

A final corrective option worth considering is formal training in statistics. Nisbett, Krantz, Jepson, and Kunda (1983) demonstrate that some judges can and do use statistical heuristics, simple judgment rules that are roughly equivalent to formal statistical principles. Furthermore, they show that training in statistics increases both the use and the effectiveness of these rules. Further investigations of the origin and extent of these statistical heuristics promise to provide important insights into the potential for avoiding biased judgments.

Given the current state of knowledge about heuristics, there is no alternative to an intuitive assessment of the prevalence of bias. Christensen-Szalanski and Beach (1984) suggest that those who conclude that severely biased judgments are widespread are victims of a "citation bias," a propensity to pay more attention to the published evidence of poor performance than the evidence showing good performance. They go on to suggest that this bias is a passing fad in the literature. While the notion of judgment researchers falling prey to one of their own biases is certainly ironic, it is not at all clear that focusing attention on systematic errors is a mere fad. For instance, Evans (1984) points out that instances of poor judgment are important since when heuristics do fail, judges are usually not aware of the failure (also see Einhorn & Hogarth, 1978; Kahneman & Tversky, 1973). Furthermore, focusing on errors is an explicit research strategy that has proven useful in other domains of cognitive psychoblogy. As Tversky and Kahneman (1983) point out, studies of "cognitive illusions" are diagnostic of basic judgment processes in the same way that studies of visual illusions are diagnostic of basic perceptual processes:

> "The focus on bias and illusion is a research strategy that exploits human error, although it neither assumes nor entails that people are perceptually or cognitively inept. In cognition, as in perception, the same mechanisms produce both valid and invalid judgments." (p. 313).

4.3. CHOICE STRATEGIES AND TASK STRUCTURE

The identification of cognitive mechanisms has not been limited to the area of judgments under uncertainty. A parallel stream of research focuses on the information processing strategies used for evaluative judgments and choices. Researchers have identified a large number of rules that decision makers use, including compensatory rules (additive utility, additive difference) and noncompensatory rules (conjunctive, disjunctive, elimination by aspects), among others (for an extensive catalog, see Svenson, 1979). In summarizing the initial work in this area, Slovic et al. (1977) noted that:

> "Researchers appear to be searching for heuristics or modes of processing information that are common to a wide domain of subjects and choice problems. However, they are finding

that the nature of the task is a prime determinant of the observed behavior." (p. 7)

4.3.1. Task Determinants

In fact, the central finding in recent judgment and choice research has been the lack of invariance of decision strategies across decision environments. In other words, information processing in decision making is highly contingent on the demands of the task (Payne, 1982). Seemingly insignificant manipulations of problem formulation cause severe shifts in observed behavior (Tversky & Kahneman, 1981). For example, consider the choice between losing 50 dollars for sure and a gamble involving either losing 200 dollars with a .25 probability or losing nothing with a .75 probability. In experimental settings, most subjects choose to take the risky gamble rather than submit to a sure loss. However, consider the situation where one is faced with a .25 probability of losing 200 dollars, and the options are either to pay a 50 dollars premium for insurance protection or to not purchase the insurance. Although this problem has the same probabilities and outcomes as the gambling situation, most subjects make the opposite decision, choosing to avoid the risk and pay for the insurance premium (Hershey & Schoemaker, 1980).

This example illustrates an important point: The influence of task structure on information processing is mediated through the decision maker's internal representation of the problem. Differences in wording and context, like the insurance/gambling problem discussed above, influence decision making through their influence on this representation (also see Simon & Hayes, 1976). Other factors that influence the problem representation include the decision maker's prior experiences and knowledge (Chase & Simon, 1973; Elstein, Shulman, & Sprafka, 1978; Larkin, McDermott, Simon, & Simon, 1980). Models of knowledge representation have yet to be integrated with decision making models, although the <u>script</u> concept may be useful for this purpose (Abelson, 1976, 1981).

Payne (1982) provides an extensive review of the impact of task structure on decision strategies. Some examples of these effects will be briefly described below. The task effects to be discussed here are: (a) task complexity (b) response mode (c) information display and (d) quality of the option set.

As task complexity increases, decision makers often cope by using less effortful choice rules. For instance, when faced with only two choice alternatives, subjects use strategies like the additive and additive difference rules that permit compensatory trade-offs among the dimensions. In more complex multialternative situations, they adopt simpler non-compensatory rules, like the conjunctive rule or elimination-by-aspects, that are easier to implement (Payne, 1976).

One of the more dramatic task effects involves the pattern of responses observed when subjects are asked to judge the attractiveness of two gambles and then choose which of the two gambles they would rather play. Lichtenstein and Slovic (1971) found that for specially constructed pairs of gambles, subjects often choose the gamble that they judge to be less attractive. This "preference reversal" phenomenon has proven to be extremely robust, replicating outside of the laboratory (Lichtenstein & Slovic, 1973) and persisting despite a variety of attempts to control for competing explanations (Grether & Plott, 1979; Slovic & Lichtenstein, 1983). Although a number of hypotheses about the psychological nature of the phenomenon have been advanced, detailed explanations of the underlying cognitive strategies have only recently been attempted. Tversky and Slovic (1984) hypothesize that the differing response modes influence the evaluation process, while Goldstein and Einhorn (1985) suggest a competing explanation involving the process of translating an evaluation into a final response.

Information display formats may influence a decision maker's ability to use that information in conjunction with certain processing strategies. This was demonstrated by Russo (1977) in a study of consumer's use of unit price information in a supermarket. When the unit price information was presented on a single posted list, shoppers used that information more often than when the prices were presented on individual shelf tags. The list made comparisons across brands of a product category less difficult, which led people to purchase less expensive brands.

The processing strategy used by a decision maker may depend upon the quality of the options available. An important point is that the relevant characteristic of the task is the perceived quality of the outcome. This often implies the existence of a reference point that is used to encode the outcomes as gains versus losses. In fact, relatively simple changes in the wording of a problem can induce decision makers to use a different reference point, changing the perceived quality of the option set, and in some instances, causing a change in choice behavior (Tversky & Kahneman, 1981). Payne, Laughhunn, and Crum (1980) were able to experimentally manipulate the reference points that decision makers used to evaluate gambles. The nature of the evaluation and choice process depended upon whether the gambles' outcomes mainly exceeded the decision maker's aspiration level, mainly fell below that level, or were both above and below that level.

4.3.2. Cognitive Effort and Strategy Selection

Task structure often influences performance because of the demands the task places on limited human information processing capabilities. Extremely complex, information-laden environments may force decision makers to adopt simple rules, like heuristics, since the complexity of the environment overwhelms the information processing system (Simon, 1955, 1956). Alternatively, the selection of decision strategies may depend, at least in part, on considerations of information processing convenience rather than necessity; strategies may be selected through a deliberate cost-benefit evaluation of accuracy (effectiveness) versus the cognitive effort the strategies require from the decision maker (Beach & Mitchell, 1978; E. J. Johnson & Payne, 1985; Russo & Dosher, 1983; Shugan, 1980). By understanding how task structure influences both effort requirements and performance characteristics of decision strategies, it may be possible to predict what strategies decision makers will use for various tasks.

This performance-effort framework is one hypothesis about how individuals cope with the metadecision problem, in which one decides how to choose. Einhorn and Hogarth (1981) suggest a more general scheme, where each decision strategy is characterized as a multidimensional object having attributes like effort requirements, memory requirements, accuracy, justifiability, and so on. The process of selecting a decision strategy then becomes an evaluation problem that requires the examination of those attributes. Each strategy is also a possible metastrategy for evaluating itself and others. Unfortunately, choosing a metastrategy then becomes an even higher level choice problem, creating the possibility of infinite regress. Payne (1982) summarizes a number of other unresolved issues related to strategy selection, including problems with the measurement of error and effort, questions about the extent to which we are aware of our strategy selection processes, and questions about how the relationship between task structure and decision strategy is learned.

4.4. DYNAMIC DECISION ENVIRONMENTS

Most of the research discussed up to this point has focused on discrete incidents of judgment and choice. The repetitive, time-dependent nature of many decision environments is usually not addressed. For instance, a study of medical diagnosis might involve presenting a clinician with some initial information about a patient. The clinician would then ask questions about the patient's history, request results of physical examinations and diagnostic tests, and try to arrive at a diagnosis. Both the accuracy of the final diagnosis and the pattern of information search could be analyzed (for examples of this approach, see Elstein et al., 1978 and B. Kleinmuntz, 1968).

More generally, the medical decision maker's task could be considered to be a component of a larger decision problem that does not simply terminate with the final diagnostic assessment. Instead, the diagnosis is used as the basis for treatment selection. Furthermore, depending upon the response to the initial treatment, a long sequence of subsequent decisions may occur. In fact, not only are there opportunities for more diagnostic information collection, but the outcome feedback from earlier treatment decisions may itself provide valuable guidance for further action.

4.4.1. Feedback in Continuous Tasks

The potential value of this outcome feedback has been discussed by Hogarth (1981). The contributions of feedback become apparent if one observes that judgment often has an incremental or continuous character. For instance, when walking a path, one does not make a priori judgments about the best path. Instead, there is a continuous monitoring of progress through a judgment-action-outcome feedback loop. While many important tasks involve similar feedback loops, the artificial tasks used to study heuristics usually lack any form of meaningful feedback. Thus, our estimates of the prevalence of biased judgments are conceivably based upon a nonrepresentative sample of task environments. The effectiveness of cognitive heuristics ought to be evaluated in tasks having this continuous character.

The role of feedback in continuous tasks is particularly important since it can dramatically affect the accuracy required of human judgment. When feedback is readily available, initial judgment errors are not necessarily serious, since feedback provides information on which to base corrective adjustments. Consider, for instance, the analogy between judgment and aiming at a target (Hogarth, 1981). It is more difficult to aim and hit a target from a distance than to continuously monitor and correct the trajectory. When viewed from this perspective, decision making success critically depends upon two major characteristics of the task: (1) the availability of feedback and (2) the opportunities for taking corrective action based upon that feedback.

The latter point follows from the close relationship between judgment and choice (Einhorn & Hogarth, 1981; Einhorn, Kleinmuntz, & Kleinmuntz, 1979). The judgment process can be conceptualized as a continuous series of mental operations that is periodically interrupted by the selection of actions. Choice is distinguished from judgment by the fact that choice commits the decision maker to action while judgment does not. The key issue is the extent to which a decision maker is committed to a particular course of action. If initial judgments lead to choices that are irreversible, then the opportunities for subsequent corrective actions are lost.

The role of feedback has not been completely overlooked in past research. For instance, research on multiple cue probability learning has focused attention on the role of feedback in the acquisition of judgmental rules (Castellan, 1977; Hammond, Stewart, Brehmer, & Steinmann, 1975; Slovic & Lichtenstein, 1971). However, the process of modifying and improving judgment rules through learning from experience is often difficult and time-consuming (Einhorn, 1980). In the short-run, decision makers may not be able to significantly improve performance through the development of new rules. On the other hand, existing judgmental strategies may be able to capitalize upon outcome feedback by catching and correcting errors through the sort of adaptive response outlined above. Viewed from this perspective, judgmental heuristics may indeed be functional despite their potential for error (Hogarth, 1981).

4.4.2. Dynamic Decision Making From a Continuous Perspective

This continuous perspective also provides a framework for interpreting the results of some previous research on dynamic decision making, which did look at sequences of time dependent decisions (for an overview, see Rapoport, 1975). This research paradigm emphasized the comparison of human performance to optimal solutions derived from mathematical models and, in contrast to the literature on discrete judgment tasks, generally found performance to be effective (e.g., Peterson & Beach, 1967). Hogarth (1981) points out that:

> "In the experimental paradigms used, the structure of tasks is such that deviations from optimal analytic solutions are not heavily penalized. Errors made at one stage can be attenuated by the end of the task. The implications of these results, however, have not been pursued with the attention they merit." (p. 198)

Slovic et al. (1977) discuss a number of reasons for the neglect of dynamic decision tasks. One possible explanation is the difficulty and expense associated with developing the required computer hardware and software. However, recent developments, particularly with respect to relatively inexpensive microcomputers, have made this technology much more accessible. A more serious deterrent may be the difficulties associated with the development of optimal solutions for these complex tasks. However, the continuous perspective outlined above provides a conceptual framework that makes these optimal solutions unnecessary. Instead, research should focus on the impact of task structure on the relative performance of different decision strategies, with a particular emphasis on the role of feedback in determining the cost of erroneous judgments or choices. Furthermore, as an alternative to normative or optimal models, the comparisons can include very simple baseline strategies based on naive principles like random trial and error.

Several recent studies have applied this continuous perspective to dynamic tasks. Hogarth and Makridakis (1981) observed the performance of teams of decision makers in a dynamic, competitive business simulation game. The human decision makers were pitted against artificial teams managed by a set of arbitrarily chosen, non-adaptive rules. The artificial teams "operated at approximately the same level as half the human teams" (p. 102), which indicates that much of the time and effort expended by the human decision makers did not lead to improved performance. In addition, the costs and benefits of rule consistency in dynamic tasks were evaluated by varying the level of consistency of the artificial teams. Results suggest that inconsistency may have strategic advantages in complex, competitive environments.

One common characteristic of studies of dynamic decision making is the construction of elaborate experimental tasks that simulate many of the complexities of realistic decision environments. The use of business simulation games is one example of this. Another example is the multistage medical decision making simulation described by D.N. Kleinmuntz and B. Kleinmuntz (1981). The decision maker is a physician who must diagnose and treat patients by interpreting both diagnostic tests and the outcome of previously administered treatments. Simulation techniques were used to evaluate the performance of three different decision strategies that varied in cognitive sophistication. One strategy was based on normative principles (Bayesian decision theory), another used heuristic principles, and the third used a simple random selection of treatments. While the more sophisticated strategies performed better, in some cases the differences were small. The results illustrate some possible trade-offs between rule complexity and performance.

D.N. Kleinmuntz (1985) extended this work by examining a wider variety of heuristic strategies in the same medical decision task. In addition, selected task characteristics were systematically varied and their influence on performance observed in a set of simulation experiments. The results confirmed the importance of feedback; task variables often considered in discrete contexts (e.g., symptom diagnosticity, disease base-rates) had less influence on performance than feedback-related aspects of the task. The results also provided insights into the costs and benefits of various cognitive heuristics.

Finally, D.N. Kleinmuntz and Thomas (1987) used the same medical decision task, but observed the performance of human decision makers. This was accomplished with microcomputer software that implements the simulated medical task (Kleinmuntz & Larsen, 1984). The subjects' performance was observed in an experiment that directly manipulated the interpretability of outcome feedback and the opportunities for corrective action. Performance did improve with better feedback and with more opportunities for corrective action. However, when the opportunities for corrective action were greatest, almost all the subjects performed worse than a simple random strategy. They could actually have performed better and expended less effort by adopting a trial and error strategy. This result raises important questions about the descriptive validity of error-effort explanations of strategy selection.

4.5. METHODS FOR EXPLORING DECISION STRATEGIES

There are four methodological approaches that have been used to explore information processing strategies in decision making. These are: (1) inference from observed judgment and choice, (2) input-output models of the relationship between available information and observed behavior, (3) observation of predecisional behavior using process tracing methods, and (4) using Monte-Carlo simulations to measure the performance characteristics of decision strategies.

4.5.1. Observed Judgment and Choice

It is often possible to learn a great deal about decision strategies simply by observing subjects' responses in carefully constructed decision problems. The studies of preference reversals and information presentation described above used this method. The goal is to use observations of the end result of decision making to support inferences about the processes involved. Studies of this type can provide extremely compelling demonstrations, but also have some serious limitations: When experimental stimuli are specially constructed for purposes of the demonstration, questions about the external validity of the stimuli are almost unavoidable. Furthermore, descriptions of the processes underlying the observed choices are often subject to competing explanations. As Russo (1977) notes, "choice processes are described after the fact. They cannot be deduced logically from the data, although they are compatible with it" (p.200). Testing the validity of the described process would involve either designing new stimuli to differentiate among competing theories, or collecting more detailed data about the process, as described below.

4.5.2. Input-Output Models

A large body of literature exists on the use of algebraic models to study how judges weight and combine information (for an extensive review, see Slovic & Lichtenstein, 1971). One of the most common approaches is correlational, where each object to be judged is represented as a collection of quantified cue dimensions. A multiple regression, usually linear, is computed using the cues as independent variables and the judge's evaluation as the dependent variable. Variations include estimating the model parameters using discriminant function analysis, analysis of variance, functional measurement, and conjoint measurement. Tests of goodness of fit and predictive validity are used to assess whether the model has "captured" the judge's strategy. Model parameters have been used to measure the importance weights assigned to various cues, to explain suboptimal information utilization, and to provide information on the source of interjudge disagreements (Hammond et al., 1975). Estimates of random error in the model provide an explicit measure of the consistency of judgment (also known as "cognitive control", Hammond & Summers, 1972). Although a number of studies have looked for nonlinearities and configural information combination, linear models have proven to be highly predictive of complex judgments in a variety of settings. (Dawes & Corrigan, 1974).

Although linear models have proven successful in predicting judgments, there has been concern expressed about the adequacy of those models for describing information processing strategies. For instance, Dawes and Corrigan (1974) note that linear models are extremely robust, making it difficult to distinguish among alternative algebraic models, particularily in the presence of fallible data. Furthermore, many different processes could be captured by the same algebraic model (Hoffman, 1960). Finally these models have focused exclusively on the information combination stage of decision making and have not been used to explicitly model processes related to information search or selective attention.

On the other hand, linear judgment models are not devoid of psychological content. Einhorn et al. (1979) argue that these models embody an important aspect of judgment behavior, the process of vicarious functioning (Brunswik, 1952). Postman and Tolman, (1959) interpret this concept as follows:

> "Cues can be used interchangeably so that different patterns of cues can lead to equivalent results. Similarly, different motor responses can result in equivalent behavioral achievements. This is the principle of vicarious functioning which is the essential underpinning of adjustment to an environment which remains partly erratic." (p. 553)

Einhorn et al. (1979) point out that the additive combination function and the random error component of the linear model accurately reflect the ambiguities that the judge faces with regard to substitutions and trade-offs between cues in a redundant environment. While the linear model may fail to capture certain detailed aspects of a judge's strategy, it does reflect the general characteristics of a judgmental metastrategy, based upon the process of vicarious functioning, that may be quite common.

4.5.3. Process Tracing Methods

Process tracing methods are data collection techniques that seek to provide detailed observation during the decision process; collecting information on the sequential aspects of the processing that leads up to a decision. Payne, Braunstein, and Carroll (1978) extensively review these methods, dividing them into two categories: (a) verbal protocol analysis, and (b) records of information search behavior.

4.5.3.1. Verbal protocols

Verbal protocols are collected by instructing subjects to "think aloud" while engaging in the decision task. The protocol is then transcribed, broken up into discrete phrases, and the phrases are then coded or content analyzed to identify cognitive operations, strategies, and knowledge states (Newell & Simon, 1972). Protocol data have been used for exploratory research, as a supplement to other data, for testing specific theoretical hypotheses about decision processes, and for the construction of computer models of decision behavior (for examples, see Payne et al., 1978). Model construction will be further discussed below.

There has been natural concern with the validity of verbal reports of information processing. Nisbett and Wilson (1977) have questioned the ability of subjects to report on higher level mental operations and note that subjects may distort their verbal reports in situations where strong norms of behavior exist. Ericsson and Simon (1980) point out that Nisbett and Wilson's argument holds mainly for retrospective reports of mental processes, particularly when reports are prompted by specific questions. Evidence shows that these difficulties can be largely avoided by collecting unprompted reports, concurrent with the information processing activity.

On the other hand, while protocols do provide a rich source of detailed information, their significance should not be overstated. Simon (1978) notes that:

> "There is very little explicit evidence on the relation of the information contained in the thinking aloud protocols to the underlying thought processes. Most experimenters who have analyzed protocols have assumed that the vocalizations correspond to a subset of the symbol structures that were temporarily present in short-term memory during the course of the problem-solving process." (p. 291).

Furthermore, not all processes are equally accessible. For instance, routine or overlearned tasks may involve automatized processes that will not produce informative verbal reports (Smith & Miller, 1978). Also, the extent to which decision makers are aware and able to report on their own metastrategies is an open question (Einhorn et al., 1979; Payne, 1982).

4.5.3.2. Information search records

One simple procedure for monitoring information utilization involves constructing experimental settings so that information is hidden until the decision maker explicitly searches for it. A common approach is to use an information display board, consisting of a matrix display of cards in envelopes (Jacoby, 1977; Payne, 1976). The rows of the matrix correspond to alternatives, the columns to attributes of the alternatives. Subjects obtain an item of information by pulling a card from an envelope. This approach is particularly appropriate for testing specific hypotheses about the sequence or the extent of information search. Using information search records together with verbal protocols can also be quite useful (Payne, 1976).

A recent development in information search studies has been the use of computer systems that overcome some of the limitations of display boards. These computer systems permit relatively effortless, unobtrusive measurement of information search. E.J. Johnson and Schkade (1985) used a microcomputer display for a study of the anchoring and adjustment heuristics in preference assessment. This permitted the use of graphical displays and, with the use of a "mouse" as a pointing device, a continuous response scale that was particularly well suited for an investigation of anchoring processes (also see E.J.

Johnson, Payne, Schkade, & Bettran, 1986). As noted earlier, D.N. Kleinmuntz and Larsen (1984) have developed a computer program that monitors information search and decision making in a dynamic decision game that resembles medical diagnosis and treatment selection. This permits investigations of information search in a task that is more complex, and hence, more realistic than the relatively simple choice environment represented by the display board matrix. Other recent examples of computer records of search are described by Brucks (1985) and Hoyer and Jacoby (1983).

An alternate technique for observing information acquisition involves recording eye-fixations (Russo & Dosher, 1983; Russo & Rosen, 1975). Analysis of the patterns of eye movements reveals information search behavior and, like other information search methods, can be analyzed to infer characteristics of the underlying strategies. Russo and Dosher (1983) note some advantages of eye fixation data: it requires less effort from the subject than manual search techniques and is not easily censored by the subject. On the other hand, the measurement apparatus is expensive, is relatively difficult to use, and may be quite obtrusive. Technical considerations may also limit the amount of information that can be presented at one time. While these problems have limited the use of the technique, it can provide valuable information on cognitive processes.

4.5.4. Simulation of Decision Strategies

A number of researchers have constructed computer models of decision strategies and then used Monte-Carlo simulations to evaluate the strategies' performance characteristics. One reason for doing this is to validate the descriptions uncovered by process tracing methods. After a subject's decision rules are inferred from the protocol, they are incorporated into a computer program (Bettman, 1979; Clarkson, 1962; B. Kleinmuntz, 1968; Newell & Simon, 1972). The computer model can then be tested by observing the extent to which the program produces the same decisions as the subject it was derived from. An alternate and stronger procedure is to use a predictive test: both the subject and the model generate decisions for new cases and the results are compared (B. Kleinmuntz, 1968). Validation can also be extended to other aspects of the model besides the final choice. This might involve matching the model's information search pattern to the subject's pattern or comparing a detailed trace of the model's processing operations to the subject's verbal protocol (Newell & Simon, 1972).

Another use for simulation techniques is to explore the consequences of various assumptions about the composition of decision strategies. Detailed computer models of strategies can be constructed and their performance observed across a variety of task conditions. This approach differs from the one mentioned above because the models of strategies are often constructed prior to observation of subjects. The simulations of dynamic medical decision making described above demonstrate this approach. D.N. Kleinmuntz and B. Kleinmuntz (1981) note some advantages: (1) it can promote the development of explicit models of processing strategies, (2) it allows the performance of heuristics to be evaluated from a normative perspective in tasks that are more complex, and hence, more realistic, than those tasks that are accessible to other techniques, and (3) the interaction between task characteristics and the performance of various strategies can be explicitly addressed. E.J. Johnson and Payne (1985) have extended the use of simulations to include the measurement of both performance and cognitive effort. Their approach involves decomposing strategies into constituent elementary information processes, simple and basic cognitive operations like comparing two quantities, adding two numbers, or reading a number. Heuristic strategies are implemented as a set of production rules (Newell & Simon, 1972) where each rule specifies a set of elementary information processes and the circumstances under which they occur. Effort is measured by the number of elementary information processes needed to make a given decision. The simulations identified a number of heuristic strategies that approximated the performance level of normative strategies while requiring substantially less effort, although the level of performance was highly contingent on characteristics of the task environment.

4.5.5. Multimethod Approaches

While a variety of methods have been suggested for investigating human decision processes, the possibilities have by no means been exhausted. For instance, there have been relatively few attempts to construct formal mathematical models of decision making at the process level. Notable exceptions include Kahneman and Tversky (1979b), Shugan (1980), and Wallsten (1980). One problem is that some of the characterizations of judgment heuristics have not been developed in enough detail to support efforts at formal modeling. To a certain extent, this observation by Slovic et al. (1977) still holds:

> "Most of the research on heuristics and biases can be considered pretheoretical. It has documented the descriptive shortcomings of the normative model and produced concepts such as representativeness and anchoring that may serve as the basis for new descriptive theories." (p. 6)

New descriptive theories that incorporate representativeness could use Tversky's model (1977) of similarity judgments as a starting point. Anchoring processes have already been incorporated into models of preference assessment (E.J. Johnson & Schkade, 1985) and probabilistic inference (Einhorn & Hogarth, 1985).

The process of formulating new descriptive theories is likely to be aided by a multimethod approach (Payne et al., 1978). Using several different approaches to observe and model the same decision process, including both statistical models and process tracing techniques, allows the strengths of one approach to offset the weaknesses of the others and can also provide convergent validation. Furthermore, concern with finding the "correct" method is misplaced, since each of the various approaches provide valid insights into the decision process, although often at different levels of detail (Einhorn et al., 1979).

4.6. IMPLICATIONS FOR HUMAN-MACHINE SYSTEMS

The research summarized in the preceding sections of this chapter have been concerned with descriptive models and theories of decision making. However, the emphasis on performance leads to a natural concern with the development and implementation of decision aids. To the extent that performance deficiencies are caused by cognitive limitations, the appropriate decision-aiding strategy is to provide tools that will help the decision maker overcome those limits. These aids may be as simple as a pencil and paper, a calculator, or instructions on the use of statistical heuristics. On the other hand, the aids may take the form of highly sophisticated computer systems. Understanding the nature and limitations of human decision strategies can provide important insights about appropriate strategies for aiding and improving decisions.

4.6.1. Problem Decomposition as a Basic Approach

Decision-aiding techniques often rely on the general principle of problem decomposition. This is a "divide and conquer" approach: A large and difficult choice problem is reduced to a set of smaller and ostensibly easier judgments. These judgments are then combined using an algorithm based on either statistical or mathematical considerations. An obvious advantage of this approach is the reduction of information processing demands. The decision maker can isolate and concentrate on each portion of the problem in turn. Furthermore, the laborious task of information combination can then be handled by a formal model, usually implemented on a computer.

The most comprehensive approach to decomposing decision problems is the set of techniques known as decision analysis (Keeney, 1982; Keeney & Raiffa, 1976). These techniques are based largely on normative theories of rational choice, and divide the problem into distinct assessments of preferences (utility or value functions) and uncertainties (subjective probability distributions). Decision analysis has been widely applied (e.g., Bell, Keeney, & Raiffa, 1977) and has also provided a focal point for researchers interested in using descriptive accounts of decision behavior to extend normative theory (e.g., Bell,

1982; Dyer & Sarin, 1982; March, 1978).

One of the major problems that has surfaced in the application of decision analysis is based upon cognitive considerations. It appears that cognitive heuristics play an important role in the assessment of uncertainty (Hogarth, 1975) as well as the assessment of preferences (Hershey, Kunreuther, & Schoemaker, 1982). In particular, the assessments of these subjective inputs seem to be prone to cognitively induced biases. Fortunately, there is at least some evidence to indicate that the results of decision analysis are robust over assessment errors, although the degree of robustness depends on characteristics of the problem (Dawes, 1979; Einhorn & Hogarth, 1975). Perhaps decision analysis and related decomposition approaches are effective because they enable decision makers to accomplish the parts of the task at which they excel, like structuring the problem and using their knowledge and expertise to identify the important variables (Dawes & Corrigan, 1974). At the same time, the decision aid handles the difficult aspects of the task, like aggregating data consistently and using all the relevant information (Einhorn, 1972).

4.6.2. Alternate Approaches

Decomposition seeks to improve decision making effectiveness by providing an explicit strategy for the decision maker to use on complex problems. Another approach is to improve decision making by changing the context rather than the process of decision making. For instance, one could restructure the display of information so that existing cognitive heuristics are more efficient. Russo (1977) did just that by reorganizing the display of price information in the supermarket. Similarly, computer-based decision support systems are capable of presenting information in a variety of formats and in differing amounts and sequences. These are design choices that should be supported by research on the impact of information displays on decision strategies.

More generally, the design of these systems can be conceptualized as a joint optimization of the task structure and the decision strategy (Russo, 1978). Computer-based systems may reorganize the information environment, or provide alternate strategies (like decomposition), or do both. Knowing what blend of these approaches is best is still an open question, requiring research on the interaction between the decision maker, the decision support system, and the decision environment.

Another general issue has been the possibility of tailoring the design of a decision aid to fit particular users. Previous attempts to match users to systems have focused on very broadly defined characterizations of "cognitive style" that have failed to provide meaningful guidelines (Huber, 1983). This is not surprising in light of the research showing the task-contingent nature of cognitive strategies, which suggests that the key issue is to match the system to the task, rather than to the individual.

To the extent that consistent individual differences have been detected, these have been related to the decision maker's degree of expertise and how that expert knowledge is organized (Larkin et al., 1980). This ties in well with another approach for assisting human decision makers, the development of knowledge-based systems. These systems use techniques drawn primarily from the field of artificial intelligence and emphasize the representation of large quantities of domain specific knowledge, along with a set of inferential rules that operate on that knowledge (Duda & Shortliffe, 1983). Also known as expert systems, their knowledge base is usually derived from human experts and the goal of the system is to mimic and, ultimately, to replace the human expert. One criterion for evaluating these systems is to compare their performance to that of the expert.

An occasionally confusing issue is the extent to which an expert system is in fact a model of the expert. While studies of cognitive processes focus on descriptive accounts of decision making and problem solving, expert systems development has a more prescriptive, engineering orientation. Thus, an expert system may be developed in consultation with a human expert, but could perform its task using strategies that are fundamentally different from the expert's.

Does this imply that descriptive studies of decision making behavior are irrelevant to the development of expert systems? P.E. Johnson (1983) argues that using descriptive studies may result in more effective systems. Experts often develop non-authentic accounts of their own expertise. Rather than relying on suspect retrospective accounts, concurrent protocol studies should be used. Although systems developed from non-authentic accounts may be quite effective in some cases, Johnson notes that authentic accounts are also helpful since:

> "In other cases, however, especially when the problem domain of interest is poorly structured, expert practitioners provide a powerful source of successful procedures for doing a task. These procedures have the additional advantage that they are fitted to the information-handling capabilities of human problem solvers". (p. 92)

4.6.3. Evaluating Decision Aiding Systems

Suppose that one has designed a sophisticated computer-based decision support system. How could one evaluate whether that system was in fact an effective aid? One approach would be to compare the quality of outcomes resulting from decisions made with and without the system. Note that the identification of appropriate outcome criteria is itself a nontrivial decision problem. Designing a well-controlled field experiment to compare aided versus unaided decisions is itself likely to be difficult and time-consuming. In fact, Einhorn and Hogarth (1981) note that most evaluations of decision aids are one-shot case studies that are incapable of resolving alternate explanations for observed changes in decision outcomes: "Moreover, the fact that clients are likely to seek aid from decision analysts (broadly defined) when things are not going well renders evaluation of pretest-post test designs lacking control groups particularly susceptible to regression effects" (p. 80).

If objective comparisons of outcomes prove impractical, the only recourse is subjective evaluation of the impact of the aid. Fischhoff (1980) has identified a number of serious problems with this sort of evaluation. For instance, a system might be advocated because of its theoretical foundations, or because it uses sophisticated mathematics. The essentially subjective nature of this evaluation is apparent when one considers that different decision makers may well take opposite positions on the desirability or practicality of theoretical or mathematical rigor.

One could in fact apply decision analytic techniques to the problem of selecting a decision aiding system. For instance, one could decompose the problem by identifying the objectives the system needs to meet and assessing the extent to which each alternative (e.g., system #1, system #2, no system at all, and so on) meets each objective. In essence, this process involves applying a formal decision aid to a metadecision problem. This leads to an interesting dilemma: Should the metadecision be made with the (untested) formal aid, or should it depend upon the same intuitive processes that are candidates for replacement?

Further research may help to resolve some of these issues. For instance, mapping out the contingent relation between task structure and the performance of cognitive strategies will indicate the situations where decision aids are likely to be needed versus those where intuitive decisions are sufficient. Furthermore, this framework should be extended to include the task determinants of performance for machine-aided strategies, as well. The designers and the users of human-machine systems already make intuitive judgments about the costs and benefits associated with various system features. Explicitly identifying the factors that make a problem a good or bad candidate for a particular decision aiding approach would be a positive step.

4.7. CONCLUSION

The study of human decision processes is currently an active and exciting research domain. Recent progress promises a more complete understanding of information processing strategies in judgment and choice and is building important links to other aspects of human cognition. In addition, decision behavior is being examined in a wider variety of environments, including domains that have a dynamic, continuous nature. Examining a broader range of task environments will improve the external validity of descriptive theories of decision making, accounting for differences across tasks and between the laboratory and the real world. In addition, theories that encompass the task-contingent nature of decision making can help in the design and implementation of decision aiding systems and, ultimately, improve the effectiveness of decision making in important problems.

ACKNOWLEDGEMENT

Preparation of this chapter was supported by the Office of Naval Research under Contract N00014-85-K-0752 and a Faculty Research Award from the College of Business Administration, University of Texas, Austin.

REFERENCES

Abelson, R.P. (1976). Script processing in attitude formation and decision making. In J.S. Carroll & J.W. Payne (Eds.), *Cognition and social behavior*. Hillsdale New Jersey: Lawrence Erlbaum Associates.

Abelson, R.P. (1981). Psychological status of the script concept. *American Psychologist, 36*, 715-729.

Beach, L.R., & Mitchell, T.R. (1978). A contingency model for the selection of decision strategies. *Academy of Management Review, 3*, 439-449.

Bell, D.E., Keeney, R.L., & Raiffa, H (Eds.). (1977). *Conflicting objectives in decisions*. New York: Wiley.

Bell, D.E. (1982). Regret in decision making under uncertainty. *Operations Research, 30*, 961-981.

Bettman, J.R. (1979). *An information processing theory of consumer choice*. Reading, Massachusetts: Addison-Wesley.

Brucks, M. (1985). The effects of product class knowledge on information search behavior. *Journal of Consumer Research, 12*, 1-16.

Brunswik, E. (1952). *Conceptual framework of psychology*. Chicago: University of Chicago Press.

Castellan, N.J., Jr. (1977). Decision making with multiple probabilistic cues. In N.J. Castellan, Jr., D.B. Pisoni, & G.R. Potts (Eds.), *Cognitive theory*. Hillsdale, New Jersey: Lawrence Erlbaum Associates. Vol.2.

Chase, W.G., & Simon, H.A. (1973). The mind's eye in chess. In W.G. Chase (Ed.), *Visual information processing*. New York: Academic Press.

Christensen-Szalanski, J.J.J., & Beach, L.R. (1984). The citation bias: Fad and fashion in the judgment and decision literature. *American Psychologist, 39*, 75-78.

Clarkson, G. (1962). *Portfolio selection: A simulation of trust investment*. Englewood Cliffs, New Jersey: Prentice-Hall.

Dawes, R.M., & Corrigan, B. (1974). Linear models in decision making. *Psychological Bulletin, 81*, 95-106.

Dawes, R.M. (1976). Shallow psychology. In J.S. Carroll & J.W. Payne (Eds.), *Cognition and social behavior*. Hillsdale, New Jersey: Lawrence Erlbaum Associates.

Dawes, R.M. (1979). The robust beauty of improper linear models in decision making. *American Psychologist, 34*, 571-582.

Duda, R., & Shortliffe, E. (1983). Expert systems research. *Science, 220*, 261-268.

Dyer, J.S., & Sarin, R.K. (1982). Relative risk aversion. *Management Science, 28*, 875-886.

Ebbesen, E.B., & Konecni, V.J. (1980). On the external validity of decision-making research: What do we know about decisions in the real world? In T.S. Wallsten (Ed.), *Cognitive processes in choice and decision behavior*. Hillsdale, New Jersey: Lawrence Erlbaum Associates.

Eddy, D.M. (1982). Probabilistic reasoning in clinical medicine: Problems and opportunities. In D. Kahneman, P. Slovic, & A. Tversky (Eds.), *Judgment under uncertainty: Heuristics and biases*. Cambridge: Cambridge University Press.

Edwards, W. (1968). Conservatism in human information processing.

In B. Kleinmuntz (Ed.), *Formal representation of human judgment*. New York: Wiley.

Einhorn, H.J. (1972). Expert measurement and mechanical combination. *Organizational Behavior and Human Performance, 7*, 86-106.

Einhorn, H.J., & Hogarth, R.M. (1975). Unit weighing schemes for decision making. *Organizational Behavior and Human Performance, 13*, 171-192.

Einhorn, H.J., & Hogarth, R.M. (1978). Confidence in judgment: Persistence of the illusion of validity. *Psychological Review, 85*, 395-416.

Einhorn, H.J., Kleinmuntz, D.N., & Kleinmuntz, B. (1979). Linear regression and process-tracing models of judgment. *Psychological Review, 86*, 465-485.

Einhorn, H.J. (1980). Learning from experience and suboptimal rules in decision making. In T.S. Wallsten (Ed.), *Cognitive processes in choice and decision behavior*. Hillsdale, New Jersey: Lawrence Erlbaum Associates.

Einhorn, H.J., & Hogarth, R.M. (1981). Behavioral decision theory: Processes of judgment and choice. *Annual Review of Psychology, 32*, 53-88.

Einhorn, H.J., & Hogarth, R.M. (1985). Ambiguity and uncertainty in probabilistic inference. *Psychological Review, 92*, 433-461.

Elstein, A.S., Shulman, L.E., & Sprafka, S.A. (1978). *Medical problem solving: An analysis of clinical reasoning*. Cambridge, Massachusetts: Harvard University Press.

Ericsson, K.A., & Simon, H.A. (1980). Verbal reports as data. *Psychological Review, 87*, 215-252.

Evans, J. St. B. T. (1984). In defense of the citation bias in the judgment literature. *American Psychologist, 39*, 1500-1501.

Fischhoff, B. (1980). Clinical decision analysis. *Operations Research, 28*, 28-43.

Fischhoff, B. (1982). Debiasing. In D. Kahneman, P. Slovic, & A. Tversky (Eds.), *Judgment under uncertainty: Heuristics and biases*. Cambridge: Cambridge University Press.

Goldstein, W.M., & Einhorn, H.J. (1985). A theory of preference reversals. Unpublished paper, University of Chicago, Graduate School of Business, Center for Decision Research, 1985.

Grether, D.M., & Plott, C.R. (1979). Economic theory of choice and the preference reversal phenomenon. *American Economic Review, 69*, 623-638.

Hammond, K.R., & Summers, D.A. (1972). Cognitive control. *Psychological Review, 79*, 58-67.

Hammond, K.R., Stewart, T.R., Brehmer, B., & Steinmann, D.O. (1975). Social judgment theory. In M. Kaplan & S. Schwartz (Eds.), *Human judgment and decision processes*. New York: Academic Press.

Hershey, J.C., & Schoemaker, P.J.H. (1980). Risk taking and problem context in the domain of losses: An expected utility analysis. *Journal of Risk and Insurance, 46*, 111-132.

Hershey, J.C., Kunreuther, H.C., & Schoemaker, P.J.H. (1982). Bias in assessment procedures for utility functions. *Management Science, 28*, 936-954.

Hoffman, P.J. (1960). The paramorphic representation of clinical judgment. *Psychological Bulletin, 57*, 116-131.

Hogarth, R.M. (1975). Cognitive processes and the assessment of subjective probability distributions. *Journal of the American Statistical Association, 70*, 271-289.

Hogarth, R.M. (1980). *Judgment and choice: The psychology of decision*. New York: Wiley.

Hogarth, R.M. (1981). Beyond discrete biases: Functional and dysfunctional aspects of judgmental heuristics. *Psychological Bulletin, 90*, 197-217.

Hogarth, R.M., & Makridakis, S. (1981). The value of decision making in a complex environment: An experimental approach. *Management Science, 27*, 93-107.

Hoyer, W.D., & Jacoby, J. (1983). Three-dimensional information acquisition: An application to contraceptive decision making. In R.P. Bagozzi & A.M. Tibout (Eds.), *Advances in consumer research*. Ann Arbor, Michigan: Association for Consumer Research. (pp. 618-623).

Huber, G.P. (1983). Cognitive style as a basis for MIS and DSS designs: Much ado about nothing. *Management Science, 29*, 567-579.

Jacoby, J. (1977). The emerging behavioral process technology in consumer decision-making research. In W.D. Perreault (Ed.), *Advances in consumer research*. Chicago: Association for Consumer Research, University of Illinois. (Vol.4) (pp. 263-265).

Johnson, P.E. (1983). What kind of expert should a system be? *Journal of Medicine and Philosophy, 8*, 77-97.

Johnson, E.J., & Payne, J.W. (1985). Effort and accuracy in choice. *Management Science, 31*, 395-414.

Johnson, E.J., Payne, J.W., Schkade, D.A., & Bettman, J.R. (1986). Monitoring information processing and decisions: The mouse lab system. Unpublished manuscript, Fuqua School of Business, Duke University.

Johnson, E.J., & Schkade, D.A. (1985). Anchoring, adjustment, and bias in utility assessment. Unpublished paper, Carnegie-Mellon University, Graduate School of Industrial Adminstration, 1985.

Kahneman, D., & Tversky, A. (1979). Intuitive prediction: Biases and corrective procedures. *TIMS Studies in Management Science, 12*, 313-327. (a).

Kahneman, D., & Tversky, A. (1979). Prospect theory: An analysis of decisions under risk. *Econometrica, 47*, 263-291. (b).

Kahneman, D., Slovic, P., & Tversky, A (Eds.). (1982). *Judgment under uncertainty: Heuristics and biases*. Cambridge: Cambridge University Press.

Kahneman, D., & Tversky, A. (1973). On the psychology of prediction. *Psychological Review, 80*, 251-273.

Keeney, R.L., & Raiffa, H. (1976). *Decisions with multiple objectives: Preferences and value tradeoffs*. New York: Wiley.

Keeney, R.L. (1982). Decision analysis: State of the field. *Operations Research, 30*, 803-838.

Kleinmuntz, B. (1968). The processing of clinical information by man and machine. In B. Kleinmuntz (Ed.), *Formal representation of human judgment*. New York: Wiley.

Kleinmuntz, D.N., & Kleinmuntz, B. (1981). Decision strategies in simulated environments. *Behavioral Science, 26*, 294-305.

Kleinmuntz, D.N., & Larsen, J. (1984). DOC: A computer program for studying decision making in a medical context. Working Paper 84/85-4-7, University of Texas at Austin, Department of Management.

Kleinmuntz, D.N. (1985). Cognitive heuristics and feedback in a dynamic decision environment. *Management Science, 31*, 680-702.

Kleinmuntz, D.N., & Thomas, J.B. (1987). The value of action and inference in dynamic decision making. *Organizational behavior and human decision processes, 39*, . in press.

Larkin, J., McDermott, J., Simon, D.P., & Simon, H.A. (1984). Expert and novice performance in solving physics problems. *Science, 208*, 1335-1342.

Lichtenstein, S., & Slovic, P. (1971). Reversals of preference between bids and choices in gambling decisions. *Journal of Experimental Psychology, 89*, 46-55.

Lichtenstein, S., & Slovic, P. (1973). Response-induced reversal of preference in gambling: An extended replication in Las Vegas. *Journal of Experimental Psychology, 101*, 16-20.

March, J.G. (1978). Bounded rationality, ambiguity, and the engineering of choice. *Bell Journal of Economics, 9*, 587-608.

Meehl, P.E. (1954). *Clinical versus statistical prediction*. Minneapolis: University of Minnesota Press.

Newell, A., & Simon, H.A. (1972). *Human problem solving*. Englewood Cliffs, New Jersey: Prentice-Hall.

Nisbett, R.E., & Wilson, T.D. (1977). Telling more than we can know: Verbal reports on mental processes. *Psychological Review, 84*, 231-259.

Nisbett, R.E., & Ross, L. (1980). *Human inference: Strategies and shortcomings of social judgment*. Englewood Cliffs, New Jersey: Prentice-Hall.

Nisbett, R.E., Krantz, D.H., Jepson, C., & Kunda, Z. (1983). The use of statistical heuristics in everyday inductive reasoning. *Psychological Review, 90*, 339-363.

Payne, J.W. (1976). Task complexity and contingent processing in decision making: An information processing and protocol analysis. *Organizational Behavior and Human Performance, 16*, 366-387.

Payne, J.W., Braunstein, M.L., & Carroll, J.S. (1978). Exploring predecisional behavior: An alternative approach to decision research. *Organizational Behavior and Human Performance, 22*, 17-44.

Payne, J.W., Laughhunn, D.J., & Crum, R. (1980). Translation of gambles and aspiration level effects in risky choice behavior. *Management Science, 26*, 1039-1060.

Payne, J.W. (1982). Contingent decision behavior. *Psychological Bulletin, 92*, 382-402.

Peterson, C.R., & Beach, L.R. (1967). Man as an intuitive statistician. *Psychological Bulletin, 68*, 29-46.

Pitz, G.F., & Sachs, N.J. (1984). Judgment and decision: Theory and application. *Annual Review of Psychology, 35*, 139-163.

Politser, P. (1981). Decision analysis and clinical judgment. *Medical Decision Making, 1*, 361-389.

Postman, L., & Tolman, E.C. (1959). Brunswik's probabilistic functionalism. In S. Koch (Ed.), *Psychology: A study of a science*. New York: McGraw-Hill.

Rapoport, A. (1975). Research paradigms for studying dynamic decision behavior. In D. Wendt & C. Vlek (Eds.), *Utility, probability, and human decision making*. Dordrecht, The Netherlands: Reidel.

Russo, J.E., & Rosen, L.D. (1975). An eye-fixation analysis of multialternative choice. *Memory & Cognition, 3*, 267-276.

Russo, J.E. (1977). The value of unit price information. *Journal of Marketing Research, 14*, 193-201.

Russo, J.E. (1978). Comments on behavioral and economic approaches to studying market behavior. In A.A. Mitchell (Ed.), *The effect of information on consumer and market behavior*. Chicago: American Marketing Association.

Russo, J.E., & Dosher, B.A. (1983). Strategies for multiattribute binary choice. *Journal of Experimental Psychology: Learning, Memory, and Cognition, 9*, 676-696.

Shugan, S.M. (1980). The cost of thinking. *Journal of Consumer Research, 7*, 99-111.

Simon, H.A. (1956). A behavioral model of rational choice. *Quarterly Journal of Economics, 69*, 99-118.

Simon, H.A. (1956). Rational choice and the structure of the environment. *Psychological Review, 63*, 129-138.

Simon, H.A., & Hayes, J.R. (1976). The understanding process: Problem isomorphs. *Cognitive Psychology, 8*, 165-190.

Simon, H.A. (1978). Information processing theory of human problem solving. In W.K. Estes (Ed.), *Handbook of learning and cognitive processes*. Hillsdale, New Jersey: Lawrence Erlbaum Associates. Vol.5.

Slovic, P., & Lichtenstein, S. (1971). Comparison of Bayesian and regression approaches to the study of information processing in judgment. *Organizational Behavior and Human Performance, 6*, 649-744.

Slovic, P., Kunreuther, H., & White, G.F. (1974). Decision processes, rationality, and adjustment to natural hazards. In G.F. White (Ed.), *Natural hazards: Local, national, and global*. Oxford: Oxford University Press.

Slovic, P., Fischhoff, B., & Lichtenstein, S. (1976). Cognitive processes and societal risk taking. In J.S. Carroll & J.W. Payne (Eds.), *Cognition and social behavior*. Hillsdale, New Jersey: Lawrence Erlbaum Associates.

Slovic, P., Fischhoff, B., & Lichtenstein, S. (1977). Behavioral decision theory. *Annual Review of Psychology, 28*, 1-39.

Slovic, P., & Lichtenstein, S. (1983). Preference reversals: A broader perspective. *American Economic Review, 73*, 596-605.

Slovic, P., Lichtenstein, S., & Fischhoff, B. (1987). Decision making. In R.C. Atkinson, R.J. Herrnstein, G. Lindzey, & R.D. Luce (Eds.), *Steven's Handbook of Experimental Psychology*.

New York: Wiley. (2nd. ed), in press.

Smith, E.R., & Miller, F.D. (1978). Limits on perceptions of cognitive processes: A reply to Nisbett and Wilson. *Psychological Review, 85*, 355-362.

Svenson, O. (1979). Process descriptions of decision making. *Organizational Behavior and Human Performance, 23*, 86-112.

Tversky, A., & Kahneman, D. (1974). Judgment under uncertainty: Heuristics and biases. *Science, 185*, 1124-1131.

Tversky, A. (1977). Features of similarity. *Psychological Review, 84*, 327-352.

Tversky, A., & Kahneman, D. (1981). The framing of decisions and the rationality of choice. *Science, 211*, 453-458.

Tversky, A., & Kahneman, D. (1983). Extensional versus intuitive reasoning: The conjunction fallacy in probability judgment. *Psychological Review, 90*, 293-315.

Tversky, A., & Slovic, P. (1984). Compatibility effects and preference reversals. Unpublished manuscript, Stanford University, Department of Psychology, 1984.

Wallsten, T.S. (1980). Processes and models to describe choice and inference behavior. In T.S. Wallsten (Ed.), *Cognitive processes in choice and decision behavior*. Hillsdale, New Jersey: Lawrence Erlbaum Associates.

HUMAN FACTORS PSYCHOLOGY, Peter A. Hancock (ed.)

5. AUTOMATIC AND CONTROLLED PROCESSING THEORY AND ITS APPLICATIONS TO HUMAN FACTORS PROBLEMS

Arthur D. Fisk, University of South Carolina
Phillip L. Ackerman, University of Minnesota
Walter Schneider, University of Pittsburgh

5.1. INTRODUCTION

We begin this chapter with a series of human performance related questions. Why is it that a person can be so intoxicated that he/she would not remember a phone number long enough to dial it but still get in his/her car and successfully (not at all safely, but successfully) make it home? Why is it that in tasks requiring people to remain alert for extended time periods (vigilance tasks) people will generally show a decline in performance, but, with subtle changes in the task these people will actually improve over time? Why is it that a person will practice one task for days without showing any improvement, but on another, seemingly more difficult task, show substantial improvement after several hours of practice?

In this chapter we will address these types of questions by providing a discussion of a theoretical and empirical perspective on human information processing that offers a theoretical framework important to human factors. This framework allows quantification, prediction, and interpretation of a broad range of human performance.

The above questions bring to mind the extreme differences often observed between the novice and the expert. Human performance in many cognitive and motor tasks shows profound changes with practice. Consider the characteristics of a tennis match, a chess game, or an auto race between a novice and an expert. The striking change that occurs with the proper kind of experience has led researchers to propose that there are two qualitatively distinct ways that humans process and/or respond to information (e.g., James, 1890; LaBerge, 1973, 1975, 1976;

Logan, 1978, 1979; Norman, 1976; Posner & Snyder, 1975; Schneider & Shiffrin, 1977; Shiffrin & Schneider, 1977). We will refer to these two forms of information processing as controlled and automatic processing.

In this chapter, we review data that indicate human performance can be characterized with these two qualitatively distinct forms of information processing, discuss determinants of skilled performance in complex tasks, and illustrate the applications of this theoretical perspective to problems facing the practice of human factors.

It is important to keep in mind that we are describing a set of information processing characteristics when we use the terms controlled and automatic processing. Controlled processing is characterized as relatively slow, serial, effortful and capacity limited. Controlled processes are under an individual's direct control (for example, rehearsing word lists or telephone numbers, planning a strategic move, etc.), are used to deal with novel or inconsistent information, and reach asymptotic performance with little or no training. Controlled processing is also used to modify long-term memory (LTM).

Automatic processing, on the other hand, is characterized as fast, parallel, fairly effortless (from the perspective of allocation of cognitive resources), and not limited by short-term memory (STM) capacity. Automatic processes are not easily altered by a person's conscious control and are developed only through extensive, consistent practice. It is our belief that automatic processes allow for the attainment of skilled performance.

The automatic/controlled processing perspective has demonstrated that the statements of "practice making perfect" or "hands-on experience produces experts" are only partially correct. Qualitative changes in performance (i.e., a change from controlled to automatic processing) occur when individuals are given extensive consistent practice (not just practice of any kind). Consistent practice occurs when the stimuli and responses are consistently mapped (CM), that is, across training trials the individual makes invariant overt (or covert) responses to stimuli (or classes of stimuli). If individuals receive varied mapping (VM) training, that is a given stimulus requires responses that change across time, automatic processing will not develop and performance will not improve or qualitatively change with practice.

There are six principles of human performance derived from automatic/controlled processing theory which are particularly pertinent to human factors specialists. Throughout this chapter, we will present data illustrating the following human performance guidelines:

* Performance improvements will occur only for situations where stimuli (or information) can be dealt with the same way from trial to trial.

* The human operator is limited, not by the number of mental operations he/she is required to perform, but by the number of inconsistent or novel cognitive (or psychomotor) operations.

* In order to alleviate high workload situations, consistent task components must be identified and, once identified, training of those components should be given to develop automatic component processes.

* Similar to number 3, to make performance reliable under environmental stressors (e.g., alcohol, fatigue, heat, noise, etc.), training should be conducted to develop automatic task components.

* For tasks requiring sustained attention (vigilance), automatic target detection should be developed prior to participating in the vigilance task; also, variably mapped information should not be presented in a continual and redundant pattern.

* When preparing training programs, instructional designers should consider the nature of the underlying processing modes (automatic or controlled) when choosing part-task training strategies.

5.2. REVIEW OF THE RESEARCH

5.2.1. Basic Laboratory Paradigm

In this section we will describe the basic laboratory paradigm and the research that empirically demonstrates the dramatic differences in performance between controlled and automatic processing.

The theory of automatic and controlled processing was put forth by Schneider and Shiffrin (1977) and Shiffrin and Schneider (1977). The characteristics of both types of processing were established by Schneider and Shiffrin in an extensive series of experiments. They utilized a paradigm referred to as a multiple frame procedure.

5.2.2. Multiple Frame Procedure

The multiple frame search procedure is a variant of that first used by Sperling, Budiansky, Spivak, and Johnson (1971) and is illustrated in Figure 1.

A trial in this search task required subjects to detect one of a set of several characters (referred to as the memory set) in a series of CRT displays. Each display included four characters arranged to form a square around a central fixation and is termed a frame. Each trial consisted of the presentation of several frames in immediate succession. The time from the onset of one frame to the onset of the next is referred to as the frame time.

In the Schneider and Shiffrin experiments, three basic independent variables were manipulated in the multiple frame procedure. In order to explore the connection between selective attention and visual search, the number of characters in each frame was varied from one to four. This variable was referred to as frame size. Frame size was held constant during all frames making up a trial. To explore the connection between selective attention and memory scanning, the number of possible targets to be held in memory and searched for during the trial was varied from one to four. This variable is denoted memory-set size.

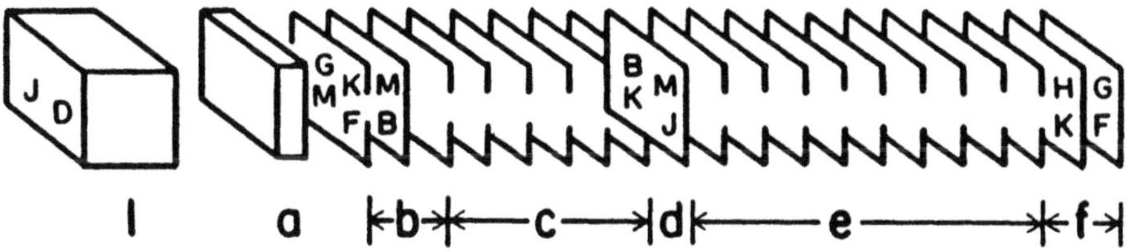

Figure 5-1: An example of a multiple frame search paradigm. Memory set size is 2 (J, D); frame size is 4. (1) Presentation of the memory set. (a) Fixation dot. (b, c) distractor frames. (d) target frame. (e, f) distractor frames.

The crucial manipulation concerned the relationship between the memory-set items and the distractor items. Previous work (e.g., Corballis, 1975; Briggs & Johnsen, 1973; Egeth, Atkinson, Gilmore, & Marcus, 1975; Estes, 1972; Kristofferson, 1972) led Schneider and Shiffrin to suspect that automatic processing would occur when the subject had a great deal of practice at searching for and detecting memory set items that were never distractors (i.e., when there was a consistent mapping (CM) between target and distractors). Conversely, Schneider and Shiffrin expected that subjects would employ controlled processing when there was a varied mapping (VM) between targets and distractors (i.e., when no consistent relationship existed between targets and distractors) or when subjects had been given only small amounts of training.

5.2.3. Basic Results of the Multiple Frame Experiment

The results of the Schneider and Shiffrin Experiment 1 are shown in Figure 2. Frame times greatly diverged among conditions such that it was necessary to graph the results in two panels, using different scalings. In the VM conditions, frame time for a given accuracy level depended strongly upon the memory-set size (M) and the frame size (F). It is apparent from these data that there is a substantial "selective attention" effect. That is, the greater the comparison load (comparison load is the product of M and F) the greater the reduction in the speed at which frames could be presented for subjects to reach a given level of accuracy. Thus, the frame time required to reach a given accuracy level is monotonically related to the product of memory-set size and frame size.

The CM conditions produced a strikingly different pattern of results. First, CM performance is much better than VM performance. Even the most difficult CM combinations of frame size and memory-set size (using an 80-msec frame time) show performance equivalent to or better than that of the 120-msec frame time of the easiest (i.e., small memory-set size and frame size) VM conditions. Also, there is almost no decrement in performance as a function of increased frame size and only a small effect of increased memory-set size on the subjects' performance. That is, it appears that attention may be successfully divided in the CM condition. The CM and VM data clearly show substantially different patterns of performance after practice.

Schneider and Shiffrin also used a single frame reaction time procedure to examine the relation between the accuracy results described above and reaction time results of previous search studies. Analogous to the accuracy data, after extensive training there was virtually no effect on latency due to increased memory set and/or frame size for the CM conditions, but a large increase in reaction time as the memory set and/or frame size increased in the VM conditions. The functions relating comparison load (memory set size x frame size) is reasonably linear with approximately a 2 to 1 negative to positive slope. The variance of the reaction times indicates little effect of memory-set size or frame size in CM conditions but again large effects in the VM conditions. In the VM condition, the data indicate a serial self-terminating search process. Fisk and Schneider (1983) have examined word and semantic category search performance. We will present the data later in this chapter to illustrate the generality of the above results to more complex information processing.

The authors' conclusion, then, was that controlled processing is used in the VM conditions and is serial in nature, while automatic processing is used in the CM conditions and is parallel in nature. Detection performance in the CM condition is virtually independent of the

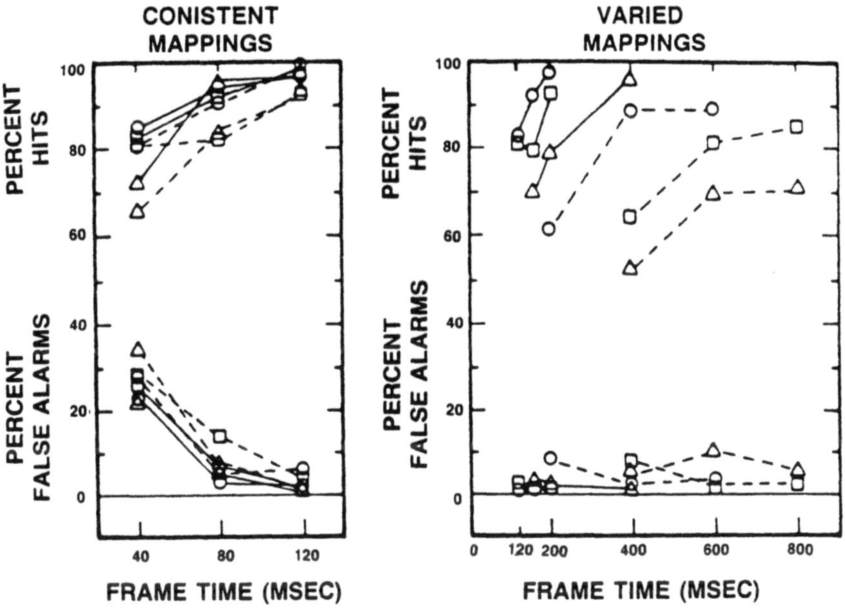

Figure 5-2: Multiple frame detection performance for CM and VM conditions. Circles represent Frame size 1; squares represent Frame size 2; triangles represent Frame size 4. The solid lines represent performance when memory set size was 1 and the dashed lines represent performance when the memory set size was 4 (from Schneider & Shiffrin, 1977).

of characters in one display, the number of characters in successive displays, and the number of characters potentially held in memory.

5.3. FURTHER QUALITATIVE DIFFERENCES

Based on the experiments discussed above, one might think that automatic processing is simply a faster version of controlled processing, with a floor or ceiling effect preventing the effects of either memory set or frame size from appearing. Such a hypothesis, however, is inconsistent with the results of experiments discussed below.

5.3.1. Reversal Experiments

Automatic processing can hinder performance when an automatic process is incompatible with the required task performance. Shiffrin and Schneider (1977) gave subjects 2,100 trials of consistent mapping letter search training. After this initial training, the roles of the memory-set items and the distractors were reversed (i.e., the memory-set items became distractors and vice versa). Under these conditions, the subjects were given 2,400 trials of CM training, which Shiffrin and Schneider termed "reversal training".

Reversing the targets and distractors resulted in severe decrements in the subjects' performance. At the start of reversal training, target detection dropped from 90% to 30% which was below the level found at the beginning of the experiment, when the subjects were naive to the task. This finding suggests that this kind of reversal produces strong negative transfer. That is, controlled processing can be hampered when the distractors are items that have been previously trained to produce automatic attention responses. The second result of particular importance is the slow rate of recovery from the effects of reversal. People can learn a new set of attention responses (and inhibit the old attention responses), but this relearning process is very difficult. In fact, it took subjects almost three times longer to reach a 90% hit rate after reversal that it did in the initial learning procedure.

5.3.2. Inability to Ignore CM Trained Items

Schneider and Shiffrin (1977) explored the degree to which stimuli could be trained to attract attention. The experimental paradigm was similar to the multiple-frame procedure previously described with the following change. One diagonal of each frame was defined to be irrelevant on all trials. Subjects were instructed to focus all of their attention upon the relevant diagonal and ignore all stimuli appearing on the irrelevant diagonal.

There are two conditions of particular interest for the present discussion; these are the conditions when memory set items for the current trial appear on the irrelevant diagonal and when memory-set items have previously been trained to induce automatic detection appear on the irrelevant diagonal. Memory-set items on the irrelevant diagonal in VM conditions are referred to as "controlled foils," and previously trained targets from CM conditions on the irrelevant diagonal are denoted "automatic foils." The main results were as follows: Search performance in the CM conditions (automatic processing) was generally unaffected by foils, either automatic or controlled. Search performance in the VM conditions (controlled search) was largely unaffected by controlled foils. However, search performance in the VM conditions (controlled search) markedly declined (20%) when an automatic foil appeared in the same frame as the target, and was even decreased to a small degree (5%) when an automatic foil appeared in the frame following the target. The decrements caused by automatic foils support the view that attention is immediately attracted by automatic foils, even when they appear on an irrelevant, to-be-ignored diagonal. Such results make it plausible to assume that automatic detection involves the learning of an automatic attention response to consistently trained memory-set items regardless of subjects' attentional intentions (i.e., an "obligatory" response).

5.3.3. Degree of Consistency

To this point consistency has been described as if it was all or none. In the previous experiments, if an item was consistently mapped it always appeared as a target (or always as a distractor) that is, the item was always attended to and never ignored or vice versa. Unfortunately, real world tasks are not so invariant, or, at least are not perceived as invariant by many novices. To address this issue, within a multiple frame letter search experiment, Schneider and Fisk (1982a) examined how varying degrees of consistency determined detection performance over extended practice. In those experiments, degree of consistency was manipulated by controlling the number of times a letter appeared as a distractor while holding constant the number of times a letter appeared as a target. There were five consistency conditions; 1) completely consistent (the CM control); 2) letters appeared as targets twice as often as distractors (67% consistency); 3) items presented equally as often as targets and as distractors (50% consistency); 4) items as targets half as often as distractors (33% consistency); and the VM control (8% consistency, i.e., letters appeared as targets 1/12 as often as distractors).

The results of that experiment are presented in Figure 3. The results are plotted as a function of practice and degree of consistency. The data show that performance improves as a quasi-multiplicative function of the degree of consistency and the amount of practice. In the perfectly consistent condition, practice resulted in substantial performance improvement. However, if the consistency was 33 percent (or less). Performance did not improve with practice.

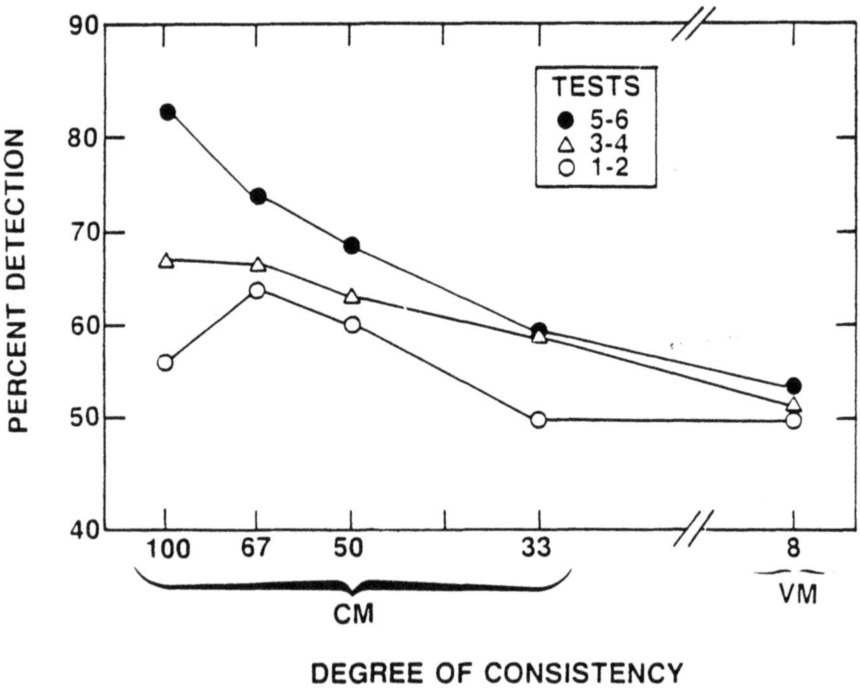

Figure 5-3: Detection accuracy as a function of degree of consistency (in percentages) at different levels of training. Each TEST refers to performance after 400 trials of practice (from Schneider & Fisk, 1982a).

5.3.4. Searching Versus Detecting

What is the nature of the benefit of consistent practice? Is this benefit due to consistently detecting a stimulus or due to consistently searching for a stimulus? If practice at consistently searching for a stimulus item is the locus of performance improvement, then consistent searches without detecting should show some performance improvement. Schneider and Fisk (1980) examined this issue by varying the ratio of positive searches to negative searches (a positive search involved correct detection of a target, a negative search required a correct rejection response). The results indicated that performance improves only as a function of successful consistent detections. A more striking result was that experience at searching for characters without detecting them actually resulted in a decrement in detection performance. The benefit of knowledge of results (see Adams, 1971) and guided training (Welford, 1976) could be interpreted, based on these data, partly as a result of allowing consistent positive executions of the consistent task components.

5.3.5. When Will Learning Occur?

A basic issue in cognitive psychology, and one that has direct implications for human factors, is the manner in which learning and information processing are interrelated. Automatic/controlled processing theory states that accurate processing can occur with little or no long-term memory modification. Fisk and Schneider (1984b) have presented data which strongly suggest that learning and controlled processing are closely linked. These data also indicate that, once developed, accurate automatic processing can occur with little or no further learning. The application of these results to the training environment is straightforward. For learning to take place, controlled processing must be utilized by the learner and this controlled processing must be task appropriate.

5.3.6. Component Consistency

The Schneider and Fisk (1980, 1982a) studies have demonstrated that development of automatic processing in visual search is a function of the number of correct detections of consistent stimulus-response patterns. Fisk and Schneider (1984a) examined whether or not the performance of a total task could benefit by automatic process development for consistent task components. That is, if the benefits of automatic processing were limited to only those tasks that were consistent from input to output then there would be little relevance in the natural environment for automatic processes. To evaluate this issue, the effects of the consistency of attending (CM or VM attending) and the consistency of motor responding (consistent or inconsistent response translation) on overall performance of the total task were jointly and independently examined. Consistency of attending and consistency of motor responding were factorially combined in a multiple frame detection paradigm. The data indicated that consistent attending produced a substantial improvement in search performance regardless of the consistency of the responding component. These results suggest that automatic component processing can improve total task performance relative to tasks without consistent components.

These data are particularly relevant to the human factors practitioner because they suggest how performance improvements due to practice may be predicted. As the number of critical consistent components increases so should the positive effects of practice (assuming some weighting function for "criticality" of the components). The more consistent components in a task, the more likely it is that individuals will benefit from practice. As a corollary to prediction of performance improvements, these results demonstrate the need to identify and train the consistent components of a task in order to obtain maximum training benefit.

5.4. ATTENTIONAL RESOURCE REQUIREMENTS OF AUTOMATIC/CONTROLLED PROCESSING

A central issue of automatic/controlled processing theory concerns the differing attentional requirements of automatic and controlled processing. Much research examining attentional-resource capacity assumes that there is, generally speaking, a fixed task-specific amount of attentional resources available for task performance (see Kahneman, 1973; Navon & Gopher, 1979; Wickens, 1980). The proposal that all processes consume resources from a limited pool (or pools) implies that there is an upper limit to human processing ability (due to limits of attentional resources) for every task. While this statement may at first appear reasonable, everyday experience contradicts it.

Our theoretical position is that consistent information processing can become effectively free from attentional resource consumption (when automatic processing is fully developed). Automatic processing may not be free from attentional resource consumption in an absolute sense. However, for most practical applications, capacity limits do not influence performance when sufficient consistent practice is provided. Conversely, controlled processes are resource consumptive. Therefore, performance is limited by the number and extent of task components requiring controlled processing.

Several experiments have demonstrated that subjects can concurrently perform two tasks without measurable deficit (Fisk & Schneider, 1983, 1984; Schneider & Fisk, 1982a, 1982b). However, for performance to be relatively deficit free, one (or both) of the two tasks must be automatically processed (see Schneider & Fisk, 1982a, 1982b, for a detailed discussion). Two tasks that require controlled processing will show substantial performance deficits. It is critical for the human factors specialist to keep in mind that combining two tasks can have a substantially different outcome depending on whether the tasks are carried out by controlled or automatic processing. Of course, for consistent tasks, the amount of practice prior to (and during) combining the two (or more) tasks is crucial to such performance findings.

5.4.1. Metaphor - How Is Capacity Limited?

The following metaphor (from Schneider, 1982) illustrates the theoretical issues of attention limits we have outlined in the above discussion. The metaphor illustrates the "capacity" limits of a large office with an adaptive communications network. Assume that controlled processing is equivalent to a public address system and automatic processing is equivalent to private message communication links. Also assume that each individual in the office can process information independent of the others (i.e., in parallel). Individuals initially communicate through the controlled processing public address system. When someone sends a message, everyone else receives the message. That is, we have a system with parallel processing but serial communication.

How would this office "perform" a visual search task? In this office, each processing unit (each person) can be inactive, be processing information, be waiting to use the public address system, or be transmitting on the public address system. We present the words "Dog," "Pine," "Apple," and "Red" to four input processors (four people). Each input processor translates, in parallel, the words into their semantic categories. The first person calls out the category "animal" over the public address system, while the others wait (in order not to garble the message). The match processor (some other person) receives the message and determines whether this is the category "fruit." Serially, the second, third and fourth processors use the public address system to call out "Tree," "Fruit," and "Color." When "Fruit" is output, the match processor detects a match, waits to get the public address system, and then outputs "Respond Yes." The response processor receives that message and the category response is made.

The predicted results are that reaction time is a linear function of the number of comparisons. As more processors must wait to make their response, the reaction time increases. This pattern of data has been presented above during our discussion of VM performance.

5.4.1.1. Office Metaphor for VM Dual Task

Suppose this office now has two simultaneous tasks to perform. In this case words are presented at a fixed rate and performance is measured by how many category words the office detects. For added workload, a concurrent digit task which is to be treated as the primary task is also presented to the office. In this case the office must determine whether the presented digits match the digits in memory.

This means that the office personnel (processing units) processing words must wait until the office personnel processing digits have finished using the public address system. The digit task increases the effective processing time of the words, resulting in more missed targets. Thus, the hypothetical single/dual task deficits will mimic the controlled processing results observed in actual dual task visual search experiments. This office might be described as a very slow, serial, capacity-limited system.

5.4.1.2. Office Metaphor with Automatic Processing

Now let us add automatic processing to this office communication network. Assume that, in addition to the public address system, the office managers can install private communication channels between frequently communicating office workers. With such a private communication linkage, one unit can communicate a specific message to another unit without using the public address system. Private lines are strengthened only after a message is transmitted by a unit and the follow-on unit also transmits a message. If a unit transmits a message and there is no follow-on message, the private line is weakened. Therefore, consistent responses to messages produce strong private lines that can support reliable communication without using the public address system. Varied response to messages results in little or no private line growth, which would necessitate the use of the public address system for reliable communication.

After developing the private communication links, how would the office operate? For a reaction time task there would be no delay caused by personnel (processing units) waiting to transmit their information. Thus, there would be no increase in reaction time with additional comparison demands. (If the private line transmission is noisy there will be some speed benefit if messages are transmitted over both the public address system and the private line. Thus, making one comparison could be faster than four comparisons, which would result in a slight comparison slope.)

How would this office perform on the dual-task problem that was presented previously? (Assume that the digit detection task requires controlled processing while the word detection task can be performed via automatic processing.) The digit tasks processing units would utilize the public address system (i.e., controlled processing). The word processing units would use their private lines and any free time on the public address system. As the private lines develop for the word units, semantic messages can be transmitted without use of the public address system and any initial capacity bottleneck is eliminated.

The use of private communication links can effectively eliminate the bottleneck caused from using the public address system. However, this does not imply that our office with private communication links is "capacity free" in an absolute sense. Clearly, as the processing load on the office increases, at some point, the processing for any one message will deteriorate.

5.4.2. Metaphor Summary

The office metaphor illustrates phenomena described in the beginning of the chapter. Performance on VM tasks is stable, slow, serial, and shows severe capacity limits. Performance on CM tasks shows substantial improvements with practice; performance is ultimately fast, parallel, and does not result in a measurable dual-task deficit when combined with a controlled process task. The metaphor also illustrates a mechanism by which practice could result in profound changes in performance and demonstrates the limits of the single undifferentiated resource view of attention (Kahneman, 1973).

5.4.3. Dual Task Experiments

Data from perceptual learning/visual search experiments point to the utility of the office metaphor. Schneider and Fisk (1982b) examined subjects' abilities for dual-task performance. In those experiments, simultaneous automatic and controlled processing versus simultaneous performance of two controlled processing tasks were compared. The simultaneous automatic/controlled processing conditions required subjects to perform two character search tasks - a VM character search (e.g., search for digits among digits) on one diagonal of the display and a CM character search (e.g., search for letters among digits) on the other diagonal. In the dual-task conditions, subjects searched on the CM diagonal for any letter and on the VM diagonal for a specific digit. In the single-task conditions, subjects searched for a target on one diagonal only. The results are presented in Figure 4A. The measures on the axis are A', which is a nonparametric analog of d' (see Craig, 1979; Norman, 1964).

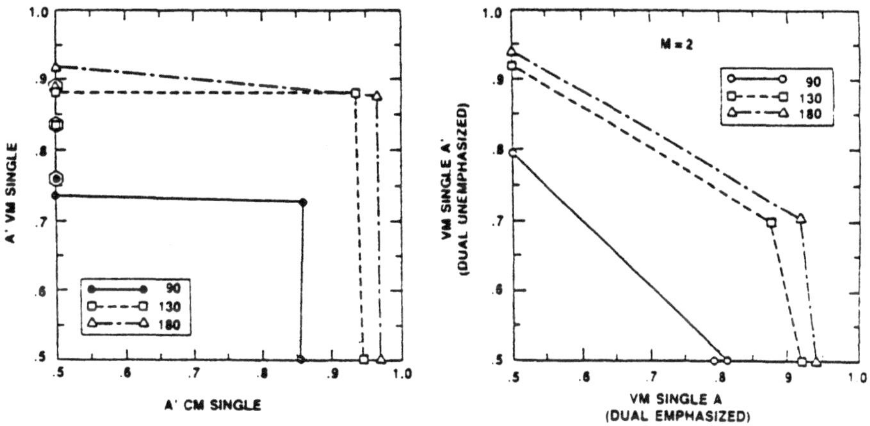

Figure 5-4: Dual task POC for CM and VM letter search tasks (4A) and two VM letter search tasks (4B). Note the rectangular POC's (4A) which indicate resource insensitivity and the triangular POC's (4B) which indicate substantial resource sensitivity (from Schneider & Fisk, 1982b).

The rectangular shape of the performance-operating curves (POC) curves in Figure 4A indicated that both tasks were carried out together without noticeable performance decrement (see Norman & Bobrow, 1976). As a measure of the dual-task trade-off, one should compare the shape and the area of the obtained POC to the shape and area of the rectangle projected from the single-task performance levels. For joint CM and VM search (Figure 4A) these data represent 98.1% of the area of the projected no-trade-off POC. This suggests that the CM and VM comparison processes do not compete for resources.

For a strong demonstration of the resource insensitivity of automatic processing, it is important to examine the same dual task combination that was described above with respect to joint controlled processing requirements. In another experiment, the subjects were required to perform controlled processing on both diagonals of the display. In that experiment there was a substantial performance decrement in the dual-task conditions (see Figure 4B). The subjects were given a substantial amount of training on the joint controlled processing task. Even with this extended training, subjects were not able to simultaneously perform both VM component tasks without performance deficit.

We have also examined dual-task performance using more complex stimuli (Fisk & Schneider, 1983). Subjects performed an automatic search for semantic category members while simultaneously performing a digit-recall task. Subjects could carry on a digit-span task and simultaneously determine whether each of 16 words were members of any one of four semantic categories with less than a two percent deficit in either the digit-span or category detection tasks. However, controlled processing search (for only four words) resulted in substantial performance deficit.

Schneider and Fisk (1982b, 1984) examined subjects' performance after extended dual-task training. Figure 5 demonstrates the differential effects of practice for CM and VM conditions. Early in practice, adding another task to the category search task resulted in a substantial performance decrement for both consistent and varied mapping search. With practice, however, CM and VM dual task performance differed markedly. Figure 5 illustrates that while the CM decrement was eliminated (1% decrement) with practice the VM decrement (61%) was relatively unchanged from the first session of training. These data lead to another human performance principle: Consistent practice can make task performance resource-insensitive. Varied mapping practice results in little, if any, reduction in resource requirements.

Others (e.g., Hirst, Spelke, Reaves, Caharack, & Neisser, 1980; Spelke, Hirst, & Neisser, 1976) have proposed that extended time-sharing training is all that is required to eliminate dual-task interference. However, our data (e.g., Figure 5) indicate that extended training is not task decrements it is necessary (but not sufficient) to provide extended training (sometimes hundreds of hours) on consistent tasks. When there is limited practice or extended training on inconsistent tasks, dual-task trade-offs will occur. (see Logan, 1979; Fisk & Schneider, 1983; Schneider & Fisk, 1984; Shiffrin & Schneider, 1977).

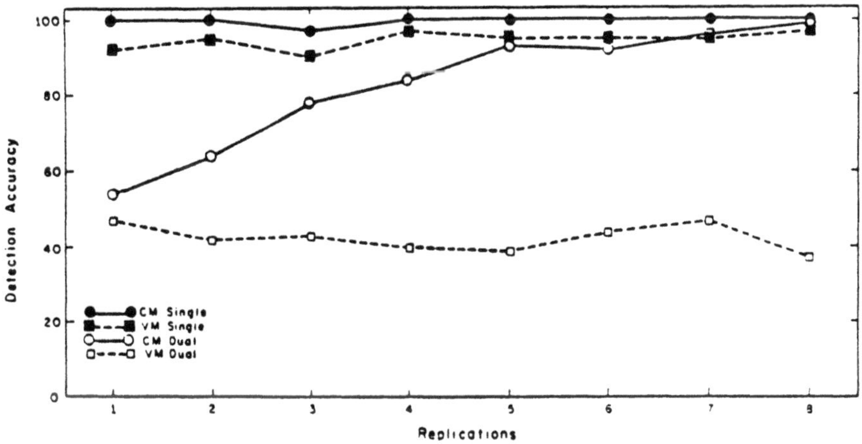

Figure 5-5: Single and dual task CM and VM semantic category search performance presented as a function of training (from Schneider & Fisk, 1984).

It is clear that, with sufficient training, automatic processing of consistent task components can occur without evidence of a need for attentional resources. These results have implications for the human factors practitioner. The implications are that: 1) acceptable multiple task performance will depend on whether the additional tasks involve primarily automatic or controlled processing; 2) multiple task performance will only improve for those components capable of being performed by use of automatic processing; and 3) properly constructed, there appears to be no inherent limit to the complexity of an automatic process.

5.5. COMPLEX STIMULI - GENERALIZABILITY OF THE VISUAL SEARCH RESEARCH

The theory of automatic and controlled processing, as with many other theories concerning cognitive processing, grew of out of research that used relatively simple stimuli (such as digits, letters, colors, etc.). In order to apply laboratory research to natural world situations, the major phenomena described by the theory must generalize to complex processing.

Eberts and Schneider (1980, 1985) trained subjects to control a second order (acceleration) system. Some of the subjects were trained with a salient cue predictor that allowed those subjects to internalize the consistencies between a given control input and resultant system output. Two other groups were trained without the benefit of the predictor. Their results provide evidence indicating that, while all groups could control the second order system, only the group trained with the salient predictor could develop an accurate internal model of the system. The accurate internal model resulted in a reduction of workload. Eberts and Schneider conclude that consistent training is necessary for development of many skills as well as the development of accurate internal models.

In another study, Eberts (1979) examined consistent and varied mapping of visual sequences of events that changed spatially and temporally. In a series of nine experiments, he showed qualitative differences between the CM and VM trained visual sequences. These differences were: 1) CM performance was best with fast stimulus presentation (optimization of apparent motion) whereas VM performance was best at very slow presentation; 2) CM trained sequences were easier to mentally rotate than VM trained sequences; and 3) there was little effect of processing load for CM sequences but substantial load effects for VM trained sequences.

We (Fisk & Schneider, 1983; Schneider & Fisk, 1984) have conducted research that examined how the major phenomena of visual search for single characters generalized to word and semantic category search. In that research we evaluated five prominent issues in the search literature: 1) linear memory set-size effect; 2) exhaustive versus self-terminating search; 3) effects of comparison load; 4) reduction of comparison slopes with practice; and 5) reduction in effort (resource cost) with practice. (Issues 1, 2 and 3 are primarily varied mapping effects while issues 4 and 5 are consistent mapping effects.)

In these word and category search experiments, subjects were presented with from one to four words or from one to four category labels as their memory set and then the subjects were presented with two words as the probe display. If either of the two words in the probe display were members of any of the presented categories in the memory set (or words if the trial was a word search trial), subjects pressed a "target present" button; if not they pushed the "target absent" button. Figure 6 shows the reaction time data plotted as a function of number of words or categories in memory. The reaction time slopes of the VM category condition were 96 msec for trials when targets were present and 202 msec for trials when targets were absent (slopes are a function of the number of comparisons, see Figure 6, left panel). These results indicate that the comparison process was serial self-terminating with a comparison time of about 200 msec. In contrast, the consistently mapped category comparison time was 2 msec per category (Figure 6, right panel).

These results show the prominent varied and consistent mapping effects seen in previous character search experiments. Therefore, there are similar cognitive mechanisms underlying search for letters, words, categories, or superordinate categories. In fact, with the exception that the word and category VM slopes are slower, the semantic category search results are equivalent to previous letter search results. The category search reaction times correlated .98 and .95 with the Briggs and Johnson (1973) and Schneider and Shiffrin (1977) letter search data, respectively.

Fisk and Schneider (1983; Schneider & Fisk, 1984) found reductions in resource cost in CM category search that parallel the earlier letter search results. They also demonstrated that the results were due to true category search, not simple element (i.e., feature or word) learning (see Schneider & Fisk, 1984). To summarize, the assumptions relating to practice effects, the importance of consistency, and resource costs have

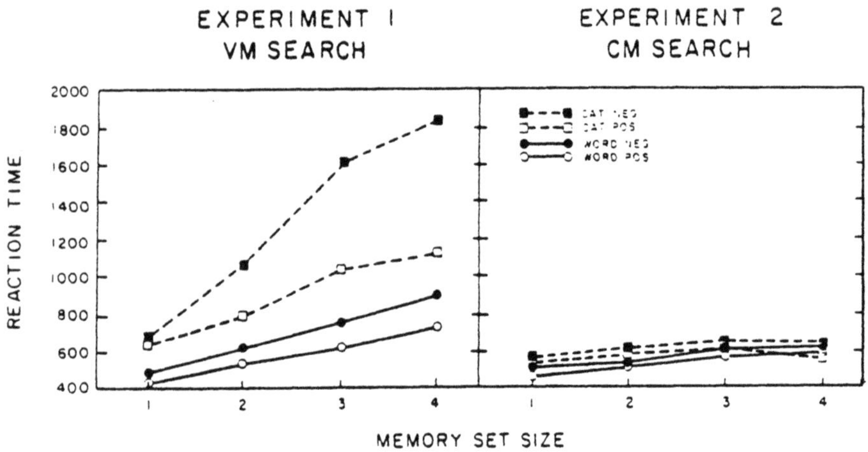

Figure 5-6: Reaction time as a function of the number of memory set items to compare for both VM and CM word and category search conditions (from Fisk & Schneider, 1983).

been confirmed for character, word and semantic category processing.

5.6. RELIABILITY OF AUTOMATIC PROCESSES

This next section begins with somewhat of a paradox. Why is it that a person who is so intoxicated that he/she cannot remember a phone number long enough to dial it can get in an automobile and successfully (not at all safely, but successfully) drive home?

An answer to this apparent paradox is that automatic processes are reliable. Research suggests that consistent task training can make performance resistant to the effects of heat stress, alcohol ingestion, and fatigue (Fisk & Schneider, 1981; 1982; Hancock, 1984; 1986; Hancock & Pierce, 1984). For example, in an experiment examining the effects of alcohol on performance, Fisk and Schneider (1982) found that six drinks had little effect on consistent mapping search performance. In contrast, varied mapping search performance dropped considerably. In that study, the effects of alcohol on task performance were examined by looking at six different tasks. Consistent and varied mapping training was combined with a compatible responding task, a variable responding task (within-task time-sharing), and a tone discrimination task (between-task time-sharing). Figure 7 shows the results of that experiment.

To summarize the data, it is clear from Figure 7 that there is a minimal effect of alcohol (.1% Blood Alcohol Concentration) on CM performance. However, there is a substantial effect of alcohol intoxication on performance of VM tasks; that is, those tasks carried out primarily with controlled processing. More importantly, substantial alcohol-related deficits occurred when subjects were required to divide attention between two controlled processing tasks.

These data do suggest that an inability to divide controlled processing resources either between or within tasks is the locus of the alcohol intoxication effect. This experiment also demonstrates a powerful theoretical framework for prediction of stressor effects. Such results suggest that processing mode (i.e., automatic vs. controlled) is more important than stage of processing, task complexity, etc. when predicting stressor effects.

Another demonstration of the reliability of automatic processing is work by Fisk and Schneider (1981) on vigilance performance. Vigilance is the ability to remain alert (or sustain attention) over extended time periods. The amount of vigilance decrement indicates subjects' failure at sustaining attention. Vigilance decrements are interpreted within the automatic/controlled processing framework by specifying that the type of processing and amount of practice can predict vigilance decrements. The vigilance experiments reported in Fisk and Schneider (1981) showed that the normal vigilance decrement was greatly attenuated when subjects could detect signals using automatic processing. Furthermore, the research suggests that maximum vigilance decrements will occur when observers must continually and redundantly allocate controlled processing resources to the task during the vigil.

Four guidelines are provided for reducing vigilance-related problems in natural settings (Fisk & Schneider, 1981, page 748): "(1) When possible, critical signals (targets) and target features should be consistently mapped as targets. (2) Training must be carried out to

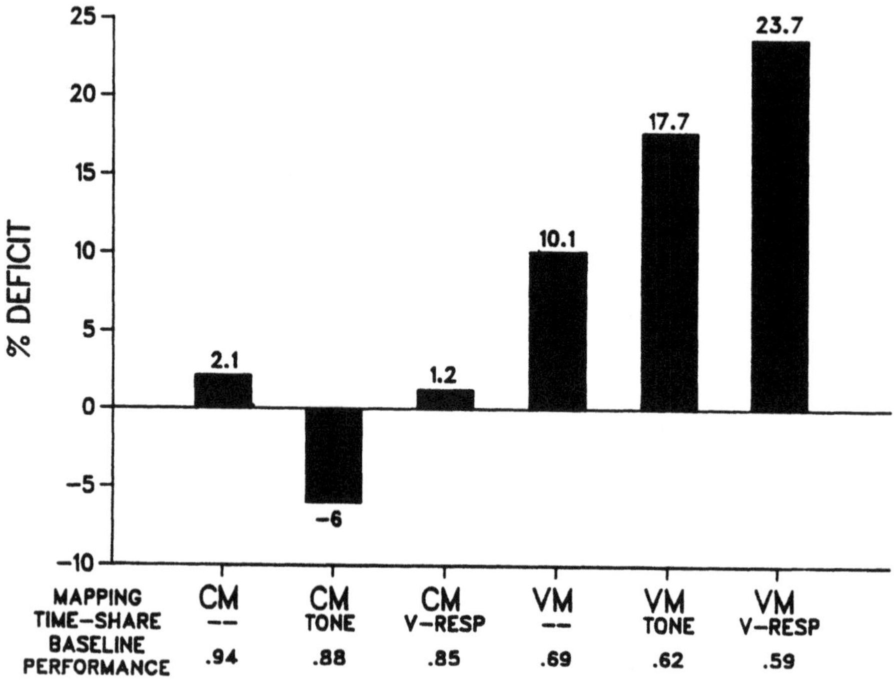

Figure 5-7: Performance decrements due to alcohol. Mapping refers to the consistency of the target letters. Time-share refers to the additional dual task either no dual task (--), a TONE discrimination task, or a variable position responding (V-RESP) to the target. Baseline refers to performance without alcohol (from Fisk & Schneider, 1982).

maximize the development of automatic processing prior to placing the observer in the task requiring vigilance. (3) If tasks require controlled processing, they should be structured to encourage the observer to reallocate controlled-processing resources to other tasks from time to time. This implies that the operator should be given enough task diversity (of mental processes) to allow reallocation of controlled processing resources. (4) Information that cannot be consistently mapped should not be presented in a continual and redundant pattern to the observer. For example, instrument design and training in situations requiring primarily controlled processing should induce the operator to sample the instruments in a varying pattern."

Further evidence of the reliability of automatic processes is provided by Hancock (1984, 1986; Hancock & Pierce, 1984). Hancock's analysis of operators' performance under heat stress indicates that little deterioration in performance will be observed when automatic processing would be primarily involved in task performance. However, adverse heat stress effects seem to occur when tasks primarily demand controlled processing. Hancock's review of the literature suggests that, in addition to heat stress, the automatic/controlled processing framework can be used to predict performance decrements for various environmental stressors (see Hancock, 1984 and 1986 for detailed discussions).

5.7. INDIVIDUAL DIFFERENCES IN AUTOMATIC/CONTROLLED PROCESSING

The characteristics of automatic and controlled processes lead human factors researchers to many inferences about the normative training environment. However, the striking differences between automatic and controlled processes have prompted research into individual differences as well. Determination of how people differ in controlled processing capacity (or efficiency) as well as the speed (or asymptote) of automatic process development may have ramifications for trainee selection facets of human factors. In addition, such research aids in the design of instructional programs by pointing to the underlying causes of performance differences between individuals of equal experience as well as expert-novice differences. A few background points are discussed below that are relevant to individual differences in automatic and controlled processing.

5.7.1. The Performance-Resource Function (PRF)

The notion of the PRF postulated by Norman and Bobrow (1975) together with automatic and controlled processing theory provides a somewhat richer language for understanding the sources of normative performance differences among tasks that often differ to a seemingly small degree. Norman and Bobrow offer two concepts relevant to task performance. The two concepts are: 1) Performance limitations based on the amount of cognitive or attentional resources devoted to a task; and 2) Performance limitations imposed by task characteristics. These concepts have been respectively termed "resource-limitations" and "data-limitations." In theory, by gathering the relevant performance information when the amount of cognitive resources devoted to operation of a task is manipulated, a derived PRF may be considered and evaluated with respect to task requirements and subject performance.

The concepts of data-limitations (which we refer to as "resource-insensitivity") and resource-limitations (or "resource-dependence") may be linked with task and performance characteristics in the automatic and controlled processing domain. To review, novel or inconsistent tasks that require controlled processing may be resource-dependent. For inconsistent tasks, no automatic processing development can occur; thus performance will be continuously resource-dependent, even after considerable amounts of practice. However, for a novel task which has the consistent characteristics necessary for development of automatic processing components, task performance will change from the controlled processing, resource-dependent stages to automatic processing development stages. As automatic processing components develop, performance becomes less resource-dependent and thus more resource-insensitive. When a task becomes automatic, fast and effortless performance is possible. This stage is characterized by stable performance and relatively unaffected by the amount of cognitive resources devoted to the task (i.e., a resource-insensitive condition).

This automatic process condition, though, represents only one situation when resource-insensitivity occurs. Other task characteristics may impose such limitations when controlled processing is required (such as ceiling or floor effects). For example, if an inconsistent task only requires 20% of a subject's controlled processing resources to attain errorless performance, then the PRF will also show a broad range of resource-insensitivity. Conversely, a task may have automatic processing components but still have a substantial number of inconsistent information processing requirements. In this case, even though improvements in performance occur with practice, the task will remain resource-dependent, based on the influence of the controlled processing components.

5.7.2. Modeling Individual Differences in Automatic/Controlled Processing

The PRF exposition also provides the descriptive language for a general ability/performance approach to individual differences. Norman and Bobrow describe resources as a portion of a subject's attentive capacity. It has been proposed that, slightly modified, the concept of resources may be used in the consideration of individual differences (Ackerman, 1984b). In fact, any subject's performance may be evaluated as a joint function of the subject's relative attentive/cognitive capacity (i.e., "relative" to the population) and the proportion of the subject's total capacity devoted by the subject to the task.

An interesting study that illustrates individual differences in automatic processing development is one by Adams (1957). As the data from the Adams study demonstrated, subjects "learn" quite different amounts over a period of task practice, depending on their initial performance levels. Adams divided his subjects into decile groups, based on a rank-ordering of performance on the first task trial. The results indicate that (for a simple and consistent psychomotor task) subjects with initial performance falling in the highest decile, on the average, showed only marginal relative improvements in performance over practice trials. Subjects in the lowest decile, though, showed substantial performance increments over time (in the Adams data, mean performance improvement for the lowest decile exceeded improvement in the highest decile by a factor of six).

It is essential to note that the group mean data do not tell the whole story. The high decile groups are relatively homogeneous from the beginning to the end of the training. The low decile groups, while relatively homogeneous at trial one, are greatly heterogeneous in performance at the last practice trial. In fact, the heterogeneity is so pronounced that many of the subjects in the lowest group actually attain scores in the top performance categories after practice. The major implication is that some subjects can make up for poor initial performance and develop automatic processes to the same degree as the subjects that are in the upper deciles.

In general, those subjects who begin a task with superior performance ultimately maintain their relative superiority in performance subsequent to practice. The converse is true for many subjects that begin the task with clearly inferior performance. However, there is a substantial asymmetry between the initial and final performance distributions for low and high performance groups. The important concerns for theories of individual differences in learning (automatic processing development) are to identify the subjects who will: a) perform well initially, and b) ultimately perform well (relative to the sample) subsequent to practice even though they fail to perform the task with distinction early in the training sequence. Merely being concerned with initial scores, final scores, or "gains" in performance obscures these differential effects that occur during the practice sequence.

The data adapted from Adams (1957) in Figure 8 show the type of transition from "ability-dependent" to "ability-insensitive" performance-ability functions that is expected in the performance-resource function approach (for consistent tasks). A reanalysis of results from Ackerman (1984a) shows a similar finding for a consistent (CM) task as well as provides the appropriate contrast for a task which requires a greater continued investment of controlled processing resources (Figures 9 and 10 respectively). As predicted, although some early performance improvement is demonstrated, the VM task shows ability-dependent performance throughout the five hours of cumulative task practice. Further discussion is provided in Ackerman (1986).

5.7.3. Resources and Processes: How Do They Relate to Human Abilities?

An issue related to determining the characteristics of resources and automatic/controlled processes concerns the specialization or differentiation of resources. A current controversy pertains to the utility of considering resources as undifferentiated (Kahneman, 1973) or specialized in separate structures (such as verbal or spatial structures; Wickens, 1980). Evidence has been reported supporting elements of both positions, indicating that each approach has some merit

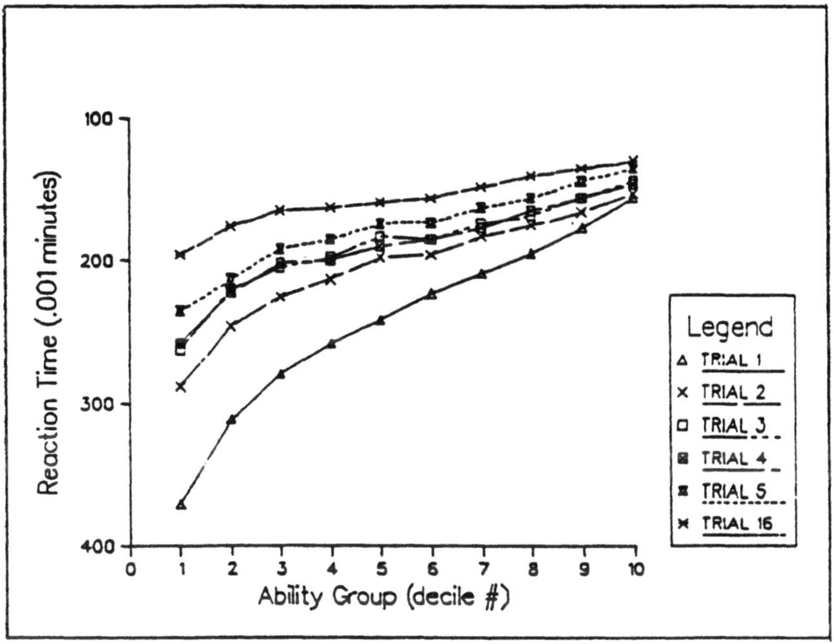

Figure 5-8: Data adapted from Adams (1957). Four unit discrimination reaction time task. Reaction time on the ordinate is reflected to demonstrate similarity to performance-resource functions. Smaller reaction time indicates "better" performance. Later trials show "ability-insensitivity".

(Navon & Gopher, 1979). To bring these theories and data into agreement with the psychometric work in intellectual abilities, it seems plausible to maintain that some compromise position is representative of the nature of cognitive resources. That is, there may be a pool of undifferentiated resources available to the individual, but there may also be more specialized structures of resources which are associated with particular task content domains. Although a classification of resource structures along the same lines as group factors in intellectual abilities goes beyond the current data, considering resources in this manner is consistent with factor analytic sources of ability data and has intuitive appeal. Such a position is advocated in the theory described below.

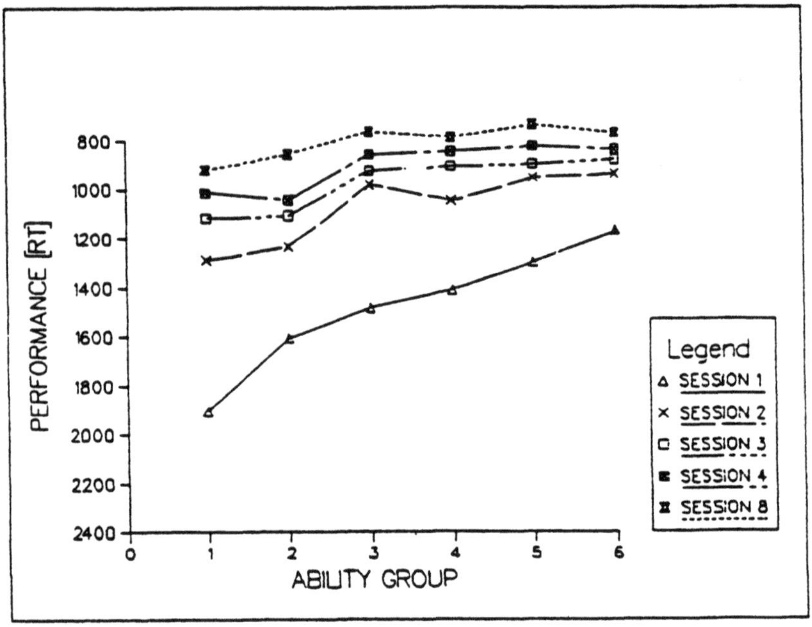

Figure 5-9: Consistent mapping verbal task performance as a function of ability group. Later sessions show "ability-insensitivity".

5.7.4. A Theory of Performance-Ability Relations

If we consider an initial domain of tasks that are novel, resource-dependent at the beginning of practice, and which allow all subjects to achieve a greater-than-chance (or greater than zero) performance level, the theory can be depicted as in Figure 11. Three main principles provide for the major effects in the theory. These are as follows:

Principle 1. Broad and general ability individual differences are equated with individual differences in amount or efficiency of attentional resources.

Principle 2. The transition from controlled to automatic processing is equated with the transition from resource to data-limited performance characteristics.

Principle 3. The ability determinants of performance are associated with the extent and type of resources required by the task.

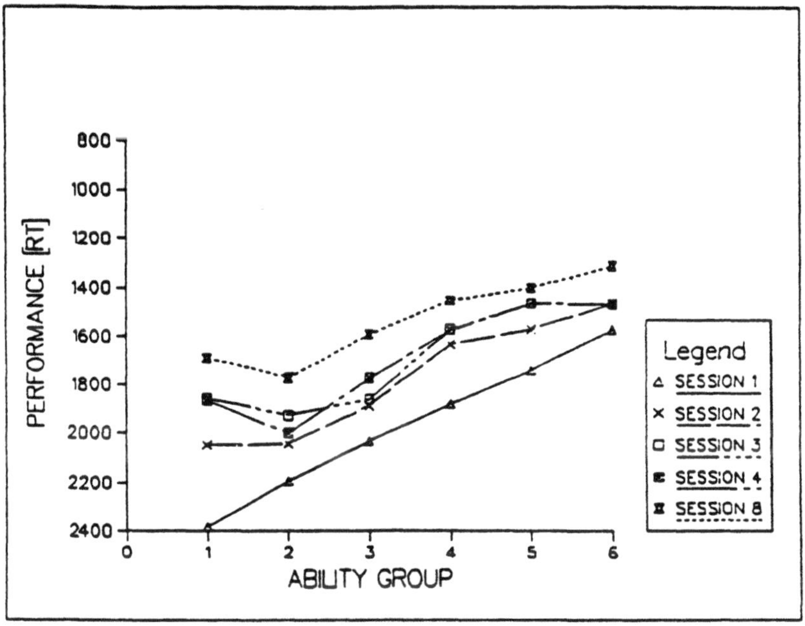

Figure 5-10: Varied mapping verbal task performance as a function of ability group. All sessions show "ability-dependence".

From the principles, the figure illustrates which types of abilities will be associated with naive and practiced task performance from either consistent or inconsistent information processing characteristics. For consistent tasks, general and content-relevant abilities will be associated with the initial performance levels (because controlled processing resources are required for processing new information). As practice proceeds, these abilities which tap the skills/processes that overlap with task automatic processing components will correlate with late, well-practiced performance. For predominantly inconsistent tasks; initial, intermediate, and late performance individual differences will be associated with both the general and content-relevant abilities (because controlled processing is required for both novel and familiar inconsistent information).

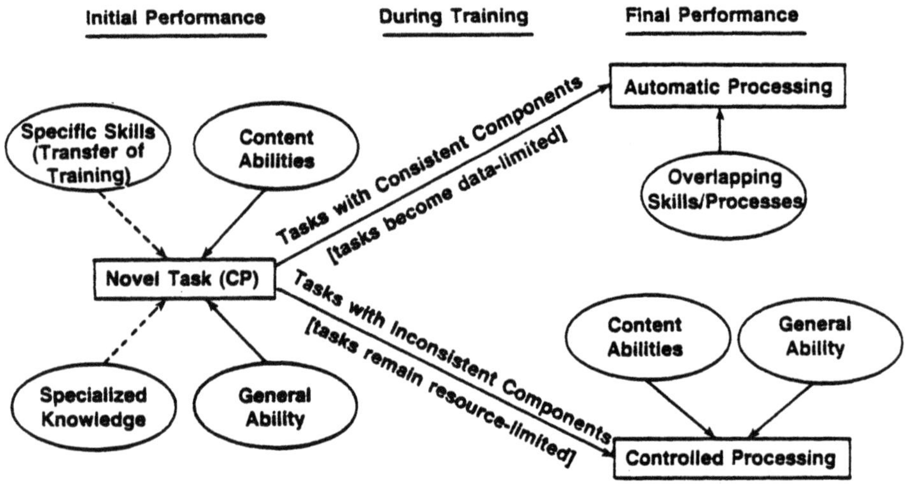

Figure 5-11: Graphic representation of the theory of performance ability relations presented in the text (from Ackerman & Schneider, 1985).

In a series of experiments examining individual differences in cognitive abilities and CM/VM performance over practice, data were acquired that demonstrate meaningful individual differences underlying automatic and controlled processes. In addition, that initial work has also demonstrated the viability of the theory discussed above (see also Ackerman, 1983a; Ackerman, 1983b). Briefly, three major theoretical predictions have been confirmed through investigations of individual differences in automatic and controlled processing. These predictions were: 1) Individual differences in performance on tasks which require greater investments of controlled processing resources were more highly correlated (than tasks which allow for development of automatic processes) with general cognitive/intellectual ability individual differences; 2) Content abilities (relevant to the task requirements) were also more highly associated with performance on tasks with increased controlled processing requirements; and 3) Overlapping task processing components give rise to identical correlations with ability determinants of such components. (For a detailed discussion, see Ackerman &

Schneider, 1985).

These data demonstrate that whether tasks require controlled processing or allow development of automatic processes is a crucial issue in determining the range and behavior of individual differences in performance. When tasks are predominantly consistent and an appropriate training procedure is utilized (i.e., the consistencies are demonstrated, sufficient task practice is given, etc.), then most population members may be capable of adequate performance (depending, of course, on a relative tolerance for small performance differences). However, when tasks are predominantly inconsistent, many applicants will be unsuitable as a result of insufficient controlled processing resources, and practice will not likely alleviate these problems.

When equipment design, employee selection, and training issues are considered, the consistency of information processing will moderate whether substantial individual differences will affect task performance at initial, intermediate, and post-training levels of practice. The human factors professional is well-advised to take such variables into account for such applications.

5.7.5. A Final Issue - Task Complexity and Mode of Processing

There is a tendency for some investigators to associate automaticity only with simplicity. Some researchers have suggested that only "simple" tasks can be automatic (e.g., Hirst, et. al., 1980, pg. 116). Those researchers, however, do not clearly define what makes a task simple or complex. Complex tasks can be performed via automatic processing, via controlled processing, or most likely, through a combination of both processes. Simple tasks can also be performed by either automatic or controlled processing. The type of processing is not determined by the complexity (or simplicity) of a task but rather by the consistency and, if the task is consistent, the amount of practice (see Fisk & Schneider, 1983 for a discussion).

5.8. CONCLUSION AND FUTURE DIRECTION

We began this chapter with a series of potentially perplexing questions. These questions involved the dramatic changes in human performance that often occur after practice or experience on some task. Throughout the chapter we have addressed the mechanisms for performance improvement - we have talked about why performance improvements occur, not just the fact that performance improvements often occur. Certainly this chapter has emphasized that the old adage "practice makes perfect" is not always correct. Only a certain kind of practice can produce a skilled performer.

The future directions for research on automatic/controlled processing will be aimed at more precisely defining the concepts of automatic and controlled processing. This will be accomplished by developing and testing precise mathematical models; further laboratory examinations of the parameters of automatic/controlled processing; continued work to determine the basic underlying principles of individual differences in learning a skill and knowledge acquisition tasks; and a pragmatic approach to the application of principles derived from automàtic/processing theory. By examining automatic/controlled processing theory in natural learning environments we will broaden our task domain and in turn have a richer database for the dual-process theoretical perspective.

REFERENCES

Ackerman, P.L. (1983). An investigation of individual differences in learning with respect to automatic and controlled processing modes. Paper presented at the Annual Meeting of the Midwestern Psychological Association, Chicago (a).

Ackerman, P.L. (1983). A theory for predicting ability/skill relations: An approach from automatic and controlled processing. *Proceedings of the Human Factors Society, 27*, 919-923. (b).

Ackerman, P.L. (1984). Multivariate evaluation of workload, resources, and performance in dual task data. *Proceedings of the Human Factors Society, 28*, 210-214. (a).

Ackerman, P.L. (1984). *A theoretical and empirical investigation of individual differences in learning: A synthesis of cognitive ability and information processing perspectives*. Doctoral dissertation, University of Illinois, Unpublished as of this date (b).

Ackerman, P.L., & Schneider, W. (1985). Individual differences in automatic and controlled information processing. In R.F. Dillon (Ed.), *Individual differences in cognition*. New York: Academic Press. Vol. 2.

Ackerman, P.L. (1986). Individual differences in information processing: An investigation of intellectual abilities and task performance during practice. *Intelligence, 10*, 101-139.

Adams, J.A. (1957). The relationship between certain measures of ability and acquisitions of a psychomotor response. *Journal of General Psychology, 56*, 121-134.

Adams, J.A. (1971). A closed loop theory of motor learning. *Journal of Motor Behavior, 3*, 111-149.

Briggs, G.E., & Johnsen, A.M. (1973). On the nature of central processing in choice reactions. *Memory & Cognition, 2*, 91-100.

Corballis, M.C. (1975). Access to memory: An analysis of recognition times. In P.M.A. Rabbit & S. Dornic (Eds.), *Attention and performance V*. New York: Academic Press.

Craig, A. (1979). Non-parametric measures of sensory efficiency for sustained monitoring tasks. *Human Factors, 21*, 69-78.

Eberts, R.E. (1979). *The automatic and controlled processing of sequences of events* (Tech. Rep. 7901). Champaign, Illinois: University of Illinois, Human Attention Research Laboratory,

Eberts, R.E., & Schneider, W. (1980). Computer assisted displays enable internalization and reduction of operator workload in higher order systems. *Proceedings of the Human*

Factors Society, 24, 59-62.

Eberts, R.E., & Schneider, W. (1985). Internalizing the system dynamics for a second-order system. *Human Factors, 27*, 371-394.

Egeth, H., Atkinson, J., Gilmore, G., & Marcus, N. (1973). Factors affecting processing mode in visual search. *Perception & Psychophysics, 13*, 394-402.

Estes, W.K. (1972). Interactions of signal and background variables in visual processing. *Perception & Psychophysics, 12*, 278-286.

Fisk, A.D., & Schneider, W. (1981). Controlled and automatic processing during tasks requiring sustained attention: A new approach to vigilance. *Human Factors, 23*, 737-750.

Fisk, A.D., & Schneider, W. (1982). Type of task practice and time sharing ,activities predicts deficits due to alcohol ingestion. *Proceedings of Human Factors Society, 26*, 926-930.

Fisk, A.D., & Schneider, W. (1983). Category and word search: Generalizing search principles to complex processing. *Journal of Experimental Psychology: Learning, Memory, and Cognition, 9*, 117-195.

Fisk, A.D., & Schneider, W. (1984). Task versus component consistency in automatic process development: Consistent attending versus consistent responding. *Bulletin of the Psychonomic Society, 22*, 330-332. (a).

Fisk, A.D., & Schneider, W. (1984). Memory as a function of attention, level of processing, and automation. *Journal of Experimental Psychology: Learning, Memory, and Cognition, 19*, 181-197. (b).

Hancock, P.A. (1984). Environmental stressors. In J.S. Warm (Ed.), *Sustained attention in human performance*. New York: Wiley.

Hancock, P.A., & Pierce, J.O. (1984). Toward an attentional theory of performance under stress: Evidence from studies of vigilance in heat and cold. In A. Mital (Ed.), *Trends in Ergonomics/Human Factors I*. Amsterdam: North-Holland.

Hancock, P.A. (1986). Sustained attention under thermal stress. *Psychological Bulletin, 99*, 263-281.

Hirst, W., Spelke, E.S., Reaves, C.C, Caharack, G., & Neisser, U. (1980). Dividing attention without alternation or automaticity. *Journal of Experimental Psychology, 109*, 98-117.

James. W. (1890). *Principles of psychology*. New York: Holt.

Kahneman, D. (1973). *Attention and effort*. Englewood Cliffs, New Jersey: Prentice-Hall.

Kristofferson, M.W. (1972). Effects of practice on character classification performance. *Canadian Journal of Psychology, 26*, 54-60.

LaBerge, D. (1973). Attention and the measurement of perceptual learning. *Memory & Cognition, 1*, 268-276.

LaBerge, D. (1975). Acquisition of automatic processing in perceptual and associative learning. In P.M.A. Rabbit & S. Dornic (Ed.), *Attention and performance V*. New York: Academic Press.

Laberge, D. (1976). Perceptual learning and attention. In W.K. Estes (Ed.), *Handbook of learning and cognitive processes*. Hillsdale, New Jersey: Lawrence Erlbaum Associates.

Logan, G.D. (1978). Attention in character-classification tasks: Evidence for the automaticity of component stages. *Journal of Experimental Psychology: General, 107*, 32-63.

Logan, G.D. (1979). On the use of a concurrent memory load to measure attention and automaticity. *Journal of*

Experimental Psychology: Human Perception and Performance, 5, 189-207.

Navon, D., & Gopher, D. (1979). On the economy of the human processing system. *Psychological Review, 86*, 214-255.

Norman, D.A. (1964). A comparison of data obtained with different false alarm rates. *Psychological Review, 71*, 243-246.

Norman, D.A. (1976). *Memory and attention: An introduction to human information processing*. New York: Wiley.

Norman, D.A., & Bobrow, D.B. (1976). On the analysis of performance operating characteristics. *Psychological Review, 83*, 508-519.

Posner, M.I., & Snyder, C.R.R. (1975). Attention and cognitive control. In R.L. Solso (Ed.), *Information processing and cognition: The Loyola symposium*. Hillsdale, New Jersey: Lawrence Erlbaum Associates.

Schneider, W., & Shiffrin, R.M. (1977). Controlled and automatic human information processing: I. Detection, search, and attention. *Psychological Review, 84*, 1-66.

Schneider W., & Fisk, A.D. (1980). *Visual search improves with detection searches, declines with nondetection searches* (Tech. Rep.). University of Illinois: Human Attention Research Laboratory,

Schneider, W. (1982). *Automatic/control processing: The limits and potential*. American Psychological Association, Washington, D.C.

Schneider, W., & Fisk, A.D. (1982). Degree of consistent training: Improvements in search performance and automatic process development. *Perception & Psychophysics, 31*, 160-168. (a).

Schneider, W., & Fisk, A.D. (1982). Dual task automatic and control processing, can it be done without cost? *Journal of Experimental Psychology: Learning, Memory, and Cognition, 8*, 261-278. (b).

Schneider, W., & Fisk, A.D. (1984). Automatic category search and its transfer. *Journal of Experimental Psychology: Learning, Memory, and Cognition, 10*, 1-15.

Shiffrin, R.M., & Schneider, W. (1977). Controlled and automatic human information processing: II. Perceptual learning, automatic attending, and a general theory. *Psychological Review, 84*, 127-190.

Spelke, E.S., Hirst, W., & Neisser, U. (1976). Skills of divided attention. *Cognition, 4*, 215-230.

Sperling, G., Budiansky, J., Spivak, J.G., & Johnson, M.C. (1971). Extremely rapid visual search: The maximum rate of scanning letters for the presence of a numeral. *Science, 174*, 307-311.

Welford, A.T. (1976). *Fundamentals of skill*. London: Methuen.

Wickens, C.D. (1980). The structure of attentional resources. In R.S. Nickerson (Ed.), *Attention and performance VIII*. Hillsdale, New Jersey: Lawrence Erlbaum Associates. (pp. 239-257).

HUMAN FACTORS PSYCHOLOGY, Peter A. Hancock (ed.)

6. DESCRIBING MOVEMENT CONTROL AT TWO LEVELS OF ABSTRACTION

Richard J. Jagacinski and Brian D. Plamondon
Department of Psychology

and

Richard A. Miller
Department of Industrial and Systems Engineering
Ohio State University
Columbus, OH 43210

ABSTRACT

Movement may be described in terms of activity modules that consist of a set of start-stop rules and associated motion generators. The motion generators determine the spatio-temporal patterning of the movement. The start-stop rules determine when a particular motion-generating process is activated and when it is interrupted or completed. These two levels of abstraction are used to describe how human subjects manually capture a moving target that uses a computer-generated escape strategy. A simulation of this two-level model mimics the episodic performance of human subjects. This multi-level style of modeling may prove useful in analyzing a wide range of skilled movements ranging from tracking to multi-limbed locomotion.

INTRODUCTION

6.1. ARGUMENTS FOR MULTIPLE LEVELS OF BEHAVIORAL ORGANIZATION

There is a growing consensus that complex motor behavior must be described at multiple levels of organization. This idea is at least as old as Bryan and Harter's (1899) work on telegraph operators. These researchers argued that with practice the units of motor organization for sending telegraphic messages changed from individual letters to words.

Similarly, the units of cognitive organization for receiving telegraphic messages progressed from letters, to words, to phrases. This ability to organize behavior hierarchically was considered the key element in achieving mastery in many types of skilled performance, including telegraphy, mathematics, music, whist, chess, and checkers. While many of these skills involve very little perceptual-motor coordination, it is significant that Bryan and Harter believed hierarchical organization was involved in the highly motoric process of sending telegraphic messages as well as in the process of receiving them.

A considerable amount of behavioral research in the twentieth century provides additional support for these ideas. For example, within the specialized realms of chess (Chase & Simon, 1973) and physics (Larkin, 1983) there is empirical evidence that performance mastery is associated with higher levels of abstraction in problem-solving and planning. One wonders whether similar demonstrations will be forthcoming within domains involving greater perceptual-motor coordination. Theoretical arguments for hierarchical and/or heterarchical organization of complex motor skills have discussed the problem of controlling the many degrees of freedom associated with moving arms and legs, fingers and toes (Bernstein, 1967; Turvey, 1977). Centralized control would be beyond human information processing capabilities. Hierarchical organization represents one method of overcoming these limitations in a manner analogous to the way chunking strategies (Miller, 1956) overcome limitations in human memory systems. A second theoretical argument for hierarchical organization of motor skills is that modular organization would permit effective recombination of well developed control processes in order to adapt rapidly to new environmental situations (Gallistel, 1980; Turvey, 1977). New skills could thus be synthesized from previously developed capabilities of intermediate complexity, rather than from extremely elementary processes (e.g., see Simon, 1969). In sum, hierarchical organization is one strategy for achieving increased levels of complexity and adaptability in a limited capacity behavioral system.

If one accepts these general arguments for hierarchical organization, then a person interested in describing the structure of behavior is faced with the problem of identifying and measuring the levels of the hierarchy. One of the more elaborate schemes of organization was presented by Powers (1973). He suggested that human behavior is organized into nine levels of feedback loops varying from control of sensory intensity to control of abstract system concepts such as government and family. Within the more delimited context of engineering psychology, Rasmussen (1983) suggested that human behavior can be usefully decomposed into three levels of organization: skill-based, rule-based, and knowledge-based behavior. Skill-based behavior is the generating of spatio-temporal trajectories by persons and machines. Feedback control, feedforward control, and the use of schemata are primary features of these motion-generating processes. People compose actions by piecing together various motions. How people decide what motions to perform and in what order to perform them is described as rule-based behavior. Environmental situations arising within the context of achieving some goal trigger rules or procedures based on previous experience. Rules are associations between equivalence classes of environmental situations and equivalence classes of motions or actions. Rules may be more or less abstract, corresponding to the broad or narrow definition of these equivalence classes. Knowledge-based behavior is a third level of organization which corresponds to the problem-solving and planning a person engages in when faced with a situation for which he or she has no appropriate rules. These three levels of behavioral organization encode environmental information at three levels of abstraction, which Rasmussen (among others) has labeled signals, signs, and symbols. Signals are quantitative spatio-temporal patterns and their various derivatives; signs are representations of familiar, environmental situations in the context of a goal or task; symbols are conceptual representations of functional relationships. These different styles of encoding information characterize the qualitative differences among skills, rules, and knowledge-based behavior.

If one uses Rasmussen's framework for reviewing work in manual control, it appears that research has been primarily concerned with the skill-based level of organization and to a lesser extent with rule-based behavior. Because manual control tasks are arising more frequently in the context of highly automated supervisory control tasks, such as piloting vehicles or controlling industrial processes, there is increased interest in interfacing skill-based descriptions with rule-based and knowledge-based descriptions. This chapter will present an example of how skill-based and rule-based descriptions can be combined to represent the way people capture a moving target. The time histories generated by people performing this task appear highly episodic. It appears that very different motion generating processes are used under different circumstances to accomplish different goals. Such data afford a good opportunity for identifying a rule-based level of behavioral organization that determines a sequence of highly varied motions. The rule-based level of organization was modeled as a set of start-stop rules which governed discrete transitions among modes of capture behavior. The skill-based level of organization was modeled as a combination of feedback control, feedforward control, and open-loop schemata described in terms of differential equations. The simulated time histories of this two-level model mimicked the episodic performance of human subjects.

6.2. ACTIVITY MODULES FOR DESCRIBING TRACKING AND OTHER SKILLED MOVEMENTS

The episodic nature of manual control was emphasized by Craik (1947) in his characterization of the human operator as an intermittent-correction servo. A number of subsequent models in which the human samples an input signal and approximates it as a sequence of simpler functions have exemplified this approach. For example, Lemay and Westcott (cited in Pew, 1974) modeled the human operator in a compensatory tracking task as linearly combining error and error rate to predict the error signal. This predicted error signal was then internally sampled every 200 ms. The human's control movement was modeled as a sequence of 200 ms episodes of acceleration and deceleration to null out the predicted error. In principle, episodic behavior could also be obtained by sampling the system error signal, approximating it as a sequence of step-ramps, and using a proportional control strategy. In contrast, smoothly continuous descriptions of tracking, such as the McRuer Crossover Model (see McRuer & Jex, 1967, for a technical review; see Jagacinski, 1977, for a tutorial review), and the continuous optimal control model (Kleinman, Baron, & Levison, 1971; see Pew & Baron, 1979, for a tutorial review), have not emphasized episodic aspects of performance. All of the above examples may be regarded as single-mode models of tracking. They are examples of motion-generating mechanisms in which the basic input-output relationship remains invariant or follows an invariant pattern.

A different aspect of episodic performance corresponds to discrete changes in the input-output relationship. Namely, multiple episodes correspond to the activation of different motion generators. This concept of episodes is more abstract than that of input approximation or output discretization and requires two types of description: rules for starting and stopping the different episodes, and a description of the processes that produce the spatio-temporal patterns within each episode. Figure 1 illustrates this concept. An activity module consists of a set of start-stop rules and an associated motion generator. When certain activation conditions occur, the process associated with the activity module begins to generate or control a motion. This motion continues until either the motion-generating process is interrupted by the activation of another activity module, or some completion condition occurs. A completion condition might consist of achieving a certain target state (e.g., bringing a system to rest, or capturing a target), creating a particular spatio-temporal pattern (e.g., performing a "Figure-8"), or simply the passage of a certain amount of time. Completion conditions correspond to achieving the goal associated with that activity module. In contrast, interruptions correspond to another goal taking precedence before the ongoing activity is completed. The motion-generating process uses a particular set of input signals (e.g., spatio-temporal patterns and their spatial and/or temporal derivatives in any of several sensory modalities) to create a set of outputs (e.g., motions of limbs or vehicles). The relationship between input signals and output signals is described in terms of some functional relationship or control law (e.g., a transfer function).

The activity module is one possible way to relate skill-based and rule-based behaviors. It represents a combination of elements developed primarily within the disciplines of control theory, artificial intelligence, and psychology. For example, the concept of associating a goal with a motion-generating mechanism such as a feedback loop was discussed by Rosenbluth, Wiener, and Bigelow (1943) in a very early paper in the development of cybernetics. Later, optimal control theory (e.g., Athans & Falb, 1966) explicitly related the goal of a movement to the parametric structure of the motion-generating process. The concept of an equivalence class of situations triggering an action is a basic assumption of stimulus-response approaches to psychology. The properties of sets of if-then rules have been formally explored as production systems in work on artificial intelligence and cognitive science (Newell & Simon, 1972; see Anderson, 1976, and Barr & Feigenbaum, 1981, for reviews). In one example particularly relevant to the present work, Goldstein and Grimson (1977) associated explicit goals with if-then production rules to simulate instrument flying of an aircraft. The productions specified selective observations of relevant instrument dials and subsequent manipulations of the appropriate controls to achieve goals such as straight-and-level flight or banking the aircraft. Norman (1981) has discussed schemata for action being first "activated" in the context of achieving some goal, and then being "triggered" by appropriate environmental conditions, perceptions, or preceding actions (see also Miller, 1985). In the present activity module (Figure 1), the activation conditions correspond to Norman's triggering conditions. Schank and Abelson (1977) developed the concept of a "planbox" in their efforts to create a computer program that would interpret natural language in terms of scripts, plans, goals, and themes. A planbox consisted of an essential action to accomplish an explicit associated goal, several types of preconditions for performing that action, and a result. All of these elements have their counterparts in the activity module of Figure 1. The notion of using an explicit test to terminate an activity was a basic part of Miller, Galanter, and Pribram's (1960) description of behavior in terms of Test-Operate-Test-Exit (TOTE) units. The TOTE unit tests for completion of an activity, and continues to perform operations until a completion is obtained (see also Reason, 1977). The activity module in Figure 1 has explicit conditions for both completion and interruption. In sum, the elements that form the activity module are by no means new. This module is one way of organizing a set of related elements governing the structure of action.

Reconsidering previous models of movement in these terms will demonstrate the potential usefulness of the activity module. In modeling manual tracking, a number of investigators (e.g., Costello,

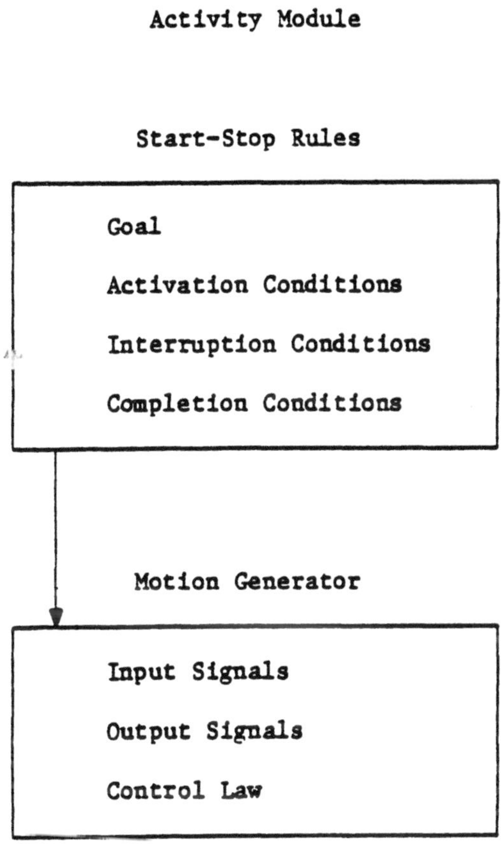

Figure 6-1: A two-level conceptualization of activity modules.

1968; Phatak & Bekey, 1969; Burnham & Bekey, 1976) partitioned the error phase plane (e.g., see Graham & McRuer, 1961) into several regions and associated a different tracking mode with each region. For example, Costello (1968) formulated a two-mode model to describe manual tracking of low-frequency random-appearing signals interspersed with sudden large, rapid, jumps (Figure 2). A single closed-loop tracking mode was sufficient to match subjects' performance in following the low-frequency portion of the signal. However, this tracking mode resulted in too much overshoot and/or ringing in response to the sudden step-like jumps. Costello therefore postulated a separate tracking mode for this second input pattern. When step-like jumps were absent, the closed-loop continuous tracking mode (consisting of a McRuer Crossover Model with lag-lead compensation and neuromuscular lag) kept tracking error (e) and error rate ($\dot{e}$) within an approximately diamond-shaped region about the origin in the error phase plane. The occurrence of a sudden step in the signal to be tracked would result in tracking error and/or error rate going outside the diamond-shaped region. This event triggered the occurrence of a time-optimal bang-bang response consisting of a large, constant, leftward (rightward) muscular force for t_s seconds followed by an equal and opposite muscular force for t_f seconds. The durations of these surges of force were chosen so that when the force pattern was filtered by the inertia of the forearm and wrist, the resulting step-like movement of the control stick would just nullify the tracking error (e_0) and error rate ($\dot{e}_0$) that existed when the surge mode was first activated.

The surge mode is open-loop; it ignores any subsequent changes in e and $\dot{e}$ during the time that the bang-bang surge is applied. There are two important implications of this open-loop strategy. First, even though the (e, $\dot{e}$) trajectory created by the surge will typically come back into the diamond-shaped phase plane region, the continuous tracking mode is not reactivated until the full surge maneuver is completed. Leaving the diamond-shaped area activates a surge, but re-entering the diamond, by itself, is not sufficient to activate the continuous tracking mode. The surge must also be completed (Figure 2). Secondly, even though the surge produces cursor movement sufficient to have nullified the initial error and error rate, subsequent changes in the signal to be tracked may result in non-zero error and error rate at the completion of the surge. If these values were outside the diamond, a second surge would presumably follow. Costello's data, however, did not illustrate this.

The details of these two tracking modes are summarized in Figure 2. The continuous tracking mode is a closed-loop strategy for nulling "small errors", where "errors" are defined in terms of both error magnitude and rate. The surge mode is an open-loop strategy for nulling "large errors", and this goal takes precedence over the nulling of small errors. The start-stop rules in this example are quite distinct from the motion-generating mechanisms. The motion generators utilize specific values of error and error rate. The start-stop rules are more abstract in that they distinguish only two broadly defined equivalence

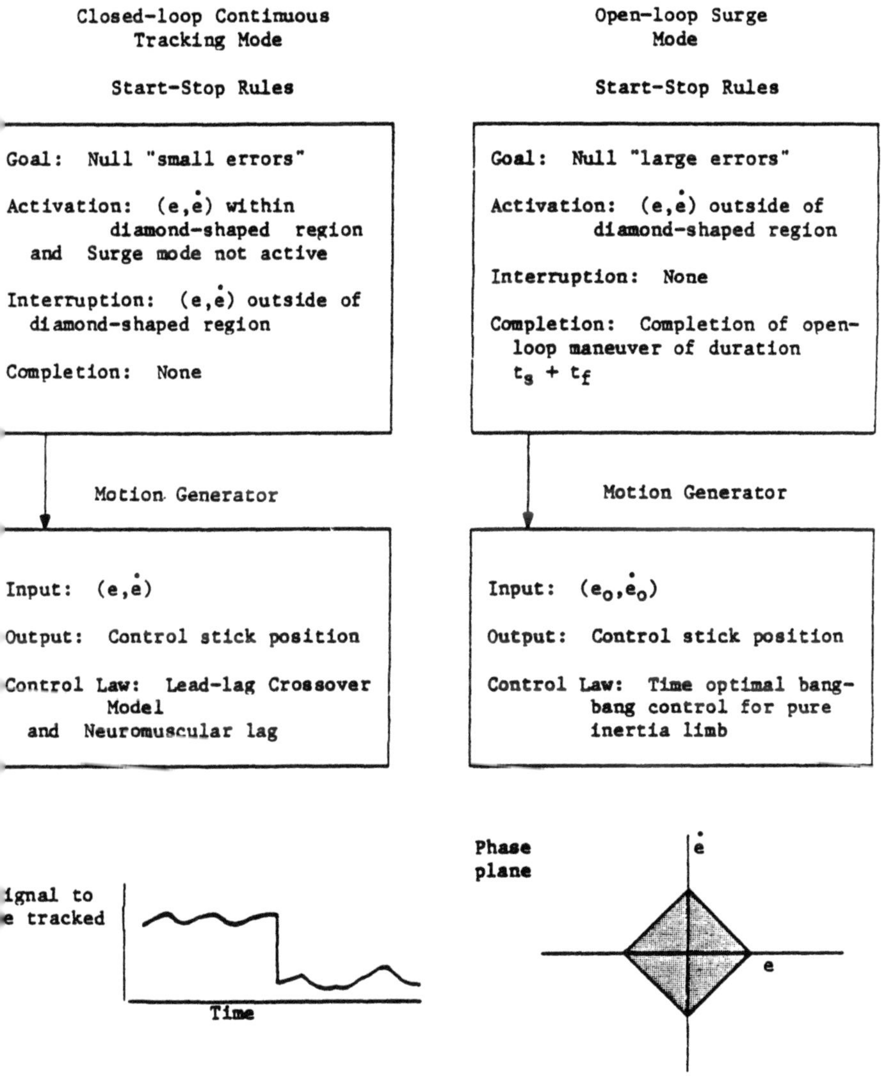

Figure 6-2: The Costello surge model characterized as two activity modules.

classes of system state, namely the diamond-shaped region (small errors) and the remainder of the phase plane (large errors). Other examples of using extended regions of the phase plane for triggering tracking mode changes involve pilots attempting to recover control of their aircraft after an equipment failure (Phatak & Bekey, 1969) and car drivers maintaining adequate headway from vehicles in front of them (Burnham & Bekey, 1976). These models could also be represented in terms of activity modules in a similar manner.

A second example of the potential usefulness of activity modules is in the description of limbed movement. One early effort to build a machine capable of walking was the "Phoney Pony" created by Frank and McGhee (McGhee, 1976). Each of the four legs of this robot had two degrees of freedom: hip rotation and knee rotation. The movements of the hip and knee were produced by motors that were switched among three possible states: forward rotation, rearward rotation, and stopped or locked in place. A state-transition diagram illustrating the movement cycle for a front leg of the Phoney Pony during a singular crawl gait is shown in Figure 3 (adapted from McGhee, 1976, Figure 10 and Table 4). States of movement are represented as rectangles in this diagram, the arrows represent transitions between states, and the events that trigger these transitions are labeled next to the arrows. State 1 is entered when the hip is fully forward (+30 degrees relative to vertical), and the knee is straightened. In States 1 and 2 the hip is moving rearward, and the knee is locked. The leg begins to support the robot body when the hip angle reaches +12 degrees. The contralateral legs (e.g., the front legs) should be 16 degrees out of phase when both of them are supporting the robot (i.e., both hip angles are between +12 degrees and -12 degrees). This condition ensures that the movement cycles of the two contralateral legs are one-half cycle out of phase. In order to ensure this relationship, there are two position checks. When one hip is 12 degrees forward, the other hip must be 4 degrees rearward, and vice-versa. If one contralateral leg gets slightly ahead of the other in its respective movement cycle, it will pause (State 2') at either +12 degrees or -4 degrees until the other leg catches up. Then the hip will resume its rearward motion (transition from State 2' back to State 2). When the hip is fully rearward (-12 degrees), State 3 is entered in which both the hip and knee are locked in place. If the other three legs are in contact with the ground, the leg can be lifted off the ground and swung forward (hip rotating forward, knee rotating backward). Since the ipsilateral rear leg was the last previous leg swung forward, it is necessary that is has begun supporting the robot again (ipsilateral hip 12 degrees forward) before the front leg can be swung forward (State 4). During the swing, the knee will reach its full rearward position (transition to State 5) before the hip is fully forward (transition to State 6). Then the knee will be straightened, and the movement cycle will begin again with State 1.

The state transition diagram represents the leg movement cycle as a stream of discrete events. In Figure 4 the movement cycle is represented as four activity modules corresponding to the extension and flexion of the hip and knee. In this representation there are basically three phases to the movement cycle. The first phase is the hip extension (rearward rotation) with the knee locked, which corresponds to States 1, 2, 2', and 3 in Figure 3. The goal of this phase is to propel the body forward. The need to pause, to synchronize with the contralateral leg, is represented as an interruption condition. This phase is completed when the hip is fully rearward. The second phase is the hip and knee flexion. Both the hip and knee activity modules are activated by the completion of Phase 1 and the ipsilateral check that ensures the three other feet are on the ground. This activity corresponds to States 4 and 5 in Figure 3. The fact that the knee flexion will be completed before the hip flexion is not explicitly represented. The third phase is the final knee extension, which corresponds to State 6 in Figure 3. Details of the motion-generation mechanisms have been suppressed in Figure 4, but they could certainly be added to provide additional detail.

Although much of the same information is contained in Figures 3 and 4, the styles of representation are different. The state transitions correspond to the start-stop rules of the activity modules; the states ignore many of the spatio-temporal details of the leg trajectory. The basic units in Figure 4 are goal-directed acts, and distinctions can be made between interruptions and completions of these acts. The style of representation in Figure 4 is in many ways similar to physiologically

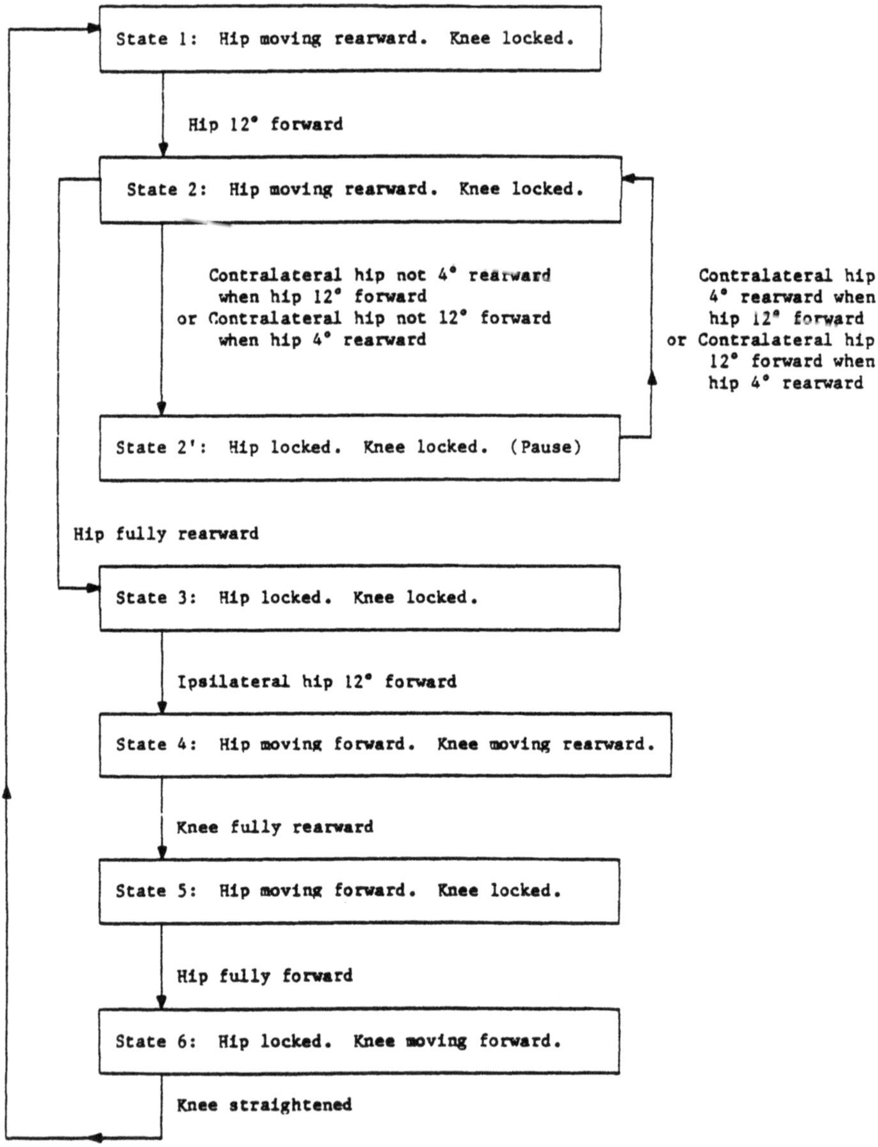

Figure 6-3: The state transition diagram for one front leg of the Phoney Pony during the singular crawl gait (after McGhee, 1976).

oriented descriptions of biological movement (e.g., see Gallistel, 1980; Grillner, 1975). Figure 4 is also in the spirit of Greene's (1982) suggestion that movement organization be represented as a collection of "demons" to distribute decision making. Indeed, this philosophy of control was behind Frank and McGhee's design of the Phoney Pony. The start-stop rules are mechanisms which control behavior generation in a manner consistent with the state transition diagram. This example of robotic movement is especially useful for comparing different styles of representation, because the generative mechanisms of this artificial system are well known. In cases of biological movement where the control mechanisms may not be known, a basic research problem is to arrive at a set of start-stop rules and associated motion generators, given the observations of state transitions and time histories. The reactive target acquisition task presented next provides one example of how to approach this problem.

6.3. A REACTIVE TARGET ACQUISITION TASK

In order to examine the inter-relationship of start-stop rules and motion-generating mechanisms, a reactive target acquisition task was constructed. This work was inspired in part by previous observations by Robinson (1965) of the smooth pursuit eye movement system. He negatively coupled the movement of a visual target to the position of the eyes. Eye movement to the right (left) by one degree caused the target to move to the left (right) by K-1 degrees. When the smooth pursuit system was activated by the appearance of a constant velocity stimulus, and K was set at values from 2 to 6, the negative coupling resulted in roughly sinusoidal oscillatory eye movements superimposed on a ramp movement. Based on the dynamic properties of the smooth pursuit system with unpredictable targets, one would expect sustained oscillations. However, the oscillations waxed and waned. Robinson interpreted this phenomenon as evidence for the eye movement system shifting into a "predictive mode" to compensate for the negative coupling. The compensation is only partially effective. Apparently, the eye-movement system needs to experience a few cycles of the sinusoidal oscillation to make effective predictions, because after the oscillations disappear, they tend to reappear again. The eye then readjusts to their regularity and again causes them to disappear. Robinson described this

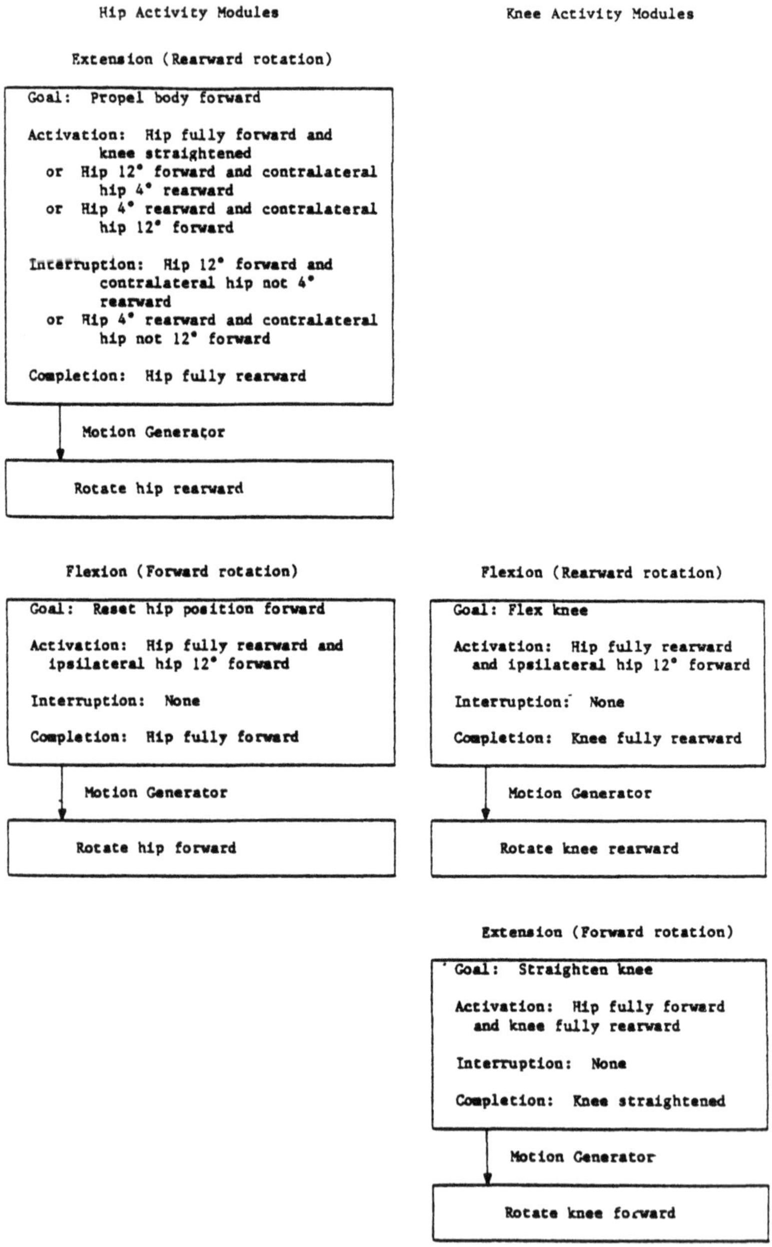

Figure 6-4: The Phoney Pony front leg movement characterized as four activity modules.

behavior as shifts between predictive and non-predictive modes of eye movement control.

The present experiment used a manual control task and generated an elusive target in a manner different from Robinson's experiment (Figure 5). Subjects captured a moving target by manipulating a control stick so as to keep a cursor over the target for 400 ms. The target reacted to the movement of the cursor with an escape strategy generated by several cascaded elements. First, a "panic function" caused the target to move away from the cursor with a velocity that increased as the cursor came closer (Figure 5). Second, the resulting movement pattern was passed through an oscillatory filter and then integrated to obtain target position (Figure 5). The filter added evasive side-to-side movement analogous to the juking maneuvers performed by a football player attempting to elude a tackler. The combination of these elements made the target movement very sensitive to slight changes in cursor movement. Therefore, even though the target behaved in a deterministic fashion, it exhibited a wide variety of movement patterns. For example, Figure 6 demonstrates the target response to three step movements of the cursor that differ only in amplitude.

Part of what makes golf and bowling such challenging sports is that the ball trajectories are so sensitive to small performance variations. In a similar manner, the sensitivity of the target response to small changes in cursor movement made the present target capture task extremely challenging. The target motion was predictable enough to encourage the subjects to generate capture strategies, but also unpredictable enough to result in many varied pursuit sequences. It was hoped that this task would result in short episodes of predictable and unpredictable target behavior, with corresponding changes in the subjects' style of pursuit.

6.4. METHOD

Eight right-handed undergraduate students served as subjects. Based on their performance in stabilizing a first-order unstable system (Jex, McDonnell, & Phatak, 1968), four subjects were chosen to continue for

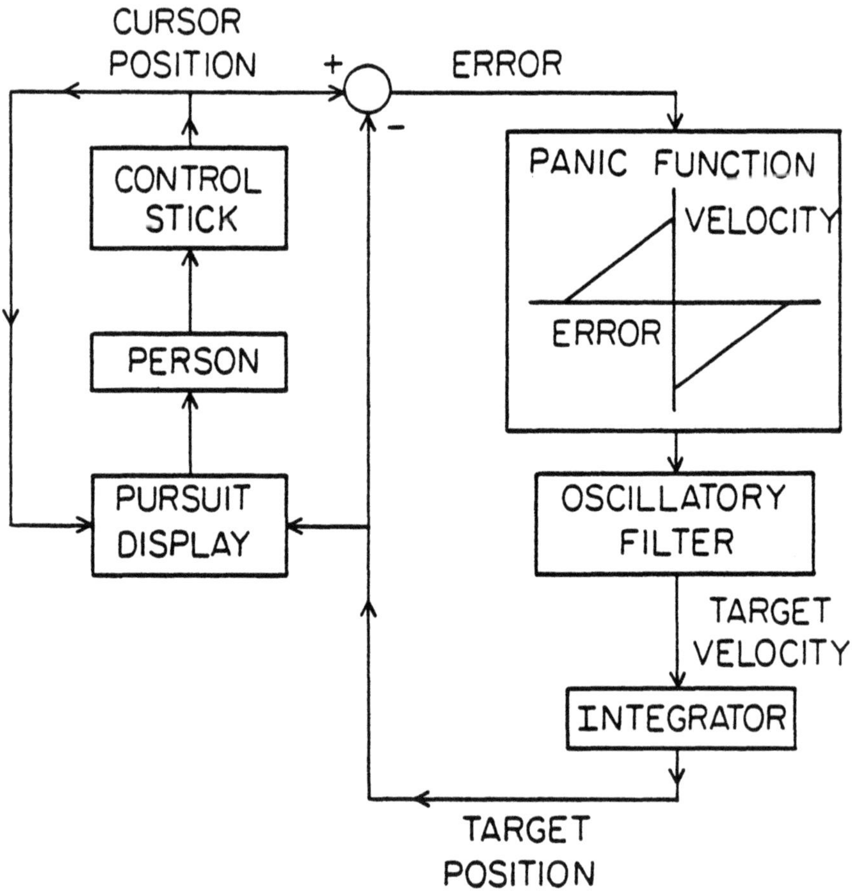

Figure 6-5: Experimental implementation of the reactive target.

ten 45-minute sessions. Subjects sat approximately 50 cm away from a 10 cm wide oscilloscope display on which they saw a target and a cursor. The target consisted of two vertical lines separated by 2 mm; the cursor was a single dot. Both target and cursor moved only in the horizontal dimension. At the beginning of a trial the cursor was centered. One second after a short warning tone, the target randomly appeared 2 cm to the right or left of center. The subjects' task was to manipulate an isometric control stick (gain = 0.35 kg per 1 degree of visual angle) so as to hold the cursor dot between the two target lines for an uninterrupted period of 400 ms. When this criterion was achieved, the target was considered "captured," and disappeared from the display. The subjects were instructed to capture the target as quickly as possible. If the target was not captured within 15 s, the target disappeared and a tone sounded, indicating that the target had "escaped." Similarly, if the target exceeded the display boundaries of 5 cm to the right and left of center, the target was also considered to have escaped. The target disappeared from the display, and as a penalty the subjects had to wait for the remainder of the full 15 s trial duration to expire.

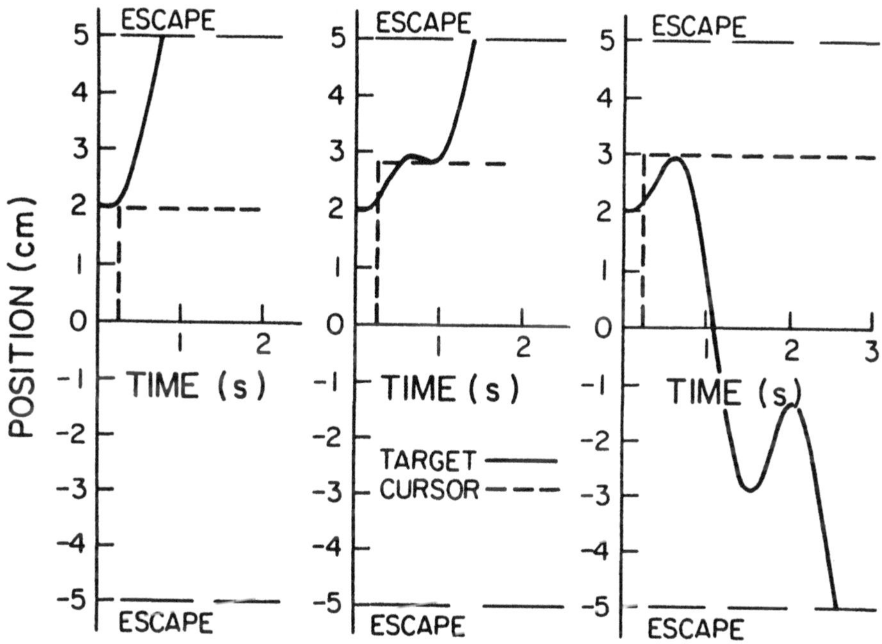

Figure 6-6: Responses of the reactive target to different size step movements of the cursor.

The target reacted to the movement of the cursor with an escape strategy represented in Figure 5. A nonlinear panic function made the target move away from the cursor with a velocity that increased as the cursor came closer. The velocity signal generated by this function had a maximum value of 4.6 cm/s for negligible error, and velocity decreased linearly to zero as the absolute value of error increased to 10 cm. The resulting velocity signal was filtered through a second-order underdamped system that made the target movement oscillatory, and then integrated to generate target position. There was a 15 cm/s saturation on velocity and a $15\omega_n$ cm/s^2 saturation on acceleration in the implementation of the underdamped system. ω_n is the undamped natural frequency of the filter. ω_n was set at either 3 or 5 rad/s, and the per unit critical damping, ζ, was set at either 0 or .25. A factorial crossing of these values produced four targets of varying degrees of evasiveness.

For each of the four targets, subjects received two practice trials followed by two 20-trial blocks. Within each block, ten randomly chosen trials started with the target on the right, and the other ten started on the left. There was a four-second interval between trials within each block. Each daily session consisted of a total of 160 data trials, 40 data trials for each target. The order of presentation of the four targets was randomized across sessions; however, subjects were informed as to which target they would receive at the beginning of each block. Subjects were instructed to capture the targets as quickly as possible, and were given feedback after each block as to the sum of their capture times over the twenty trials. Whenever the target escaped, a capture time of 15 s was recorded for that trial. There was thus a strong penalty for an escape. Subjects were also given daily feedback on their total capture time across all 160 trials, and a bonus of five dollars was offered to the subject with the lowest total capture time for Sessions 9 and 10.

6.5. RESULTS

6.5.1. State Definitions

Mean capture times on Sessions 9 and 10 ranged from 2.9 s for Subject 1, to 3.8 s for Subject 4 on successful trials. For all four subjects, mean capture times increased monotonically across targets in the following order: (ω_n=3 rad/s, ζ=.25), shortest capture time; (ω_n=3 rad/s, ζ= 0); (ω_n=5 rad/s, ζ=.25); (ω_n=5 rad/s, ζ = 0), longest capture time. The number of each type of target captured by Subjects 1 and 4, and the mean capture times on successful trials are shown in Figure 7.

Subject 4 performed more poorly than Subject 1 on both of these measures of performance.

A single trial for Subject 1 capturing the most difficult target, Target IV, is shown in Figure 8. Qualitatively, this time history appears to contain a sequence of short episodes of very different types of pursuit behavior. After a reaction time interval of approximately 300 ms (RT segment, Figure 8), the cursor moves very rapidly toward the target to reduce the initial large distance from the target (first A segment, Figure 8). Once the cursor nears the target, the cursor begins to follow the target closely and mimic the target trajectory (first F segment, Figure 8). After several changes of direction, the discrepancy between the target and cursor builds up, and the cursor no longer mimics the target trajectory (segment P, Figure 8). Rather, the cursor moves much more slowly than the target, coming close to the target only at its upper turnaround points. The cursor then begins to follow the target closely again (second F segment, Figure 8) until the target approaches the 5 cm escape boundary. The cursor then exhibits a quick pulse that has the effect of reversing the target movement due to the panic function (second A segment, Figure 8). Finally, the cursor again begins to follow the target closely, and the target is captured (third F segment, Figure 8).

The boundaries of the episodes indicated in Figure 8 were determined by a computer program that basically identified three patterns:

* A - "Fast Acquisition" - Cursor velocity is much greater than target velocity.

* F - "Close Following" - Cursor velocity is approximately equal to target velocity.

* P - "Predictive Mode" - Cursor velocity is much less than target velocity.

The distinction between a fast acquisition as in the first A segment in Figure 8 and close following is similar to the two modes in Costello's

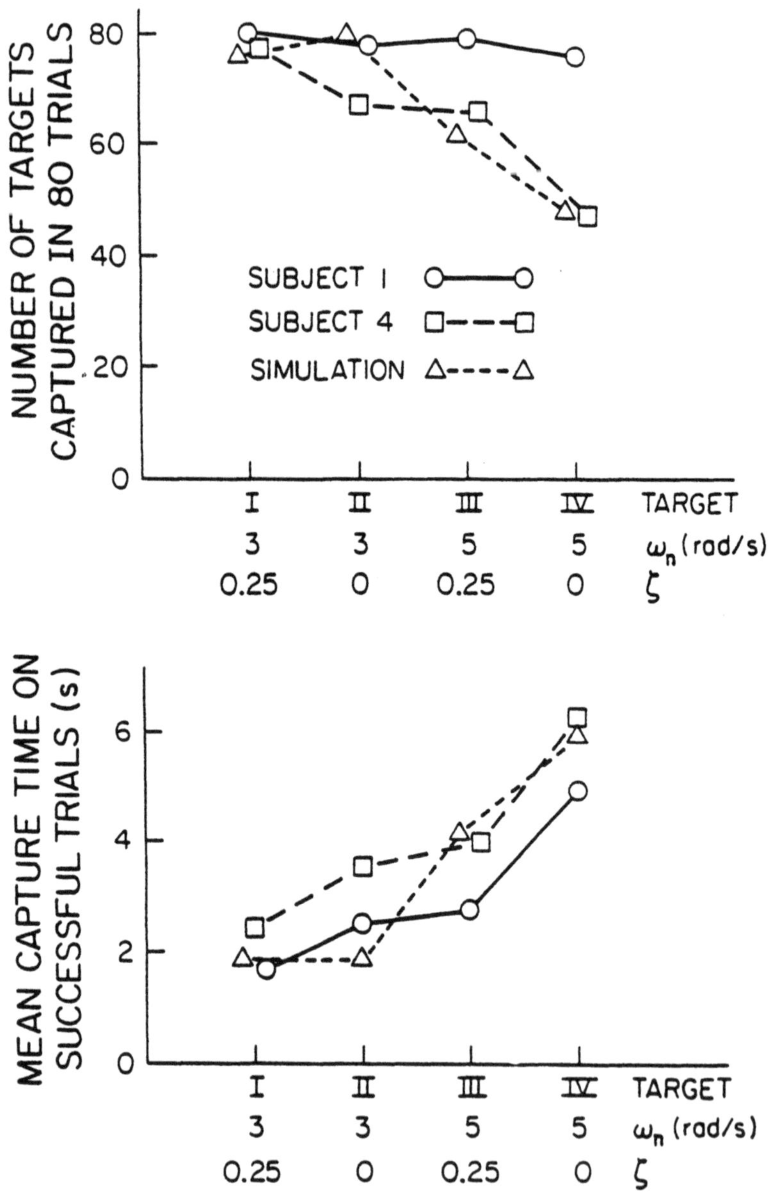

Figure 6-7: Number of targets captured and mean capture time on successful trials for Subject 1 (circles), Subject 4 (squares), and and activity module simulation (triangles).

(1968) surge model. Large errors are corrected proportionately more rapidly than small errors. The second A segment in Figure 8 keeps the target in bounds rather than reducing a large discrepancy. This type of response might better be labeled "herding." More will be made of this distinction later in this chapter. The predictive mode is quite different from both fast acquisition and close following. The subject seems to know that the target is eventually going to turn around and oscillate back toward the cursor, and therefore cursor movement is relatively slow in anticipation of these reversals. This behavior seems to involve more long-range prediction of target behavior.

The time history of each trial in Sessions 9 and 10 was represented as a sequence of A, F, and P behaviors identified by a computer program. This program used a three-stage pattern recognition procedure. The first stage approximated the continuous time history as a sequence of constant velocity segments. The break points between segments were determined by extending each segment until the mean squared error between the segment and 100 Hz samples of the actual time history exceeded 1.83 cm^2. The second stage made an initial assignment of A, F, and P behavior based on the absolute value of the ratio of target velocity to cursor velocity at .01 s intervals. Ratios less than .4 were identified as A behavior, ratios greater than 3.0 were identified as P behavior, and ratios between .4 and 3.0 were classified as F behavior. However, none of the classifications was assigned if the absolute values of both target and cursor velocities were less than .5 cm/s, or the target was turning around or exhibiting a short pause. The target to cursor velocity ratio identification scheme was considered unreliable in these instances, so an indeterminate 0 classification was temporarily assigned. Mode 0 was also assigned to segments of F behavior in which the target and cursor velocity were of opposite sign, segments of A behavior in which the target velocity was less than 5 cm/s, and segments of A, F, and P behavior less than 150 ms in duration.

The third stage of the pattern recognition procedure used more global cues to classify the initially indeterminate 0 regions and to further check the initial A, F, and P identifications. The association of close following (F) with tracking error less than .65 cm was also a major feature of these rules.

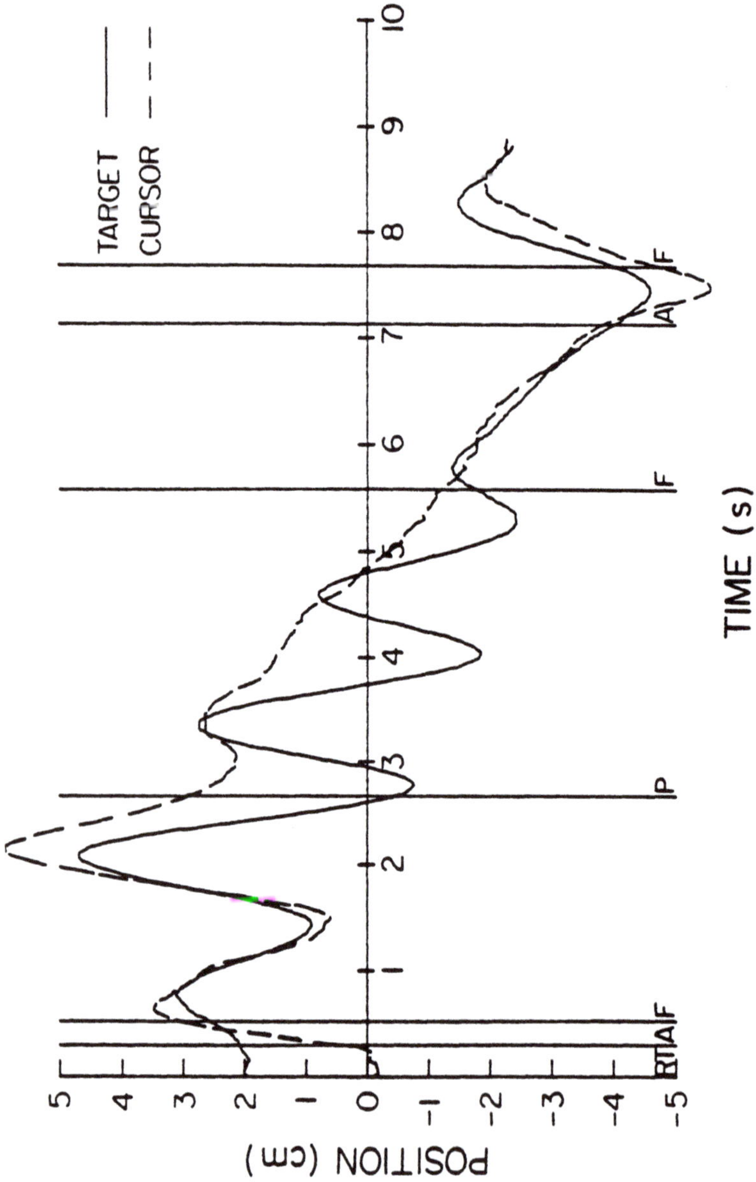

Figure 6-8: Time history of Subject 1 capturing the most difficult target, Target IV ($\omega_n = 5$ rad/s, $\zeta = 0$).

* 0 segments that were at least .5 s in duration and had tracking error continuously less than .65 cm were classified as F.

* For F-0-F, A-0-A, and P-0-P sequences, the 0 mode was given the same identification as its immediate neighbors based on the assumption of continuity.

* For A-0-F, F-0-A, P-0-F, and F-0-P sequences, that part of the 0 segment immediately adjoining the F segment and having tracking error less than .65 cm was identified as F. The remainder of the 0 segment was identified as being a continuation of the other neighboring mode.

* A-0-(End-of-Trial) and P-0-(End-of-Trial) sequences were treated as though they were A-0-F and P-0-F sequences (see above), unless the last data point in the trial had an error greater than .65 cm. In the latter case, a P mode was assigned to as much of the continuous latter part of the 0 segment as had tracking error in excess of .65 cm. The remainder of the 0 segment was taken to be a continuation of the preceding A or P mode.

* For A-0-P and P-0-A sequences, that part of the 0 segment immediately adjoining the A segment and having target velocity continuously greater than 5 cm/s was identified as A. The remainder of the 0 segment was identified as P.

* P segments not having at least one data point with tracking error greater than .65 cm were reclassified as F segments.

The result of the three-stage pattern recognition procedure was that each trial's time history was described as a sequence of A, P, and F behaviors after the initial reaction time (RT) interval. Examples of trial-to-trial variability of these descriptions are shown in Figures 9 and 10. For ease of comparison, several trials that began below the origin have been drawn mirror-imaged on the position axis. This level of description is considerably more abstract than the original time histories.

All of the intricate detail of target and cursor movement are coded in only three nominal categories of activity, A, F, and P.

6.5.2. Markov Descriptions

For each target, an individual subject's 80 trials from Sessions 9 and 10 were coded as sequences of A, F, and P behaviors. These 80 trials were then summarized by treating A, F, and P as states and each state transition as an event. The number of times each possible transition (A to F, A to P, etc.) occurred as the first, second, third, etc. event in a trial was summed over trials. The result of this summation is shown in the top part of Figure 11 for Subject 1 capturing the most difficult target. Circles represent states, arrows represent transitions between states, and the numbers next to the arrows indicate the number of times that that transition occurred. The subscripts on the state identification letters indicate the event number that led to that state occupancy, e.g., A_1 occurred after the first event, P_3 after the third. Multiple instances of a particular class of states (e.g., F_2 and F_4) were merged if a chi-square test indicated that their distributions of exit transitions to the other classes of states (e.g., A and P) were not significantly different. The merging of two states was accomplished by pooling their entry and exit states. For example, if F_2 and F_4 were merged, the state transition diagram at the bottom of Figure 11 would result. Subsequent tests would then be performed to see if A_3 and A_5, P_3 and P_5, F_2 and F_6, etc. could be merged. This merging process was designed to test whether or not different occurrences of the same activity type were essentially equivalent, based on their exit state distributions. If two states were merged, it means that the same type of activity occurred at different places in the event sequence. More elaborate tests of equivalence could have been devised by incorporating comparisons of time in state distributions (Miller, Jagacinski, Nalavade, & Johnson, 1982) and metric features of the trajectories.

The results of the merging process for Subjects 1 (best) and 4 (worst) are shown in Figure 12 for each of the four targets. The number in each circle is the mean duration of that state in seconds. The number in parenthesis below each arrow is the number of times a transition occurred. The number above each arrow is the estimated

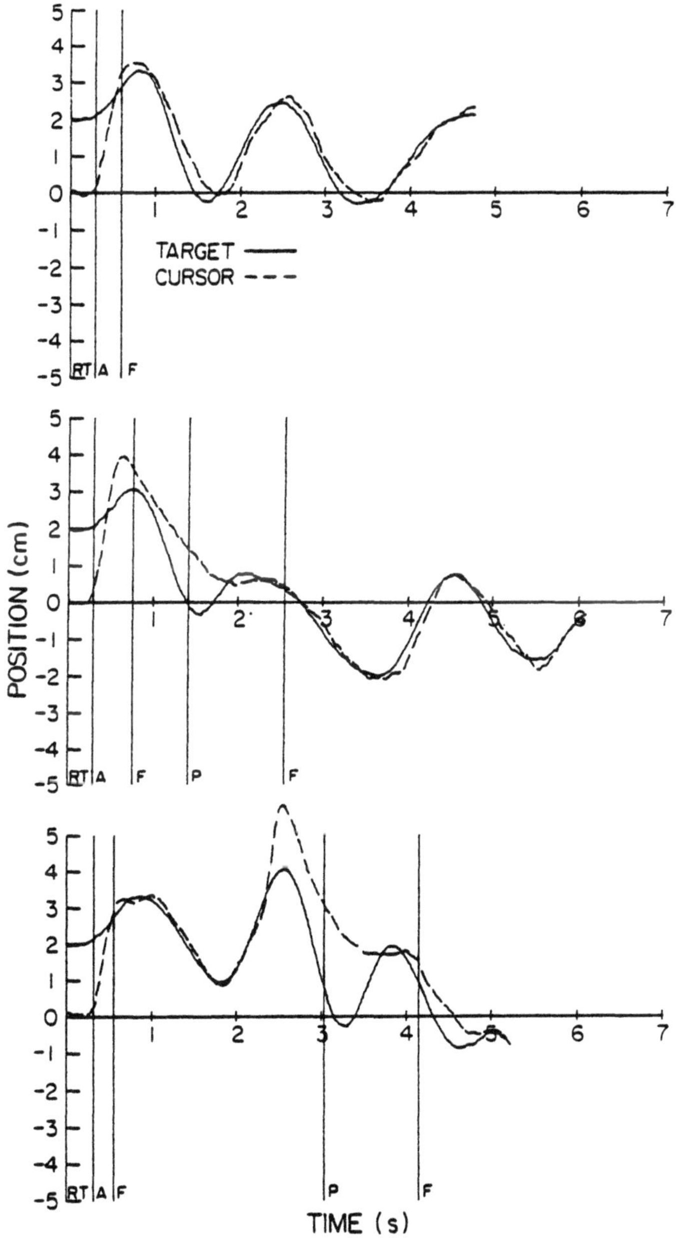

Figure 6-9: Several varied time histories of Subject 1 capturing the most difficult target, Target IV (ω_n=5 rad/s, ζ=0).

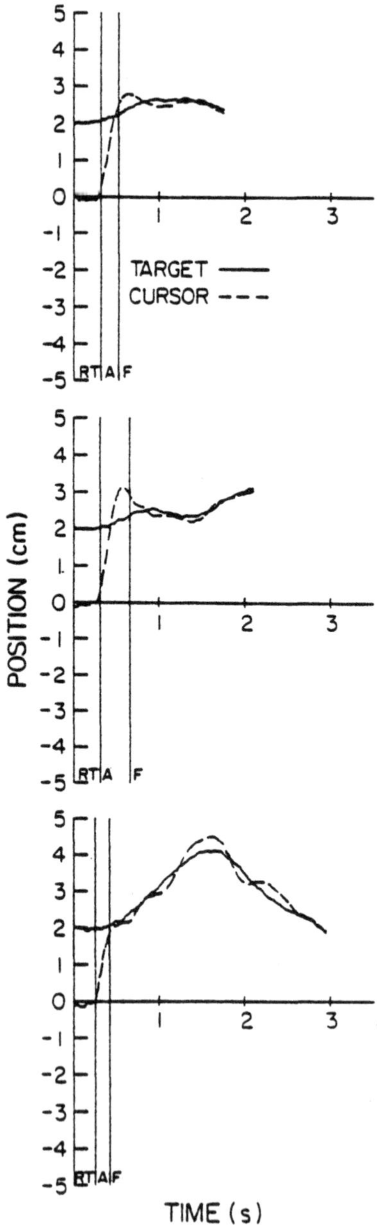

Figure 6-10: Several varied time histories of Subject 1 capturing the easiest target, Target I ($\omega_n = 3$ rad/s, $\zeta = .25$).

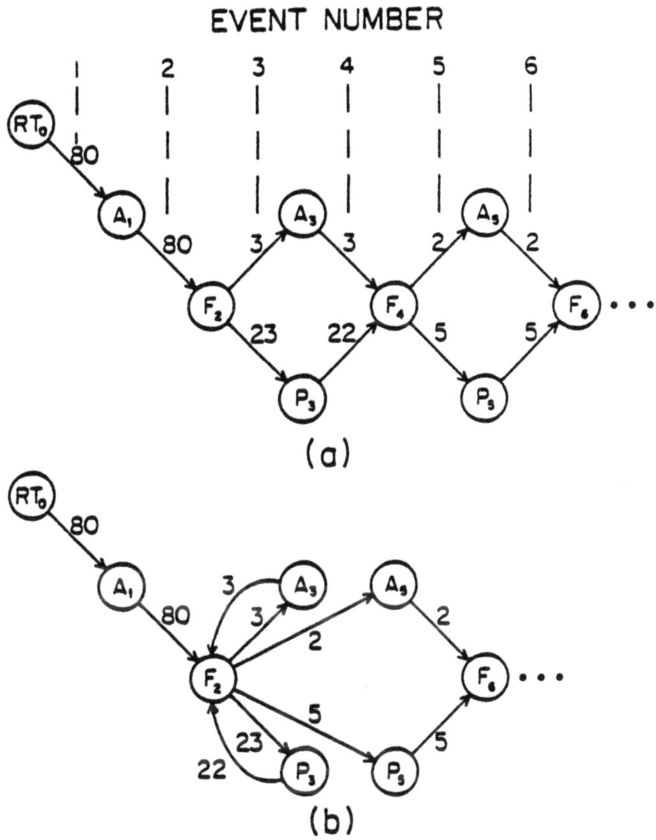

Figure 6-11: (a) Summation of mode transition (events) across 80 trials for Subject 1 capturing the most difficult target. (b) State F_4 merged with state F_2 by pooling their entry and exit states.

probability of going to a particular new state given that a transition occurred from the old state. Transitions which occurred less than 8 times in 80 trials are not shown in Figure 12. The corresponding number of targets captured and the mean capture time on successful trials for Subjects 1 and 4 are shown in Figure 7.

At thi abstract level of representation, the sequences of behavior exhibited by these subjects are quite similar. For Subject 1, the reaction time interval is followed by a fast acquisition (A), which is in turn followed by close following (F) (Figure 12). For the easier targets, Targets I and II, this sequence generally leads to a successful capture. For Target III there are some F to A transitions indicating herding behavior. For the most difficult target, Target IV, there occur additional transitions to the P mode. Subject 4's markov diagrams are very similar with the addition of some transitions to the P mode for Target II and some F to A transitions (mostly herding) for the most difficult target. Across targets, Subject 4 tended to have longer RT (.38 s vs. .29 s) and F mode durations (2.49 s vs. 1.55 s) and shorter P mode durations (.61 s vs. .86 s) than Subject 1 ($p<.01$ for all comparisons). These markov diagrams are strikingly similar given the large differences in subjects' performance. For example, for the most difficult target, Target IV, there was a 36% difference in targets captured (95% for Subject 1, 59% for Subject 4). Important differences may therefore exist in the lower level details of how some of these major categories of activity were implemented by the two subjects.

One aspect of performance missing from these diagrams is the states of the cursor and target when the mode transitions occurred. Phase plane diagrams of cursor, target, and error revealed striking individual differences between Subjects 1 and 4 when transitioning into the P mode (Figure 13). Subject 1 transitioned into the P mode primarily when the error was increasing (a well defined linear locus in the first and third quadrants of the error phase plane), and target velocity was usually greater than 5 cm/s. Subject 4 had a more diffuse spread of points in the first and third quadrants of the error phase plane, and no well defined pattern in the target phase plane. Cursor velocity (not shown) was less than 1 cm/s for 28% of Subject 4's entries into the P mode, indicating that some of the activity classified as "predictive" may have simply been pausing. In contrast, Subject 1 tended to generate ramp-like cursor movements during the P mode, and cursor velocity was never less than 1 cm/s at entry into the P mode.

6.5.3. Activity Module Simulation

Based on the previous analyses, a model of capture performance was constructed. The purpose of this exercise was to provide more generative descriptions of the A, F, and P modes of activity and test

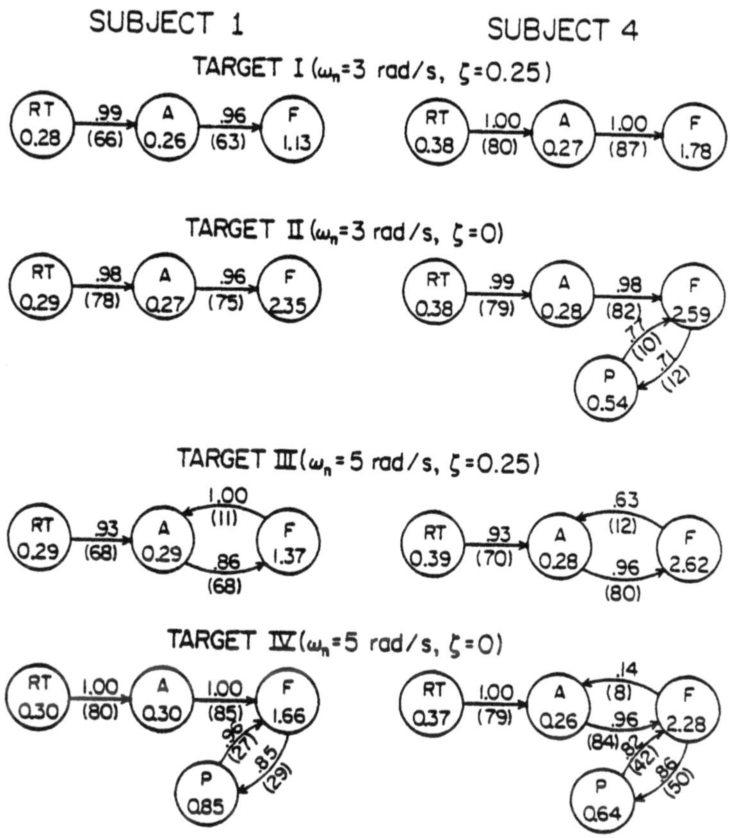

Figure 6-12: Markov diagrams of the state transition probabilities (above each arrow), the number of transitions (in parentheses), and the mean duration of each state in seconds (within the circles).

whether the two-level control structure suggested by the previous analyses could indeed capture targets in a manner similar to the human subjects.

A goal structure for the task is shown in Figure 14. The structure is read from top to bottom with each subordinate goal displayed below the superordinate goal to which it contributes. The subordinate goal must

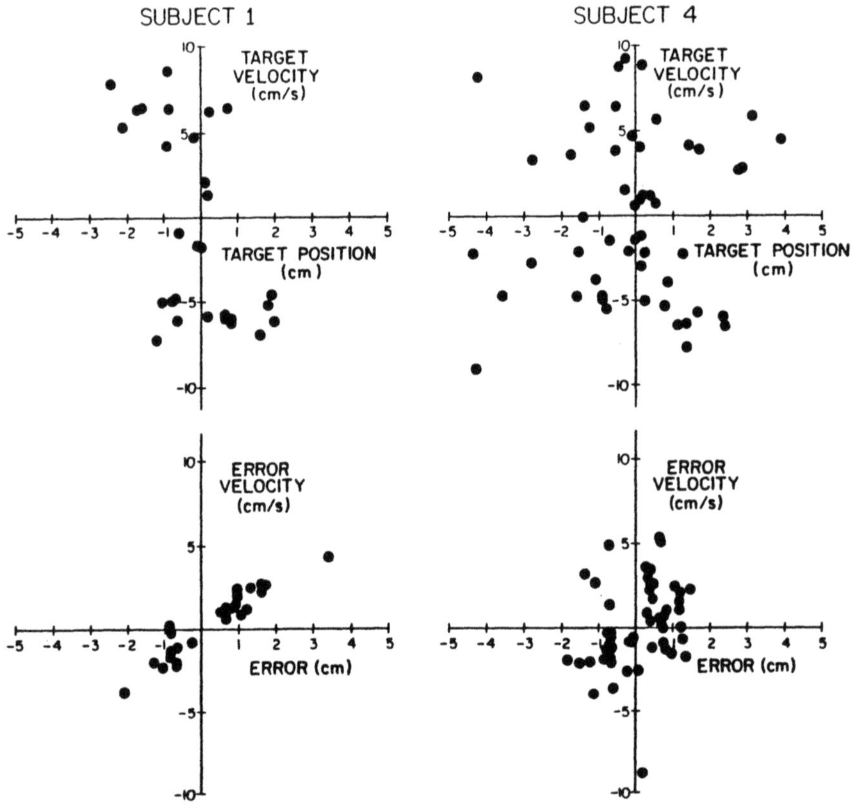

Figure 6-13: Phase plane plots of the target and error when subjects entered the predictive (P) mode while attempting to capture the most difficult target, Target IV.

be accomplished if the superordinate is to be accomplished. For example, the overall goal in the task is to capture targets, but this goal cannot be achieved without close following. Similarly, small errors cannot be nulled out via close following unless large errors are first eliminated. The diagram implies that different activities are used to reduce large errors and small errors. If large, fast, target oscillations occur, they must be eliminated by predictive mode activity before close following can be effective. If the target nears the escape boundary, herding will occur, and will be followed by either predictive mode activity and/or close following.

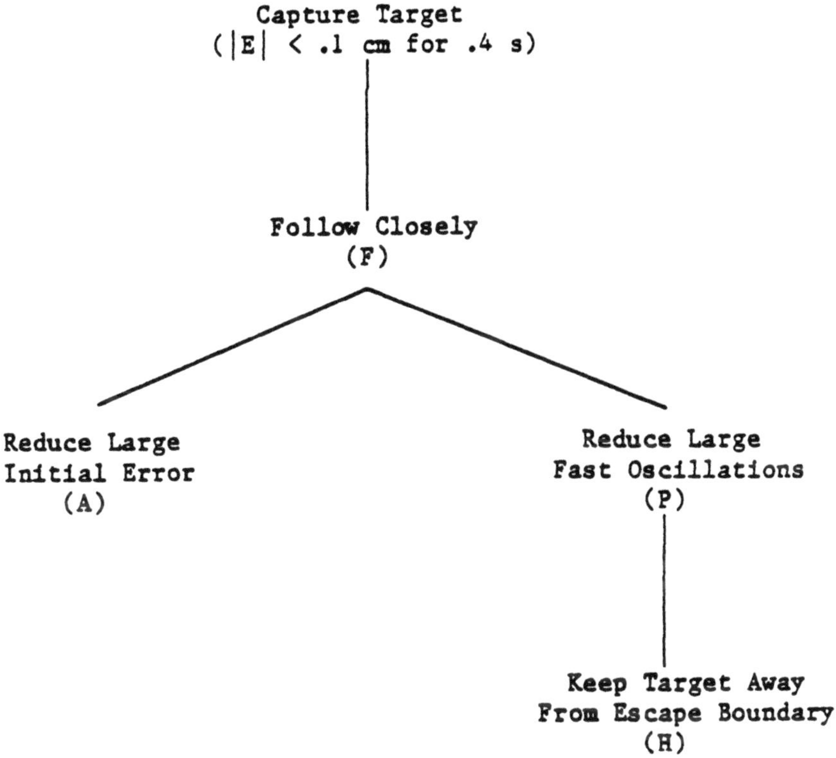

Figure 6-14: Goal structure for the target acquisition task.

Four activity modules (Figure 15) were designed to accomplish each sub-goal. Each module has a set of rules for its activation, completion, and/or interruption by other modules. These rules control the transitions among modes of activity, as represented in the markov diagrams. The fourth module arose from treating herding and the reduction of the initial large tracking error at the beginning of a trial as two separate modes, denoted as H and A, respectively. Each module has an associated goal and a priority to reflect the urgency of the goal. Preventing an escape (herding) has the highest priority, and reducing large, fast oscillations via the predictive mode, has second priority. Staying close to the target to achieve capture (close following) cannot be successful if the target is about to escape, or if it is oscillating wildly.

This goal was therefore given third priority. The fourth goal, reducing the large initial error, applies only at the beginning of trials. It was therefore given fourth priority in the simulation even though it functions as a prerequisite for close following.

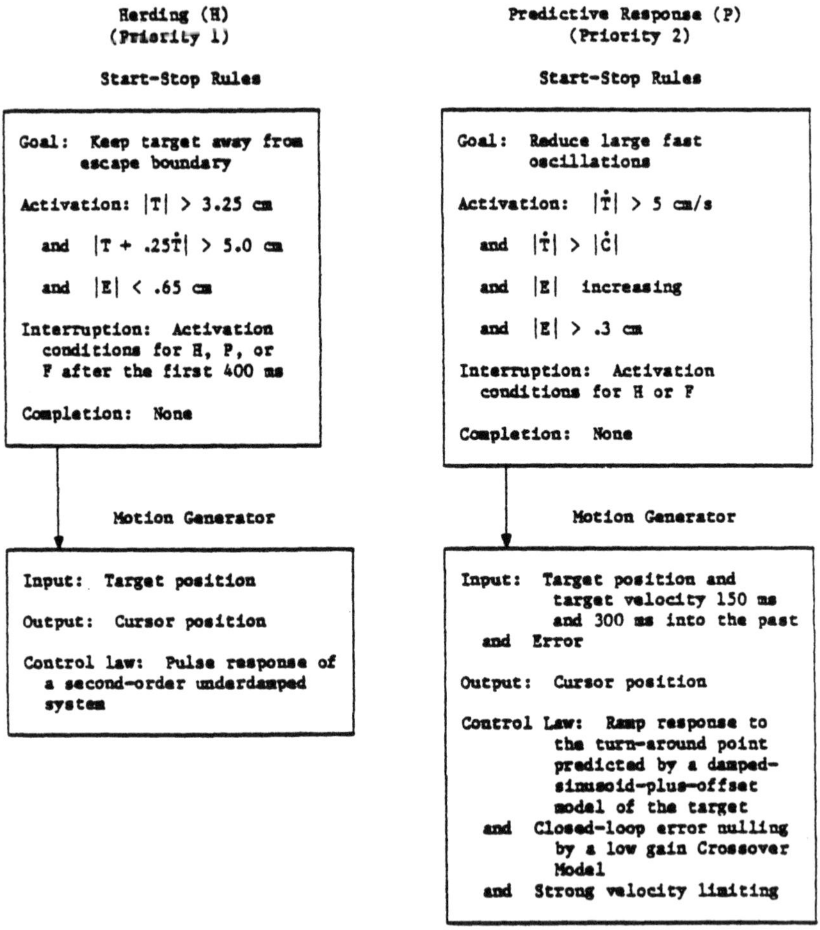

Figure 6-15: An activity module model of the target capture process.

The activation conditions for the P and F modules were based on the phase plane patterns for Subject 1. Very few herding responses were detected by the computer pattern recognition scheme previously described, so the entry conditions for the herding maneuver were not derived from subjects' data. Once activated, a module remained active until it either achieved its goal or was interrupted by another activity module. If the activation conditions for more than one module were satisfied simultaneously, the highest priority module took precedence.

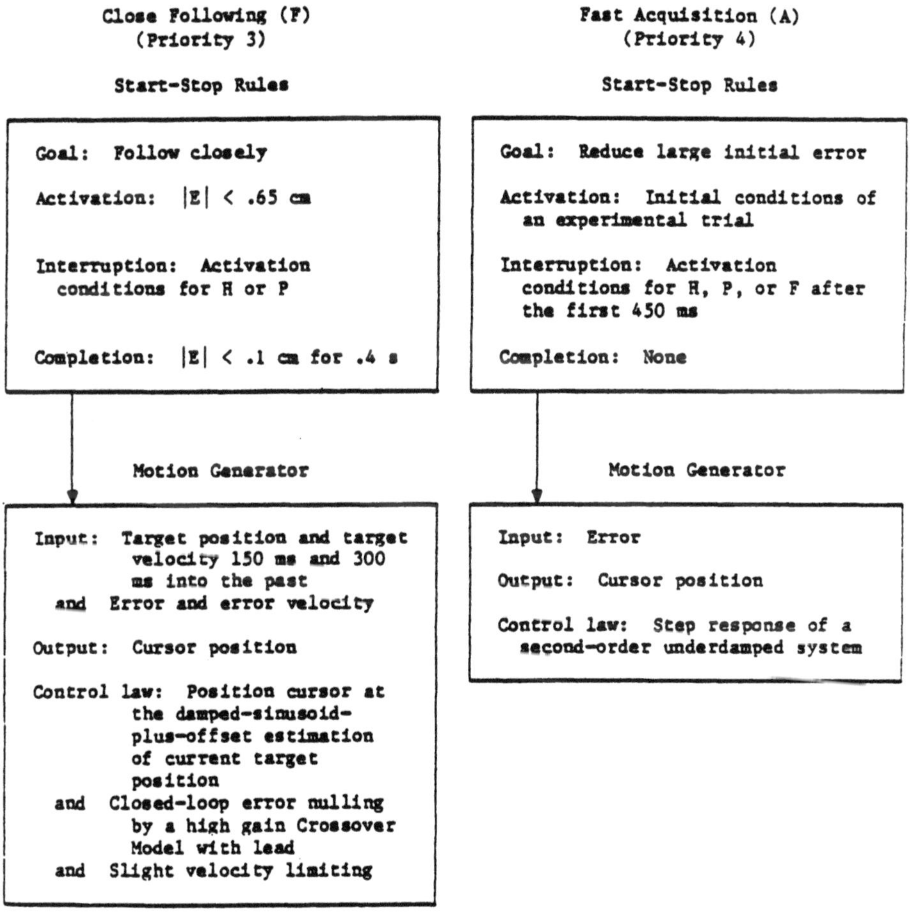

Figure 6-15: Continued.

The second level of each module is its motion-generating mechanism. The initial acquisition response, A, was generated from the step response of a second-order underdamped system. After a 250 ms reaction time, the initial response began with low damping and a high undamped natural frequency, to generate a fast rise time similar to that exhibited by Subject 1. The damping was then increased and the undamped natural frequency was decreased to shape the overshoot aspect of the response. This preprogrammed response was protected from interruption by other tracking modes for 450 ms. The herding maneuver, H, was similarly generated from the pulse response of a second-order system with high undamped natural frequency and low damping. This preprogrammed response was protected from interruption for 400 ms.

The trajectories for the predictive response, P, were generated from a combination of three mechanisms: a predictive element, a closed-loop error-nulling element and a velocity limiter. The predictive element used a damped sinusoid plus an offset as the model of target motion. Target position and target velocity 150 ms and 300 ms into the past were used to continuously estimate the target model parameters for amplitude, frequency, phase, and offset. The damping constant and the range over which the model searched for a frequency estimate were fixed as a priori knowledge of the target. In the predictive mode, the cursor did not attempt to keep up with the target. The subject anticipated that the fleeing target was going to turn around and start coming back, and then turn around again in an oscillatory manner. The predictive element therefore continuously predicted the position and time of the spatially nearer turnaround, and generated a cursor velocity sufficient to intercept the target at this turnaround (see the P segment in Figure 8). This predictive behavior was combined with a closed-loop error-nulling element in the form of a simplified McRuer Crossover Model with low gain and 150 ms time delay. A velocity limiter approximated neuromuscular smoothing.

The close following or F mode used the same three elements as the predictive mode, but modified their interaction. The predictive element used the damped sinusoidal model to estimate present target position based on target position and target velocity 150 ms and 300 ms into the past. The change in cursor position necessary to match this estimated target position was weighted by a factor reflecting how accurately the damped sinusoidal model had recently estimated past target positions.

This predictive element was combined with a high gain simplified McRuer Crossover Model with 150 ms time delay and high-frequency lead. The output of this closed-loop error-nulling element was weighted less when the predictive element was doing well at estimating target position. Finally, the third element was a less severe velocity limiter than was used in the P mode to approximate neuromuscular smoothing.

The set of four activity modules is a deterministic model. Given the constant initial condition at the beginnings of trials, only a single time history would be generated for each of the four targets. Subjects' data, however, exhibited considerable trial-to-trial variability even after ten days of practice (Figures 9 and 10). To introduce trial-to-trial variability into the model, the initial acquisition response was stochastically varied by sampling from normal distributions to determine the undamped natural frequency and the damping during the overshoot aspect of this response, and the period of protection from interruption by other activity modules. In order to avoid sharp transients in the cursor movement when mode transitions occurred, a constant velocity extrapolation of the cursor trajectory was averaged in a weighted manner with the cursor position generated by the new mode over an interval of 200 ms following the mode switch. The weighting on the constant velocity extrapolation decreased exponentially as the weighting on the cursor position generated by the new mode increased exponentially. The time constant of this exponential blending function was stochastically varied at each mode transition by sampling from a normal distribution. Any of the other tracking modes could also have been varied stochastically. However, the present stochastic variations were sufficient to generate an interesting variety of time histories.

Once a particular activity module was activated, its motion generator behaved in a deterministic manner. However, due to variations in the states of the target and cursor at the time a module was activated and variations in the 200 ms exponential blending function, the same motion generator produced many different trajectories. These different trajectories created triggering conditions for the other activity modules. The markov diagrams lack the detail of the time histories, and therefore represent transitions to more than one subsequent state probabilistically. The probability estimates in Figure 12 are estimates of how often the

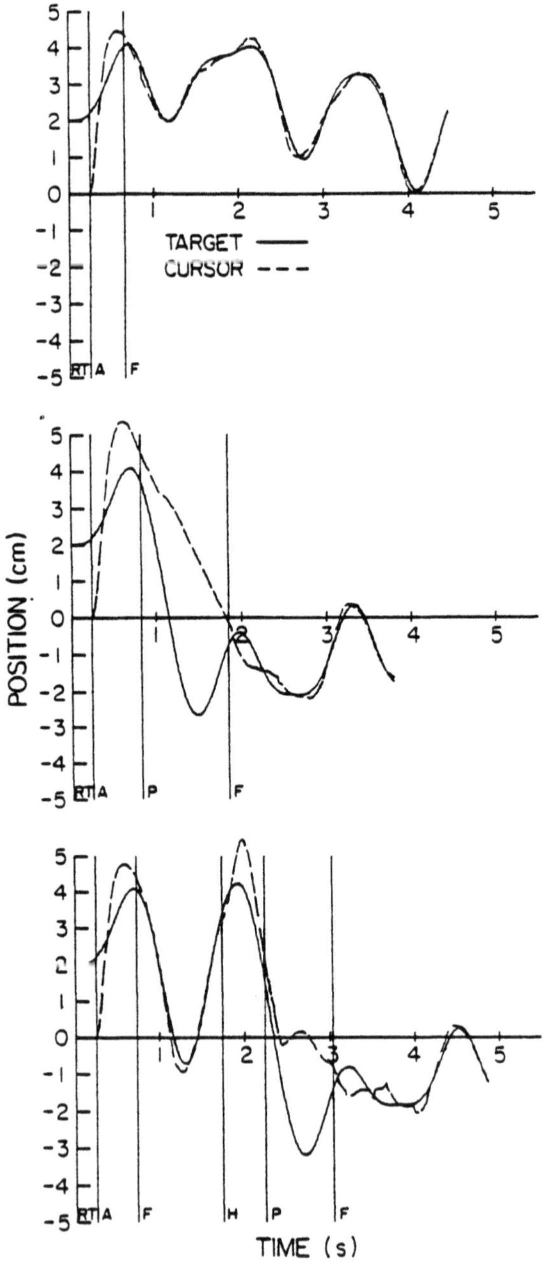

Figure 6-16: Time histories of the activity module model capturing the most difficult target, Target IV.

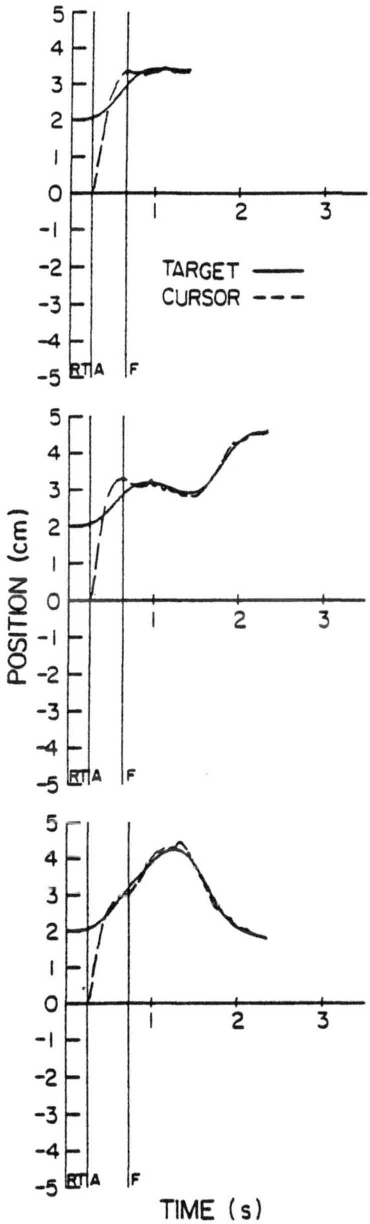

Figure 6-17: Time histories of the activity module model capturing the easiest target, Target I.

triggering conditions for the various subsequent states occurred in the detailed time histories. Sample time histories of the model capturing the most difficult and easiest targets are shown in Figures 16 and 17. Parameters in the motion generators other than time delays were adjusted for each target to match subjects' data. The mode transition rules were not varied across targets.

The performance of this multi-level model has to be judged at multiple levels of detail. In terms of gross performance measures, the number of targets captured and the mean capture time on successful trials more closely resembled the performance of Subject 1 for the easier two targets, and Subject 4 for the more difficult targets (Figure 7). The model's performance is thus in the same general range as the experimental subjects' performance, but it is only as good as the worst subject for the more difficult targets. The trials shown in Figures 16 and 17 have been chosen to demonstrate qualitative similarities with Subject 1's trajectories (Figures 9 and 10). The overall episodic quality of the simulation trajectories is very similar to that of Subject 1. Furthermore, the trajectory shapes for the individual tracking modes bear a good qualitative resemblance to the human subject's performance. One aspect of Subject 1's performance not exhibited by the simulation is predictive modes with multiple turnarounds as in Figure 8. More detailed quantitative comparisons were not performed.

At a higher level of abstraction, markov diagrams summarizing the state transitions exhibited by the model are shown in Figure 18. In order to permit easier comparison with the subjects' diagrams (Figure 12), only four classes of states, RT, A, F, and P were used in constructing the markov diagrams for the simulation. The initial fast acquisitions and herding maneuvers were treated as the same class of state (A). Like Subjects 1 and 4, the simulation's first two transitions were from RT to A and then primarily from A to F. For the easiest target, Target I, the simulation exhibited a few (9) herds. Subject 4 also exhibited a few herding maneuvers (7), but these were not drawn in Figure 12, due to the criterion of drawing only transitions that occurred at least 8 times in 80 trials. The simulation's state transitions resembled Subject 1 for Target II ($\omega_n = 3$ rad/s, $\zeta = 0$). For Target III ($\omega_n = 5$ rad/s, $\zeta = .25$) the simulation exhibited a few transitions to the predictive mode that generally led to the target escaping. The most striking difference from the subjects' performance occurred for the most difficult target, Target IV. Here the simulation exhibited two distinct close-following states. The close-following state occurring earlier in the trials, F_1, was followed most often by herding (A_2). The close-following state that occurred later in the trials, F_2, was followed most often by the predictive mode (P) (Figure 18). Given this difference in the exit state distributions, these two F states were not merged. Additionally, there were some herding to predictive mode transitions that did not appear in any of the subjects' markov diagrams. The frequencies of transitions between F_2 and P in Figure 18 are inflated by a rapid "chattering" back and forth between the two modes. It occurred for brief durations on about one-third of the trials for the most difficult target. Omission of these chatter transitions did not alter the basic structure of the markov diagram.

In evaluating the similarities and differences between the markov diagrams for the subjects and the activity module model, it is important to note that the mode transitions for the simulation were immediately available from the computerized record of the simulation. In contrast, the mode transitions for the subjects were identified from the previously described three-stage pattern recognition procedure. Inspection of the subjects' time histories suggests that the pattern recognition procedure was not very sensitive to herding maneuvers. For example, in Figure 8 there appears to be a misclassified herding maneuver at the end of the first close following episode. A similar example appears in Figure 9 (bottom trial). Further evidence for this insensitivity to herding was demonstrated by passing the time histories generated by the activity module model for the most difficult target though the three-stage pattern recognition procedure. The markov diagram for this reclassified data omitted all herding, as well as most of the simulation's initial acquisitions. A comparison of the middle and bottom trajectories of Figures 9 and 16 suggests that there are also differences in the identification of the beginnings and endings of the predictive mode by the activity module simulation and the three-stage pattern recognition scheme used for analyzing the subjects' data. These results indicate the need to judge the similarities and differences in the markov diagrams for the subjects and the simulation with some caution. A new pattern recognition scheme based on the start-stop rules of the simulation might do a better job of identifying mode changes in the subjects' data.

The markov diagrams for the simulation are also useful in suggesting how the simulation might be improved. Presently, in the F mode the simulation chases after the target too steadfastly without taking into account that it is nearing the escape boundary. Then, especially for the most difficult target, the herding mode (A_2) interrupts the close following behavior. The resulting transient delays the capture when the herding maneuver is successful, and it is not always successful in preventing escapes. Attempts at improving the F mode for the most difficult target by classical control lag and lead compensation procedures did reduce mean squared tracking error, but, surprisingly, decreased the number of captures. This result was apparently due to the increasingly forceful effects of the nonlinearity when tracking error was small. Heuristic knowledge about the nonlinearity appears necessary to increase the proportion of times the model captures the most difficult target.

6.6. DISCUSSION

The present study has demonstrated the potential of combining if-then rules and motion generating mechanisms to describe the episodic nature of target capture behavior. The present authors believe these different levels of describing behavior are examples of what Rasmussen (1983) has referred to as rule-based and skill-based behaviors. In more complex environmental situations a third level of organization corresponding to problem-solving aspects of knowledge-based manipulations might be added to the present model to account for learning and adaptation in novel situations.

Although the activity module model should be considered only a first approximation to the subjects' behavior, the performance of the model lends additional credence to this multi-level analysis of behavior. The decomposition provided by the different tracking modes proved useful in developing a simulation to match human performance. An alternative would be to work at only one level of abstraction, and attempt to represent all of the varied aspects of the target capture

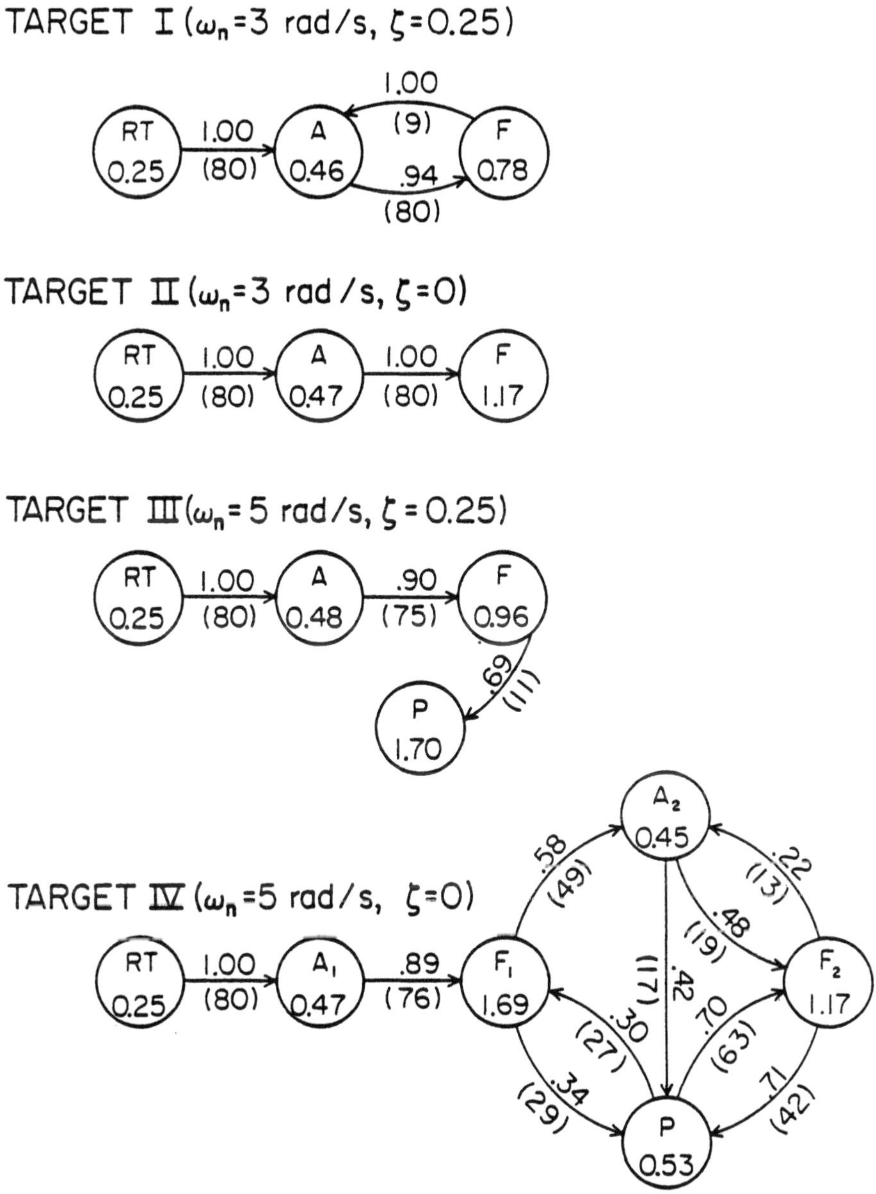

Figure 6-18: Markov diagrams of the state transition probabilities (above each arrow), the number of transitions (in parentheses), and the mean duration of each state in seconds (within the circles) for the activity module model.

behavior in a single linear or nonlinear differential equation. This approach probably would have been very difficult given the nature of the time histories exemplified by Figures 8-10. The assumption behind the present approach is that skilled behavior is organized at multiple levels, and that descriptions of these different levels will require distinctly different representations. In the present example, these levels correspond to task specific goals, the rule-based level that matches the appropriate goal with the environmental conditions at hand, and the motion generator level that produces the motion appropriate to achieve the goal.

The present simulation has also demonstrated the usefulness of a simplified predictive element for successfully capturing a higher-order nonlinear target. Although the form of the simplified target model (a damped sinusoid plus offset) was not uniquely identified from the subjects' time histories, earlier versions of the simulation suggested that some kind of predictive mechanism was essential for achieving the tracking accuracy required in this demanding task. Closed-loop error nulling did not appear sufficient. The damped sinusoidal predictive element in the P and F modes did not explicitly represent the nonlinear velocity generating escape mechanism that preceded the oscillatory filter, nor was the time history of past tracking error modeled as an input to the damped sinusoidal approximation. In order to bring the level of performance of the model up to that of Subject 1, it would probably be necessary to provide the model with some additional heuristic knowledge about the nonlinear panic function. It remains to be seen whether heuristic knowledge is sufficient, or if full veridical knowledge of the target is necessary to have the model perform as well as Subject 1. The usefulness of approximate prediction has been noted by a number of investigators, including Kelley (cited in Sheridan & Ferrell, 1974), Murril (1967), and Herzog (1968) and is supported by the limited success of the present model. Additional work on incorporating more globally defined states and global pattern recognition capability might also improve the present model without resorting to full veridical knowledge of the target.

The close following (F) and predictive (P) motion generators used the same basic elements of damped-sinusoidal prediction, closed-loop error nulling and velocity limiting, but the two generators differed in the way these elements interacted (Figure 15). This recombination of the same basic elements captures the spirit of what Greene (1982), Gallistel (1980), Turvey (1977), and others have termed coordination. Although the present model has this property, there may be other ways of representing the motion-generating mechanisms for these two modes. The present authors do not claim that the present representation is unique.

The tracking modes used in this activity module model appear to be closely related to distinct styles of tracking noted by previous investigators. For example, Costello's (1968) two-mode model for nulling large and small errors is similar to the distinction between the fast acquisition (A) and close following (F) modes in the present study. The subjects' behavior in the predictive (P) mode is somewhat analogous to crossover regression (McRuer & Jex, 1967), in which subjects do not attempt to follow high frequency characteristics of the input signal. Subjects' ability to predict sinusoidal patterns in manual control tasks is also well documented (Magdaleno, Jex, & Johnson, 1970; Pew, 1974), and Robinson (1965) described predictive and non-predictive modes of eye movement with sinusoidal target motion. Parallels such as these increase the credibility of the present mode definitions. Nevertheless, considerably more work is necessary to establish their behavioral independence as distinct modes of tracking. It is necessary to find independent variables that can alter each mode individually without altering the other modes. For example, Subject 1 only used the P mode to any appreciable degree for the most difficult target. If the other modes were not altered in structurally significant ways by this manipulation of the target values of ω_n and ζ, one would have greater confidence that the P mode was behaviorally independent from the other tracking modes. Similarly, the addition of high frequency noise to the target might affect the close following (F) mode without significantly altering the fast acquisition, herding, and predictive modes. More work needs to be done on this important issue.

In summary, the present work argues for the usefulness of combining rule-based descriptions and differential equation descriptions of target capture behavior. By explicitly representing multiple levels of organization of tracking behavior as in the present study, it may be easier to incorporate tracking into more general behavioral models. In more complex tasks involving both supervisory and active control, rule-based behavior may in turn be developed by still more abstract behavioral organization involving problem solving and learning. The rule-based aspects of active tracking can thus provide a means for interfacing it to other kinds of behavior. An interesting combination of rule-based procedures and dynamic tracking performance for a realistic aircraft landing is discussed by Baron, Muralidharan, Lancraft, and Zacharias (1980). Rule-based descriptions may also be important in describing multi-axis tracking. For example, in two-axis stabilization and hover tasks, Onstott and Faulkner (1978) found evidence for discrete episodes in which one axis was controlled at a time. Descriptions of this performance require specification of the decision rules for switching between axes.

The present work suggests that the Rasmussen (1983) definition of "skill" and the usual human performance definition are not exactly the same. Tracking ability is usually viewed as a skill, but both rule-based and skill-based levels in the Rasmussen sense were used to characterize performance in the present tracking task. One may argue that a rule-based approach might apply to any finite-state decision process which is "worked out". That is, the action or alternative is defined as a function of environmental and problem conditions. As shown by the model developed in the present work, skilled behavior may require such processes to adapt goals or implementing mechanisms. Furthermore, in some cases, skill may require the ability to learn and perform in novel situations necessitating a knowledge-based level of behavior.

It is the aim of this multi-level approach to behavioral modeling to provide a more structured picture of person-machine interaction and to contribute toward greater theoretical development in this field. Multi-level analysis forces an investigator to organize behavioral measurements at multiple levels of abstraction. This effort can potentially provide increased diagnostic insight into how a complex behavior is decomposable into functional parts. Such decompositions can be more diagnostic than grosser summary performance measures in revealing system bottlenecks, and in suggesting improved system design. If tracking is considered as just one example of how people interact with specialized devices, then the types of analyses used in this paper may be applicable to many different person-machine interactions. Recent examples of this modular style of analysis applied to an anti-aircraft artillery team and to an industrial routing task have been presented by Miller (1985). Similar ideas have been discussed by Scanlan and Agin (1978) in a complex target detection task, and by Kieras and Polson (1985) in the context of a computer text editor. The authors hope that the present effort will contribute toward the development of behavioral models at multiple levels of abstraction.

ACKNOWLEDGEMENT

This work was supported by the Air Force Office of Scientific Research, Grant AFOSR-78-3697 and by NASA, Grant NAG 2-195. The project monitor for the latter grant was E. James Hartzell. Portions of this report are based on the Masters' Thesis of the second author (Plamondon, 1982). Parts of this work were presented at the Twenty-Fourth Annual Meeting of the Psychonomic Society, San Diego, California, November, 1983, and at the Twentieth Annual Conference on Manual Control, San Jose, California, June, 1984.

REFERENCES

Anderson, J.R. (1976). *Language, memory, and thought*. Hillsdale, New Jersey: Lawrence Erlbaum Associates.

Athans, M., & Falb, P.L. (1966). *Optimal control*. New York: McGraw-Hill.

Baron, S., Muralidharan, R., Lancraft, R., and Zacharias, G. (1980). *PROCRU: A model for analyzing crew procedures in approach to landing*. (Tech. Rep. NASA No. CR 152397). Moffett Field, California: Ames Research Center,

Barr, A., & Feigenbaum, E.A. (Eds.). (1981). *Handbook of artificial intelligence*. Stanford, California: Heuris Tech Press.

Bernstein, N.A. (1967). *The co-ordination and regulation of movements*. London: Pergamon Press.

Bryan, W.L., & Harter, N. (1899). Studies on the telegraphic language: The acquisition of a hierarchy of habits. *Psychological Review, 6*, 345-375.

Burnham, G.O., & Bekey, G.A. (1976). A heuristic finite-state model of the human driver in a car-following situation. *IEEE Transactions on Systems, Man and Cybernetics, 6*, 554-562.

Chase, W.G., & Simon, H.A. (1973). Perception in chess. *Cognitive Psychology, 4*, 55-81.

Costello, R.G. (1968). The surge model of the well-trained human operator in simple manual control. *IEEE Transactions on Man-Machine Systems, 9*, 2-9.

Craik, K.J.W. (1947). Theory of the human operator in control systems: 1. The operator as a engineering system. *British Journal of Psychology, 38*, 56-61.

Gallistel, C.R. (1980). *The organization of action: A new synthesis*. Hillsdale, New Jersey: Wiley.

Goldstein, I.P., & Grimson, E. (1977). *Annotated production systems: A model for skill acquisition*. Cambridge, Massachusetts: M.I.T. Artificial Intelligence Laboratory. (M.I.T. AI Memo 407).

Graham, D., & McRuer, D. (1971). *Analysis of nonlinear control systems*. New York: Dover.

Greene, P.H. (1982). Why is it easy to control your arms? *Journal of Motor Behavior, 14*, 260-286.

Grillner, S. (1975). Locomotion in vertebrates: Central mechanisms and reflex interaction. *Physiological Review, 55*, 247-304.

Herzog, J.H. (1968). Manual control using the matched manipulator control technique. *IEEE Transactions on Man-Machine Systems, 9*, 56-60.

Jagacinski, R.J. (1977). A qualitative look at feedback control theory as a style of describing behavior. *Human Factors, 19*, 331-347.

Jex, H.R., McDonnell, J.D., & Phatak, A.V. (1966). A critical tracking task for manual control research. *IEEE Transactions on Human Factors in Electronics*, 7, 138-145.

Kieras, D.E., & Polson, P.G. (1985). An approach to the formal analysis of user complexity. *International Journal of Man-Machine Studies, 22*, 365-394.

Kleinman, D.L., Baron, S., & Levison, W.H. (1971). A control theoretic approach to manned-vehicle systems analysis. *IEEE Transactions on Automatic Control, 16*, 824-832.

Larkin, J.H. (1983). The role of problem representation in physics. In D. Gentner & A.L. Stevens (Eds.), *Mental models*. Hillsdale, New Jersey: Lawrence Erlbaum Associates. (pp. 75-98).

Magdaleno, R.E., Jex, H.R., & Johnson, W.A. (1970). Tracking quasi-predictable displays. *Proceedings of the Fifth Annual NASA-University Conference on Manual Control, NASA SP-215*, 391-428.

McGhee, R.B. (1976). Robot locomotion. In R.M. Herman, S. Grillner, P.S.G. Stein, & D.G. Stuart (Eds.), *Neural control of locomotion*. New York: Plenum Press. (pp. 237-264).

McRuer, D.T., & Jex, H.R. (1967). A review of quasi-linear pilot models. *IEEE Transactions on Human Factors in Electronics, 8*, 231-249.

Miller, G.A. (1956). The magical number seven, plus or minus two: Some limits on our capacity to process information. *Psychological Review, 63*, 81-97.

Miller, G.A., Galanter, E., & Pribram, K.H. (1960). *Plans and the structure of behavior*. New York: Holt, Rinehart, & Winston.

Miller, R.A., Jagacinski, R.J., Nalavade, R.B., & Johnson, W.W. (1982). A finite-state description of coordination in a two-handed target acquisition task. *IEEE Transactions on Systems, Man, and Cybernetics, 12*, 529-538.

Miller, R.A. (1985). A system approach to modeling discrete control performance. In W.B. Rouse (Ed.), *Advances in man-machine systems research*. Greenwich, Connecticut: JAI Press. (pp.177-248).

Murril, P.J. (1967). *Automatic control of processes*. Scranton, Pennsylvania: International Textbook Company.

Newell, A., & Simon, H.A. (1972). *Human problem solving*. Englewood Cliffs, New Jersey: Prentice-Hall.

Norman, D.A. (1981). Categorization of action slips. *Psychological Review, 88*, 1-15.

Onstott, E.D., & Faulkner, W.H. (1978). *Prediction, evaluation, and specification of closed loop and multiaxis flying qualities (AFFDL-TR-78-3)*. Wright-Patterson Air Force Base, Ohio: Air Force Systems Command.

Pew, R.W. (1974). Human perceptual-motor performance. In B.H. Kantowitz (Ed.), *Human information processing: Tutorials in performance and cognition*. Hillsdale, New Jersey: Lawrence Erlbaum Associates. (pp. 1-39).

Pew, R.W., & Baron, S. (1978). The components of an information processing theory of skilled performance based on an optimal control perspective. In G.E. Stelmach (Ed.), *Information processing in motor control and learning*. New York: Academic Press. (pp. 71-78).

Phatak, A.V., & Bekey, G.A. (1969). Model of the adaptive behavior of the human operator in response to a sudden change in the control situation. *IEEE Transactions on Man-Machine Systems, 10*, 72-80.

Plamondon, B.D. (1982). *A finite state analysis of reactive target tracking*. Unpublished Masters' Thesis,

Ohio State University, Columbus, Ohio.

Powers, W.T. (1973). *Behavior: The control of perception*. Chicago: Aldine.

Rasmussen, J. (1983). Skills, rules, and knowledge: Signals, signs, and symbols, and other distinctions in human performance models. *IEEE Transactions on Systems, Man, and Cybernetics, 13*, 257-266.

Reason, J.T. (1977). Skill and error in everyday life. In M.J.A. Howe (Ed.), *Adult learning: Psychological research and applications*. New York: Wiley. (pp. 21-44).

Robinson, D.A. (1965). The mechanics of human smooth pursuit eye movement. *Journal of Physiology, 180*, 569-591.

Rosenbluth, A., Wiener, N., & Bigelow, J. (1943). Behavior, purpose and teleology. *Philosophy of Science, 10*, 18-24.

Scanlan, L.A., & Agin, A.K. (1978). *A behavioral model of target acquisition in realistic terrain* (Tech. Rep. No. P78-70R). Hughes Aircraft Company, Culver City, California,

Schank, R., & Abelson, R. (1977). *Scripts, plans, goals, and understanding*. Hillsdale, New Jersey: Lawrence Erlbaum Associates.

Sheridan, T.B., & Ferrell, W.R. (1974). *Man-Machine systems*. Cambridge, Massachussetts: M.I.T. Press.

Simon, H.A. (1969). *The science of the artificial*. Cambridge, Massachussetts: M.I.T Press.

Turvey, M.T. (1977). Preliminaries to a theory of action with reference to vision. In R. Shaw & J. Bransford (Eds.), *Perceiving, acting, and knowing*. Hillsdale, New Jersey: Lawrence Erlbaum Associates. (pp. 211-265).

7. HUMAN COMPUTER INTERACTION

Ray Eberts
School of Industrial Engineering
Purdue University
West Lafayette, IN 47907

7.1. INTRODUCTION

Interacting with a computer is becoming more and more difficult to avoid. We pay our bills to computers, and sometimes they threaten to send collection agencies if we are delinquent. We often bank by interacting with computerized tellers. The office is becoming more and more computerized, from the copying machines to computerized mail and word processors. We are bombarded with advertisements about computers and computer products, and we are told that our children will not succeed in college unless they have a computer at home. At this writing, it is estimated that over 10% of us have computers at home. A recent estimate stated that there are around 600,000 computers in our public schools; this is about 20 times more than there were in 1980. We are told that some computers are user-friendly and others are intelligent; some may even be both. We are told that computers are revolutionizing the world and increasing productivity. Interacting with a computer is, however, difficult and frustrating. Most people would probably agree that interacting with a computer is more difficult than interacting with another human being. Human factors specialists and psychologists are working to make human-computer interaction easier, so that computers can become more effective. This chapter outlines four approaches that have been taken to make this interaction more effective. The role the human factors specialist plays for each of these approaches will be outlined.

Human-computer interaction is a challenging topic. The price of computer hardware has been decreasing rapidly. To use an analogy, if the cost of a Rolls Royce had followed the path of computers, we would now be able to buy a Silver Cloud for approximately two dollars.

Because hardware is becoming less expensive, we are no longer limited by what can be done for the interface, but we can start to think about what should be done. As an example, the Apple Macintosh incorporates an innovative interface (e.g., windows, pop-down menus, graphics, and direct-manipulation modes), sells for under 2,000 dollars, and is touted as easy to use. One version of the Macintosh contains 128K of memory, of which 40K is used to control the operating system and the interface, thus making it easy to use. Needless to say, past computers that only had, for example, 16K of memory at a comparable price, could not incorporate the features that the Macintosh has. Thus we are less limited in the kinds of things that can be incorporated in the computer design; decisions must be made about the most effective interface.

Human-computer interaction is an important topic. More and more of the work force is becoming exposed to computers. Thus, productivity is tied into how well workers interact with the computer. Besides the importance of the computer in manufacturing, the computer is vitally important as the work force becomes more service-oriented and we process more information than goods. White collar workers account for over 54 million people in the labor force today; this is expected to grow to over 62 million by 1990. That will be 55 percent of the total work force. Although advances have been made in productivity growth for blue collar workers in the last couple of decades, white collar productivity has not experienced the same kinds of growth. Sometimes, computers can make tasks more difficult and time consuming. As examples, Hansen, Doring, and Whitlock (1978) found that a computerized class examination took students as much as 100% longer to finish than did the typical paper-and-pencil test. In analyzing the task, they found that much of the extra time was spent in trying to figure out how to use the machine. Similarly, Kozar and Dickson (1978) found that students took more time making decisions if the relevant information was presented on a computer screen than if it was presented on paper. The process of interacting with computers needs to be improved.

Some improvement in human-computer interaction has been made. As a concrete example of the advances, I will consider one human-computer interaction task; computer assisted instruction (CAI). One of the earliest applications of computers interacting with humans has been in CAI. Much of the CAI work has been research-oriented, so measurements have been kept of various parameters of the production of the task. One such measurement has been the number of hours taken to author an hour of computerized instruction. Table 1 contains examples of authoring times over the past 20 years. As can be seen, a large decrease in interaction times occurred when a switch was made from FORTRAN, a general all-purpose language, to an authoring language designed especially for CAI applications. Another improvement occurred when the computer system changed from a batch-oriented system (sending a complete job to the computer and receiving feedback only on the complete program) to an on-line interactive system (line-by-line feedback can be provided). A third improvement occurred when the authoring language was equipped with menu displays and prompts that help tell the user what to do. For this particular example, the time needed to complete a task has been decreased by almost 300 times by making the computer system easier to use. I should mention that the eight hours to author an hour of instruction may be a little low when compared to the others in the table. This authoring time was reported by the manufacturer of the language and is one of the shortest times that I have seen (see Table 3 in Eberts & Brock, 1987). Advances have been made, however, in the ease and time taken to interact with computer systems. Similar examples of improvement could probably be seen in other human-computer tasks.

TABLE 1. ADVANCES IN INTERACTION EFFICIENCY

Language	Dates Available	Batch or On-Line	Hours/ Hour	References
FORTRAN	Before July 1967	Batch	2286	Avner, 1979
Authoring	Before Oct. 1968	Batch	615	Avner, 1979
Authoring	After 1968	On-Line	151	Avner, 1979
Authoring with menus and prompts	1984	On-Line	8	Fairweather and O'Neal, 1984

The problem of human-computer interaction is both challenging and important. Several different methods have been used by human factors specialists, psychologists, and computer scientists to attack this problem. In my review of this area, I have categorized the methods used into four general approaches. This will form the basis of this chapter. I have called the four approaches the empirical approach, the anthropomorphic approach, the cognitive approach, and the predictive-modeling approach. When categorizing the methods into these four approaches, the divisions between the approaches are not always clear; some studies may fall into a number of different approaches. I have, however, found this categorization scheme beneficial for organizing the wide range of material on this topic and for identifying ways to apply the different approaches to new problems.

In the next sections, each of the four approaches to human-computer interaction will be discussed. For each of the approaches, the problem of human-computer interaction is seen in a different light. For the empirical approach, it is seen as a problem because we do not know which of the various interactive methods are the best. For the anthropomorphic approach, it is seen as a problem in that computers are not as easy as humans in terms of ease of interaction. For the cognitive approach, it is seen as a problem in that humans perceive a computerized task to be a problem, and we must allow the computer to help us with that problem-solving in some way. For the predictive-modelling approach, it is seen as a problem because we need the tools to predict which method will be the best. In the following discussion, I will talk about the methodology used, examples, advantages, and disadvantages of the various approaches.

7.2. THE EMPIRICAL APPROACH

Many different choices have to be made for human-computer interaction tasks. When teaching programming, instructors have to decide between flowcharts or no flowcharts, whether to reduce emphasis on GOTO statements, or whether to emphasize that program modules should be indented. When designing a system that queries the user, the designer can choose between menu displays, fill-in-the-blank, parametric, or direct manipulation approaches. In database searches, the designer must choose between logical OR or AND searches or searching by paging through the database. The designer has a wide choice of input devices such as keyboard, voice, mouse, or touch display, to name a few. All of these methods and devices have already been developed. The problem is to choose which one is best. By incorporating the empirical approach, the researcher finds that the correct choice depends on the results of experimentation.

The methodology of this approach should be familiar to researchers who perform behavioral studies. First, the items to be tested are identified. The items used will usually correspond to the different levels of an independent variable (the variable being manipulated by the experimenter). Next, a task, that corresponds closely to the real world task but is controllable in a laboratory situation, is identified. The choice of the task determines the kinds of dependent variables that will be used and measured. Finally, the experiment is carried out in a well-controlled environment so that factors other than the independent variable do not vary from condition to condition. The results are then analyzed to determine the statistical significance of the results. The methodology of the approach has been outlined by Shneiderman (1982), Moher and Schneider (1982), and Embley (1978). It is interesting to note that all of these researchers have computer science backgrounds.

Tables 2, 3, and 4 contain a listing of some of the important articles on programming, information display, and input devices that empirically test the different methods, modes, and devices available. In each of the tables, the particular reference is broken down by the independent variables, the dependent variable, and the results. The tables do not provide a complete list of the empirical studies done, but they can be used to get an idea of the approach and some of the issues that have been tested. The results provided in the tables are summaries and could be slightly misleading if interpreted literally; the full article should be consulted for more information.

Characteristics of this approach can be outlined by examining the articles from the tables in more detail. The characteristics are not necessarily adhered to by all the articles; exceptions do occur. By looking at the general characteristics of the studies, the generalizability of the studies can be determined. As an example, if the study uses only novice subjects, the study can be generalized only to novice computer users.

First, the independent variable is generally the method, mode, or device available. As an example, the Duchnicky and Kolers (1983) experiment has display mode as an independent variable, with the levels being either 40- or 80-column displays. These experimental designs usually lend themselves to one-way analysis of variance or t-test analyses. The significance levels used for the analyses are usually set at .05, as they are for a rule of thumb in much of the behavioral research.

There is also a tendency to use less reliable levels; as an example, Shneiderman (1982) says that the .074 significance level seen in his experiment is a statistically significant result. As in most behavioral research, the dependent variables of time and errors are usually measured. Many of the experiments include preference ratings as part of the experimental design.

Another characteristic of this approach is that, for the most part, novices are used as subjects in the experiments. Thus, the results of these experiments are meant to apply mostly to first-time computer system users. There are, of course, some exceptions (e.g., Rosson, 1983, examined expert/novice differences in the use of text editors). If the research is done in industrial settings, the subjects are usually from the secretarial staff; if done in an academic setting, the subjects are usually from the student population.

A final characteristic of this research is that little or no training time is provided. A typical study lasts for an hour, a maximum of two hours in some cases. Generally, changes over time are not studied. In some cases, the changes that may occur over time are inferred by looking at differences between experienced and inexperienced groups of subjects. In the next sections, each of the three table topics will be examined in more detail.

7.2.1. Programming

One of the most costly and time-consuming human-computer interaction tasks is programming. At this time, 90% of the cost of a computer is the cost of the software. This is a complete reversal from about a decade ago, when 90% of the cost was for the hardware.

Creating, debugging, and documenting a program is a slow process. As an example, so-called superprogrammers can produce only about 50-200 lines of code a day. Typical applications programmers can only produce about 10-15 lines of code per day. Table 2 summarizes some of the research devoted to making this process faster and less costly.

TABLE 2. EMPIRICAL STUDIES ON PROGRAMMING

Independent Variables	Dependent Variable	Result	References
Batch or time-shared facilities	Code and debug	Time-shared better	Grant & Sackman, 1977
GOTO	Write program	No differences	Lucas & Kaplan, 1976
IF-THEN or GOTO	Comprehension	IF-THEN better	Smith & Dunsmore, 1982
IF-THEN, GOTO, WHILE-DO	Understanding	GOTO better	Smith & Dunsmore, 1982
IF-THEN-ELSE, GOTO	Time to complete, errors	IF-THEN-ELSE better	Sime et al., 1973
IF-THEN-ELSE, GOTO	Comprehension	No difference	Miller, 1975
Single entry, single exit or control	Comprehension	Weak support for single entry	Love, 1977
Structured, unstructured	Comprehension, modification, debugging	Little or no differences	Sheppard et al., 1979
Indented, not indented	Comprehension	No differences	Di Persio et al., 1980
Paragraphing	Comprehension	No differences	Love, 1977
Typed or typeless languages	Errors	Typed better	Gannon, 1976
Flowcharts or no flowcharts	Composition, comprehension, debugging, and modification tasks	Flowchart not useful	Shneiderman et al., 1977
Listing, flowchart, or Bowles structure	Debugging	No differences	Gilmore & Smith, 1984
Flowchart or program design language	Comprehension and quality of programs	Little differences	Ramsey et al., 1983
Debugging aids	Errors	Aids not useful	Gould & Drongowski, 1974

Besides making software development more efficient, another impetus has come from the teaching of programming in computer science courses. This teaching has been greatly influenced by books such as The Elements of Programming Style by Kernighan and Plauger (1978) and FORTRAN With Style by Ledgard and Chmura (1978), which both emphasize the structured approach in programming. Some of the characteristics of structured programming are: that GOTO statements should be avoided, indentation can be used to make the program easier to read, and the code should be written in modules that have a single entry and a single exit point. As can be seen from Table

2, structured programming is not always supported by the empirical results In some cases, GOTO statements are easier to comprehend than other kinds of data structures (Smith & Dunsmore, 1982), only weak support is found for the module concept (Love, 1977), and indentation and paragraphing are not supported (Di Persio, Isbister, & Shneiderman, 1980; Love, 1977). This lack of empirical support could be due to the methodology used rather than the failure of structured programming. The structure in a program is most beneficial when the programmer writes a program, does not use it for several months, and then has to go back to it and rewrite it. As mentioned in the introduction, the experimental design for the empirical approach does not usually incorporate the long-term studies needed to show the advantages of structured programming.

Other areas are also studied in programming. One issue has been whether or not flow charts are useful for program construction (Gilmore & Smith, 1984; Shneiderman, Mayer, McKay, & Heller, 1977). Generally, flowcharts have been found to have little utility, and thus, the tendency in computer science is to deemphasize them, As another example, the usefulness of debugging aids (Gould & Drongowski, 1974) has also been studied, with little or no support demonstrated for the usefulness of these kinds of aids.

7.2.2. Information Display

The information displayed is often the only view the user has of how the computer operates. To make the interaction easier, the information displayed must be carefully chosen. Some of the many choices available for displaying information are presented in Table 3. Various display methods have been studied, such as the kinds of messages to provide, the mode of presentation of the information (voice, graphics, or text), and the amount of information displayed.

Making generalizations from this empirical work is difficult. However, a few trends seem to stand out. First, the messages provided to the users should be as specific as possible, to make the system easier to understand (Grudin & Barnard, 1984; Shneiderman, 1982). The messages should also be as courteous as possible, though it is interesting

TABLE 3. EMPIRICAL STUDIES ON INFORMATION DISPLAY

Independent variables	Dependent variables	Results	References
10,15, or 60 cps Baud	Understanding, retention	10-15 cps best	Bevan, 1981
Baud rate, output volume, or output variability	Answer questions	Baud rate little effect; variability increase affected errors	Miller, 1977
Scrolling rate	Eye movements	Static best	Kolers et al., 1981
Color of letters and background	Scanning time and accuracy	Many interactions between foreground and background colors	Ohlsson et al., 1981
Color or shape	Time, errors in search	Color easier	Zwaga & Duijnhouwer, 1984
Window size	Reading rate	Little difference	Duchnicky & Kolers, 1983
Moving window or scrolling	Time, moves	Moving window best	Bury et al., 1982
Color or monochrome graphics	Time, errors	No effect	Tullis, 1981
Amount of visual feedback	Time, errors	No effect	Rosinski et al., 1980
40 or 80 column displays	Reading rate	80 column 40% faster	Duchnicky & Kolers, 1983
Amount of text displayed	Time	Little effect	Neal & Darnell, 1984
Tables, graphs, or combination	Comprehension	Table increased comprehension; combination slow but accurate	Powers et al., 1984
Relational or hierarchical database models	Comprehension, problem solving and memorization	Hierarchical better	Brosey & Shneiderman, 1978
Graphics or linear input for databases	Time, accuracy	Graphics better	Bury & Boyle, 1982
Menus vs. other input	Time, errors	Menus more accurate; no effect on times	Gade et al., 1981
Map or index of menu items	Time	Map better	Billingsley, 1982
Constant or non-constant position for menus	Time	Constant better	Teitelbaum & Granda, 1983
Graphics or text	Time, errors	Graphics better in terms of time	Tullis, 1981
Graphics or voice output for emergencies	Accuracy, speed or dual task	Graphics better	Robinson & Eberts, in press
Graphics, text, text and graphics, or voice commands	Time, errors	Text, text and graphics best	Stern, 1984
? or simple error message	Errors	Simple message better	Shneiderman, 1982
Hostile, neutral, or courteous messages	Accuracy	Courteous better than neutral; no other effects	Shneiderman, 1982
Specific or abbreviated command names	Time, errors, strategies	Specific better	Grudin & Barnard, 1984
Specificity of error messages	Debugging	Increased specificity better	Shneiderman, 1982

to note that Shneiderman (1982) found that hostile messages were as accurately responded to as the courteous messages. Second, regarding the mode of presentation, graphics have been found to be useful (Bury

TABLE 3. EMPIRICAL STUDIES ON INFORMATION DISPLAY (cont.)

Independent variables	Dependent variables	Results	References
HELP message design Time, errors T{ Best if user initiates	Cohill & Williges, 1982		
HELP message design	Time, errors	On-line best	Magers, 1983
Command or form/fill-in dialogue	Performance	Form fill-in best	Ogden & Boyle, 1982
Organized or random menus	Accuracy	Organized better	Liebelt et al., 1982
Screen-oriented or contextual cursor movement	Time, errors	Screen-oriented better	Hicks & Keller, 1982

& Boyle, 1982; Robinson & Eberts, 1985; Tullis, 1981), with one exception. Stern (1984) found that text alone or text with graphics was better than graphics alone. Color can also be useful for highlighting various items on the display and allowing a search for that item to occur fairly rapidly (Zwaga & Duijnhouwer, 1984). Third, much work has been performed on the usefulness of menu displays. Menu displays help decrease errors (Gade, Fields, Maisano, & Marshall, 1981) and can be improved by placing menu items in constant positions (Teitelbaum & Granda, 1983), and by providing a map of the menu structure if menus are embedded within other menus (Billingsley, 1982). Finally, work has been done on the amount of information that should be displayed. Generally, the larger the display screen, the faster the reading rate (Duchnicky & Kolers, 1983). However, both the window size (Duchnicky & Kolers, 1983) and the amount of information displayed on a screen (Neal & Darnell, 1984) have little effect on performance of the task.

7.2.3. Input Devices

Several different input devices are available for use. A typical input device is the keyboard that can include function keys and special cursor control buttons. Other input devices include the joystick, the mouse, the lightpen, and the trackball. The touch screen, voice input, and data tablet are input devices incorporated in the past couple of years. Two of the more exotic input devices in the research state are the recording of eye movements (see Calhoun, Arbak, & Boff, 1984) and the recording of brain waves (see Israel, Wickens, Chesney, & Donchin, 1980, for a slightly different application of brain wave research). In this area, empirical studies have been done on various input devices to find the characteristics that will make them more effective.

Table 4 contains a summary of some of the research performed on input devices. I had a difficult time finding studies that directly compare the different devices. One exception is the extensive study of voice as an input device. The research on voice has been contradictory; Mountford and North (1980) found that voice was better than keyboard entry and Gould and Alfaro (1983) found that voice was worse than typing or handwriting. Another study (Aretz, 1983) contained a tradeoff on the dual task employed, so that interpreting the data was difficult. Table 4 also outlines some of the research on how to improve keyboard characteristics (Emmons & Hirsch, 1982; Roe, Muto, & Blake, 1984; Suther & McTyre, 1982) and on the mouse design (Price & Cordova, 1983).

7.2.4. Advantages

TABLE 4. EMPIRICAL STUDIES ON INPUT DEVICES

Independent variables	Dependent variables	Results	References
Keyboard slope	Typing	No differences	Suther & McTyre, 1982
Keyboard height	Typing	No differences	Emmons & Hirsch, 1982
Tactile feedback on key push	Time, errors	Feedback effective	Roe et al., 1984
Auditory feedback on key push	Time, errors	Feedback effective	Roe, et al., 1984
Voice or keyboard entry	Tracking	Voice better	Mountford & North, 1980
Voice or manual entry	Time, errors, dual task	Manual better on response time; voice better on primary task	Aretz, 1983
Voice, typing, handwriting	Time, errors	Voice worst	Gould & Alfaro, 1983
Multiple mouse buttons	Time	Multiple buttons better than multiple clicks	Price & Cordova, 1983

The empirical approach offers an alternative to intuition in determining which design is best. Since humans will be using the computer systems, the design that is easiest for the human to use should be the one incorporated in the design. The effectiveness of the empirical approach can be gauged by whether intuition is unconfirmed through the empirical studies. This has occurred in human-computer interaction studies; Table 2 contained examples which showed that structured programming may not be as useful as once thought. Also, the use of flow charts has decreased due to the empirical evidence that they offer little or no advantages over other methods.

7.2.5. Disadvantages

Those with backgrounds in behavioral research are certainly aware of some of the pitfalls of carelessly designed empirical research. The experimenter must be careful to eliminate possible confounds in the study. A confound occurs when an uncontrolled factor is allowed to vary along with the levels of the independent variable. Confounds can be divided into two broad categories: design-related and implementation related. The kinds of confounds that can occur, such as the groups not being comparable, are outlined in Campbell and Stanley (1966) and in Cook and Campbell (1976). Experiments which rely on preference ratings can be particularly problematic. The subjects would probably have a tendency to best rate those aspects of the task that are the newest and flashiest even though they may not be the easiest to use. Also, the unintentional demand characteristics of the experimenter could affect the results; the results may tend to come out the way the experimenter wants, because the experimenter provides cues to the subjects about the responses desired (see Rosenthal, 1963).

As an example of some of the methodological problems that can occur with this approach, Sheil (1981) did an extensive literature review on empirical research for computer programming tasks. He concluded that a large percentage of the studies had some methodological problem associated with them that made the applicability of the results questionable.

Another disadvantage is due to the kinds of experimental designs that are used. As mentioned previously, a majority of the studies use novice subjects in one- or two-hour experiments. The generality of this research can be questioned. It is certainly appropriate to apply this research to interactions, such as the use of bank teller machines, which must be learned on the spot to be effective. In interactions which require long hours of learning a computer language; the results of these studies may be relevant only to the user at the beginning of learning and not at the end of learning. The point is, application of research cannot be done blindly; characteristics of the experimental design must be taken into account to try to predict the applicability of the results.

It is possible for this kind of empirical research to be misapplied by someone who does not take into account the characteristics of the experimental design. A similar problem occurs when a practitioner comes upon a particular problem, aware that the empirical approach is used quite often in human-computer interface designs, and thinks that the appropriate experiment to solve the problem must have already been done; an over-reliance on the empirical approach could occur. The practitioner may spend valuable time searching through the empirical literature to find the relevant experiment, rather than considering other available approaches.

The final disadvantage of the empirical approach is the lack of theoretical guidance in the interface design. The experiments listed in Tables 2-4 are performed, for the most part, to meet a specific need at a particular time; they are need-driven instead of theory-driven. In many cases this is appropriate. To fill in missing links, such as how a novice transitions to an expert, guidance from a theoretical orientation is needed. Needed experiments to cover all the applications cannot be performed. Without a theoretical perspective, when a new device or display mode is developed, the experiments must be performed all over again to determine how the new method compares to the old, see the Chapter by Kantowitz in the volume for a comparable discussion. In other words, the human factors of computer interfaces becomes a follower instead of a leader; it has little or no predictive value without a theoretical orientation. Many of the methods for the other approaches take a theoretical perspective.

The empirical approach has been good at placing the choice of interface designs on an empirical basis instead of an intuitive basis. In many cases, the experimental results have disconfirmed what was intuitively thought to be the best design. For specific applications of technology or methods already created, the empirical approach is quite useful. When implementing results from this approach, however, the practitioner must take into account the experimental design that produced the results so that the results are not over-generalized. The theoretical orientation of the next two approaches is needed to help create effective systems (instead of comparing existing systems) and to fill in the missing holes that a purely empirical approach leaves.

7.3. THE ANTHROPOMORPHIC APPROACH

Anthropomorphism can be defined as the application of human qualities to nonhuman entities. Under this approach, the human factors specialist uses the process of human-human communication as a model for human-computer interaction. The logic is that humans can effectively communicate, for the most part, between themselves, so the problem with human-computer communication must be that the computer does not act enough like a human to carry on an effective interaction. Therefore, if the computer is provided with human-like qualities, the interaction can be more effective. Several qualities can be applied to the computer: natural language, voice communication, help messages, tutoring, and friendliness.

"User-friendly" is a term which is increasingly used to describe computer products. Defining the term is difficult; determining if one system is more friendly than another is even more difficult. Defined within the context of the anthropomorphic approach, the meaning of the term can become clearer. In essence, the term means that the computer will interact with the user much the same way one human would interact with another human. In particular, the interaction will occur easily, communication will occur naturally, mistakes and errors will be accepted and mutually fixed, and assistance will be given when in trouble. Several glaring and common examples of user-unfriendly systems exist. In such systems, errors can be difficult or impossible to recover from without clearing everything and starting from the beginning. An unfriendly system requires the use of cryptic commands, such as "cp" for copy, instead of more natural commands. In an unfriendly system, error messages are vague (e.g., "illegal entry") and contain no advice on why the error occurred or how to recover from the error. Finally, in an unfriendly system, common sense is not used. I know of one computerized billing system which sent the customer a bill for -10 dollars; this error could be avoided by including a very simple check for overpayment of the bill in the computer program. A system is user unfriendly if it requires responses or utilizes commands that are avoided when humans interact with each other.

Another term used quite often with this approach is "natural". Again, much speculation has occurred in the literature about what is considered to be natural. Table 5 summarizes some of this literature by providing a brief discussion of the concept and a reference for where more detail can be found for that concept. From the table, naturalness is seen to occur if the human and computer act as partners (Press, 1971), if the natural language interaction is "graceful" (Hayes & Reddy, 1983), if the computer has some ability to critique the users' plans (Langlotz & Shortliffe, 1983), and if the computer can be introspective by providing the user with its underlying "thinking" or an image of its underlying process (Fitter, 1979).

Application of the human-human communication analogy to human-computer interaction is not fully achievable at this time. The success of this anthropomorphic approach depends on advances in technology and in artificial intelligence. Many kinds of human-human interactions cannot occur for human-computer interaction. The computer only "knows" what it receives from the user through input devices; it can do no sensing on its own. Many of the important human-human interaction cues cannot be picked up by the computer. Humans pick up visual cues, such as loss of eye contact, that may indicate stress. Voice inflections may indicate irritability to another human, who may alter normal communications so that the stress or irritability level can be reduced. A computer does not have this ability and so cannot be completely user-friendly.

The methodology for the anthropomorphic approach occurs in one of two ways. In the first, humans interacting with each other are used as a model and means for gathering information. The data from this kind

TABLE 5. "NATURALNESS" IN HUMAN COMPUTER INTERACTION

Concept	Reference
Should provide an image of the underlying processes to the user	Fitter, 1979
Normal communication between individuals is examined to apply to the interface	Kennedy, 1974
Formal communication theory applied to human computer communication	Oberquelle et al., 1983
Constructive nature of human-human communication applied to man-computer interaction	Thomas, 1978
Human and computer are seen as partners	Press, 1971
Interface that is neither too under-determining nor too over-determining	Thimbleby, 1980
Computer accepts, analyzes, and critiques users' plans	Langlotz & Shortliffe, 1983
System made more natural by being self-adaptive	Innocent, 1982
Graceful natural language interaction	Hayes & Reddy, 1983
Natural artificial languages	Perlman, 1984

of research can specify the appropriate parameters of the computer system to be designed. This methodology will be discussed in the next section under human-human interaction as a source of data. The second method is more intuitive. For this method, the human-computer task is compared with a human-human interaction task to look for mismatches. The mismatches are then attacked under separate headings. One of the mismatches that occurs is lack of friendly assistance with the task. Each of these items will be discussed under the heading of the particular application area.

7.3.1. Human-Human Interaction As A Source Of Data

In this methodology, humans interacting with each other in natural settings are used to specify the characteristics needed in an effective computer system. As an example of the use of this method in a very natural setting, Pollack, Hirschberg, and Weber (1982) recorded and analyzed the dialogues that occurred from a radio call-in show. These dialogues were used to determine how an expert system would need to respond to queries; it was very useful in pointing out the difficulty of the task. They found that the callers rarely came to the radio host with a well-formulated question. Instead, the caller and the host often had to determine what the question was and spend some time defining terms. The obvious conclusion is that the computer, to be natural and perform like a human, must be able to do more than answer questions; it may have to help formulate the question.

In a controlled laboratory setting, Ochsman and Chapanis (1974) tested human-human communication to try to determine the best human-computer communication modes. In this experiment, they tested human communication when solving problems in ten different modes: typewriting, handwriting, video, voice, and various combinations of these modes. In observing the problem-solving performance, Ochsman and Chapanis concluded that problems were most efficiently solved if voice was one of the communication channels used. The implication is, thus, that human-computer communication can be better if voice inputs are allowed.

As another example of this methodology, Kelly and Chapanis (1977) used the human-human interaction model as a source of data for determining the size of vocabulary needed for a computer to interact with a human in natural language to mutually solve a problem. Two-person teams were required to interact with each other to solve a problem. The experimenters recorded the words that the teams used to try to solve the problems. Next, new two-person teams were given the problems and told that they could only use the 300 words utilized most often by the previous teams. In comparing the teams from the restricted word environment to the previous teams from the unrestricted environment, little deficit in problem solving performance occurred.

Kelly and Chapanis concluded that a computer system with a 300-word vocabulary would be adequate to help humans solve problems.

As a final example of this methodology, Hendler and Michaelis (1983) used two-person teams to compare unlimited-grammar, natural-language interactions to limited grammar, natural language in the interactions of the team members. They found that in the first hour of the two-hour experiment, the team members using the unlimited grammar solved problems better than the team members using the limited grammar. In the second hour, however, no difference in problem-solving ability occurred between the two groups. They concluded that limited-grammar, natural-language systems could be implemented in computer systems, if the users are allowed enough time to get used to the limitations.

These studies have shown that the human-human communication model can be used to collect data on the kinds of features needed in a computer system to make interaction with the computer more natural. The underlying assumption of this methodology is that naturalness in interacting with computers, where natural is used metaphorically from human-human interaction, is desirable. That is, given the chance, humans would want to interact with computers as they interact with people.

7.3.2. Technology Of "Naturalness"

Naturalness is complicated. Implementing naturalness in computer systems is dependent on advances in technology. Two of these technologies, natural language processing and voice interaction, will be discussed in this section.

Human natural language processing involves the complex interplay between the meaning of the words, the context in which the words appear, and background knowledge. For effective communication, one communicator must have some feeling for the knowledge structures that the receiver of the information has so that the message can be altered to be understandable by that person. Often, understanding natural language is described as a process that is both bottom-up, driven from the individual meanings of the words, and top-down, the meanings are derived from the context that they appear in. A simple computerized dictionary containing the meanings of words is not enough to understand natural language.

Many reviews of natural language processing exist (e.g., Barr & Feigenbaum, 1981; Burton & Brown, 1979a). Briefly, the natural language processor tries to break down the input by "parsing" it into a form which can be interpreted by the computer. The parsing can be done by using syntactic rules (such as whether the words are verbs or nouns), semantic rules (interpreting the context and meaning of the word), or pragmatic rules (for example, categorizing the words as agents or objects, see Schank & Riesbeck, 1981). Natural language processing can be done by the present technology only if the domain is limited in some way. As an example, SAM (Cullingford, 1981) is quite proficient for understanding news stories appearing in newspapers but is not applicable to other domains. Burton and Brown (1979a) describe a natural language processor, for a computer assisted instruction course, which is quite good at interpreting natural language within its particular domain; it is not applicable to other domains. In conclusion, natural language processors are possible within a limited domain, but may require extensive development time and use a large amount of computer memory if implemented.

The technology of voice recognition devices also points to the complexity of human speech understanding. The devices that are available fall within two main classes: speaker independent and speaker dependent. For the speaker-independent devices, the system is able to understand any speaker of either sex with any dialect and pronunciation. Speaker-independent systems have been advertised with vocabularies of less than 20 words with a recognition rate of around 85%. Speaker-dependent systems can recognize a particular voice only after being "trained" by having the speaker go through a vocabulary usually from two to eight times. Inexpensive speaker-dependent systems are currently available which have vocabularies of more than 50 words. Speech recognition of phrases and sentences is tied into natural language processing and is difficult at this time. Large advances are being made in this area with both accuracy and vocabulary size increasing.

A discussion of both of these systems indicates how naturalness is coupled with advances in technology. The technology of naturalness could change very quickly in the future. At this time, however, the ability to incorporate natural language processing and complex speech recognition devices into the human-computer interface is limiting and may use up memory space that could be devoted to other enhancements to the interface.

7.3.3. Experiments On "Naturalness"

Achieving natural systems is a goal of the anthropomorphic approach. Should computers be natural? A few experiments have addressed this issue from a position of the design of command names. The following discussion shows that, counter-intuitively the experimental evidence has not been supportive of naturalness in computer systems.

One set of experiments (Scapin, 1981) compared command names that were natural with command names which used computer terminology, and thus were computer-oriented. In a free-recall memory task, inexperienced subjects could more easily recall the functions of the computer-oriented words than the functions of the natural words. The natural words may have had other connotations to the subjects besides the computer application, and thus, differentiating the computer context from the natural one may have been difficult. The computer-oriented words may have had less interference from other meanings.

In another experiment, Dumais and Landauer (1981) investigated the possibility of having novices choose the command names that were the most natural to them. The experimenters found that using commands is more complicated than trying to recall what they are; the specificity of the name must be taken into account. The novices, not being familiar with the task, chose natural nonspecific names. As the subjects acquired expertise, the nonspecific command names could be misleading. Naturalness may not be a good substitute for specificity.

7.3.4. The Influence Of CAI

Applications in computer assisted instruction (CAI) have emphasized trying to make the interaction with the human, or student in this case, as easy and natural as possible. In a situation such as this, the students' attention must be focused on learning the material instead of trying to learn to interact with the computer. The Hansen et al. (1978) experiment, mentioned previously, showed the problems that could occur when the student must devote considerable energy and attention to interacting with the computer. Thus, CAI applications have used natural language processing to make that interaction more natural (e.g., Barr, Beard, & Atkinson, 1976; Burton & Brown, 1979a). A related emphasis models the kinds of interactions that a teacher has with students, so this interaction can be done by the computer. This research has resulted in advances in computerized tutoring and in development of student models.

An application of tutoring that has received attention is the development of tutoring rules which can be used for effective CAI lessons. Burton and Brown (1979b) describe many tutoring rules which they have developed in their design and implementation of the WEST program to teach children arithmetic skills. The tutoring rules that they designed are based upon two major problems: deciding when to interrupt the student's problem-solving activity, and what to say once it has been interrupted. In another application of how tutoring rules can be implemented, O'Shea (1978) designed a CAI lesson to teach quadratic equations where the tutoring rules were incorporated in a production rule system. Based upon interactions with the student, the rules in the system can be changed and tutoring techniques modified along the way. All of these tutoring programs use a model of the teacher as the goal for computerizing these techniques, so that tutoring can be performed with little human intervention.

The tutoring that is done by computers in CAI is very often designed to detect and correct the misconceptions that students may have about certain situations. As an example, Stevens, Collins, and Goldin (1979) analyzed the kinds of misconceptions that students have about weather systems with the goal of designing a CAI lesson to "debug" these misconceptions. The misconceptions that students have are often incorporated in a student model that the computer keeps of each individual student. Five general methods are used to determine what a student knows. The first two methods used, topic marking and context models, are relatively simple approaches. For the topic-marking approach, the system keeps track of the information with which the student has been presented. For the context model, the extent of the student's knowledge is interpreted in terms of the dialogue and questions that the student asks the system.

The bug and overlay approaches compare the student's knowledge with the knowledge of an expert. For the bug approach, student knowledge is characterized in terms of the bugs or misconceptions the student has about the subject when compared with a variety of bugs until the best match is found. For the overlay approach, student knowledge is characterized as a subset of an expert's knowledge. Once the student demonstrates the correct facts or rules contained in the expert's knowledge, those facts are considered as having been acquired by the student.

The fifth method, generative modeling, is a relatively new approach. Instead of looking at the particular facts acquired, the student's knowledge is characterized in terms of the plans used to solve a particular problem. The computer instruction is then organized according to these perceived plans: factual misconceptions are corrected and feedback is tailored to the particular student's conceptions.

The characteristics of some CAI systems outlined in this section could be applied to other computer tasks. This would allow the computer to follow the tutoring rules to determine when help is needed, and to determine what kind of help should be provided. The student model could be implemented in a user model which would allow the computer to individualize the interaction or the kind of help that should be provided to different users.

7.3.5. Advantages

The human-human interaction model can be an effective model for deciding the kinds of features to be included in the human-computer interaction interface. Computers can be made more user-friendly by incorporating the naturalness of human-human communication in the design of this interface. Determining the kinds of communication modes which are most natural can either be done by observing humans in natural settings or by analyzing how humans interact in a laboratory experiment. There is little excuse for designing a computer system where the error messages are cryptic, the system is unforgiving of mistakes, and simple "intuition" checks are not used. Most of these features take only a few lines of code for implementation on any interface design.

This approach also has the advantage that it can be used to create and design computer interfaces instead of following the lead of other disciplines, as has occurred with the empirical approach. By performing a task analysis looking at the mismatches between human-human interaction and human-computer interaction, suggestions can be made for the improvement of the interface. The experimental methodology of Chapanis and his co-workers (Kelly & Chapanis, 1977; Ochsman & Chapanis, 1974) can also be used to test communication methods before they are implemented in the design of a computer system.

A logical extension of this approach can also be made. Academic fields that do research in human-human interaction, such as social psychology, can be applied to human-computer interaction to advance the field. An example of this is the application of Kelly's (1955) construct theory from psychoanalysis to the automated acquisition of expert knowledge in the design of an expert system (Boose, 1984). Other social psychology theories, such as the theory of telling (Lewis & Cook, 1969), which analyzes one-way communication, and a theory of conversation (Pask, Scott, & Kallikourdis, 1973), could be applied to computer systems under the anthropomorphic approach.

7.3.6. Disadvantages

The anthropomorphic approach is overly dependent on advances in technology. As previously emphasized, natural is complicated. Some advances in natural language processing and voice recognition have been made, but these features may be difficult to implement in some computer systems.

Another problem is that natural and friendly may not always be the best design. The experiments that looked at natural command names found that naturalness should not be substituted for specificity. Some users, especially expert users, may not want systems that are overly friendly and wordy. These user-friendly systems may be inefficient for performing a task that a user is very familiar with. Researchers have been investigating methods to allow the level of interaction to adapt to the level of the user (e.g., Badre, 1984; Mozeico, 1982; Schneider, 1982). In these studies, the interaction mode would be different for an expert and a novice.

The concept of natural language interaction has been criticized recently. An experiment by Small and Weldon (1983) compared natural language interaction to a specialized query language in data-base retrieval searches, and found that subjects worked faster with the specialized query language. Other researchers and practitioners have argued that natural language may be inappropriate for a computer system. As Shneiderman (1982) stated, "When people want a discussion they go to other humans, when they want precision and speed they go to a computer." Other reasons why natural language may be inappropriate for computers, according to Shneiderman (1982) are that the overhead of creating and maintaining a natural language interface could more appropriately be applied to other tasks. In addition, users may form unrealistic expectations of the computer's power. The latter has been validated experimentally. Jenkins (1984) found that operators of a nuclear power plant thought the expert system controlling the plant "knew" more than it actually did. By using natural language, the computer may appear to be smarter than it actually is.

TABLE 6. COGNITIVE APPROACHES

Cognitive Theory	Applications
Analogical Reasoning	Programming (Mayer, 1975) Windows (Teitelman, 1979) Icons (Bewley et al., 1983) Teaching of novice computer users (Rumelhart & Norman, 1981)
Spatial Reasoning	High spatial skills useful for text editing (Gomez et al., 1983) Programs represented spatially (Molzberger, 1983) Maps of hierarchical menus (Billingsley, 1982) Contextual presentation of graphics (Robinson & Eberts, in press) Information highlighting (Knapp et al., 1982) Animation (Gould & Finzer, 1982) Steamer (Hollan, 1984; Hollan et al., 1980; Stevens et al., 1983; Williams et al., 1981) Internalization of graphics (Eberts, 1984; Eberts & Schneider, in press)
Problem Solving	GOMS (Card et al., 1983)
Scripts, plans and goals	Text editing (Robertson & Black, 1983) Understanding of computer programs (Soloway et al., 1982; Ehrlich et al., 1982; Soloway & Black, 1983)
Mental models	Programming (Mayer, 1975) Control task (Eberts & Schneider, in press) Complex plant (Hollan et al., 1980)
Attentional Resources	Display information (Robinson et al., 1985) Manual and voice responses (McLeod, 1977; Wickens, 1980; Wickens et al., 1983)

7.4. THE COGNITIVE APPROACH

For the cognitive approach, theories in cognitive science and cognitive psychology are applied to the human-computer interface to make the processing of information by both the human and the computer easier and more efficient. The cognitive theories state how humans perceive, store, and retrieve information from short and long-term memory, manipulate that information to make decisions and solve problems, and carry out responses. The overall goal of the cognitive approach can be stated pictorially in Figure 1 (adapted from Norman, 1983). Three concepts are important. First, the conceptual model is a description of the computer system in engineering terms so that it is accurate, consistent and complete. The mental model is the model that the user forms of how the computer system works; this mental model guides the user's actions and behaviors. The mental model is built up through interactions with the display representation which provides a picture of the conceptual representation. The goal is to try to choose the information to represent on the display so that the mental model can, like the conceptual representation, be accurate, consistent, and complete.

The cognitive approach views the human as being adaptive, flexible, and actively involved in interacting with the environment to try to solve problems or make decisions. The cognitive approach has been concerned with applying specific theories to the human-computer interaction. Theories which have been applied include those on analogical reasoning, spatial reasoning, problem solving, attentional models, and goals, plans, and scripts. These theoretical approaches and applications of the theory are summarized in Table 6 and will be considered in more detail in the next sections.

7.4.1. Analogical Reasoning

Analogies are an important kind of learning that is used quite often in teaching. In teaching, the instructor chooses some concrete situation that the student is familiar with and presents new information in terms of how it relates to the old familiar information. One of the most popular examples of using analogies in teaching is to say that light acts like particles in some situations and like waves in other situations. Neither is totally correct but it does help the student develop the concept of light by using familiar information built up through interacting with particles or waves. The use of analogies was an important part of Rumelhart and Norman's theory (1981) of human learning.

Gentner and Gentner (1983) found that the kinds of analogies used by subjects in experiments determined the kinds of mistakes that were made in problem solving tasks. They found that people very often form an analogical model of electricity based upon either a water flow analogy or a teeming crowd analogy. By analyzing the analogies, Gentner and Gentner determined that the water flow analogy would be better for predicting the effects of a battery on an electrical circuit, and the teeming crowd analogy would be better in predicting the effects of a resistor on the circuit. The results indicated that the predictions were fairly accurate and that the analogy used can affect how people think about a task. Related to analogies is learning by example. Again, new knowledge is created by giving subjects a concrete example of its use. Anderson, Greeno, Kline, and Neves (1981) found that learning by example was an effective approach for learning algebra skills.

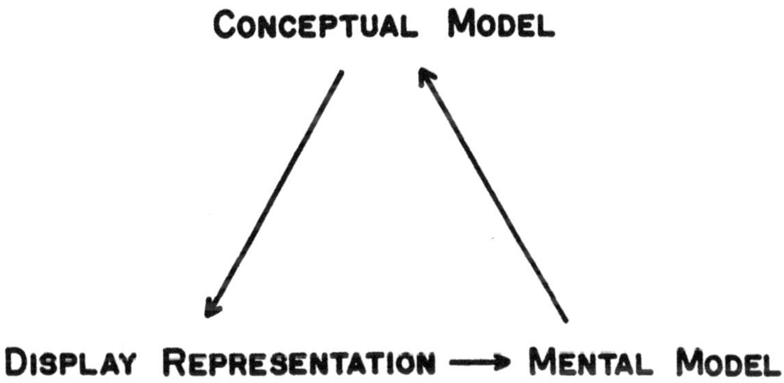

Figure 7-1: Relationship between the conceptual model, the display representation, and the mental model. The goal is to choose the display representation so that the mental model can closely correspond to the conceptual model.

In human-computer interaction tasks, Mayer (1975) has used analogies to teach computer programming skills to novices. In his experiments, he told the subjects that computer input was similar to a ticket window, output was similar to a message pad, control structures were similar to a shopping list with a pointer, and computer memory was similar to an eight space erasable blackboard. Analogies helped the subjects produce programming code for novel tasks; tasks different from those on which they were trained. Additionally, he found that the subjects with lower Student Aptitude Test (SAT) scores were benefited more than those with high SAT scores; apparently the high scorers naturally use analogies.

Analogies are also used when displaying information. The use of windows on computer screens (see Teitelman, 1979) can be called an analogy. A window implies that the user is only allowed to see a small part of the world that is out there; similarly a window on a computer screen only allows the user to see a small part of the information that is available. In the design of the Xerox Star interface, analogies were used quite often (Bewley, Roberts, Schroit, & Verplank, 1983). As an example, a wastepaper basket is pictured on the screen, which can be pointed to if the user wants to delete a file; deleting a file is similar to discarding trash in a wastepaper basket. Many of the characteristics of Star were later incorporated into the Apple Macintosh.

As another example of the use of analogies, Rumelhart and Norman (1981) used analogies to teach novices how to use an editing system. They first told the subjects that the editor was analogous to a secretary, in that they give the secretary commands and text to type. This caused a problem, however, in that the subjects thought that the editor should be as intelligent as a secretary and be able to differentiate between the commands and the text. To get around this problem, Rumelhart and Norman used a second analogy, the tape recorder analogy, to indicate to the subjects that the editor is put into record mode and faithfully records everything until it is told to stop by issuing the appropriate command. This analogy was useful for getting around the first problem but introduced another problem by not being able to account for the delete command. As a third analogy, Rumelhart and Norman told the subjects that using the editor was similar to filing information in a filing cabinet, in that cards could be inserted or removed. Rumelhart and Norman concluded that analogies could be useful for teaching novices about computer systems.

The Rumelhart and Norman (1981) experiment does point out, however, one possible problem with the use of analogies: finding a perfect match between the old information and the new information may be difficult. If a perfect match is not found, then the user could inappropriately overextend the analogy as the subjects did when ascribing intelligence to the text editor. Similarly, the wastepaper basket analogy could be overextended to make the user believe that the discarded file could be retrieved until the end of the day. An appropriate interface would allow the user to recover discarded files until the end of the day when, supposedly, the "garbage" is removed.

7.4.2. Spatial Reasoning

Manipulation of images is required to solve many kinds of problems. As examples, Bower (1972) found that groups of subjects told to use imagery are better than a no-imagery control group in solving certain kinds of problems. In a three-term series problem (e.g., Al is taller than Bob, Bob is taller than Carl, who is the tallest?), Huttenlocher (1968) found that, to solve the problem, subjects would often visualize the size of the people and then decide from that visual image which one was taller. In other kinds of problems, where subjects are asked to answer relatively simple questions, research has indicated that imagery is used to determine which object is larger (Moyer, 1973); imagery is used to determine if objects in different orientations are rotated or reflected from a standard (Cooper, 1975; Cooper & Shepard, 1974); and images can be scanned to solve problems such as the most efficient route to get from one place to another.

Some recent experiments indicate that having the ability to make use of the manipulation of spatial information may make human-computer interaction tasks easier to perform. In particular, Gomez, Egan, Wheeler, Sharman, and Grunchacz (1983) found that those subjects who scored high on spatial ability tests could learn how to use line and screen editors better than those who scored low on spatial ability. In another area, Molzberger (1983) found that computer programmers often form a mental picture of the program before they actually start to develop a code. Supposedly, high spatial ability would also make this task easier.

A goal for human-computer interaction is to make the task easy enough so that anyone can do the task regardless of the kinds of abilities that people have when they come into the task. Along the lines of this reasoning, then, has been the use of computer graphics to make the task easier. Many times, an assumption is made that graphics will always make the task easier. As can be seen from Table 3, this is not always the case. The Tullis (1981) and Bury and Boyle (1982) experiments showed that graphics were better on some measures; the Stern (1984) and Powers, Lashley, Sanchez, and Shneiderman (1984) experiments showed that the other presentation modes were better. Graphics may only be appropriate in certain situations; Billingsley (1982) showed that having a map of the hierarchical menu structure was useful for keeping one's place in the menus. Additionally, good design principles must be used when incorporating graphics into a situation. As an example, Robinson and Eberts (1985) designed the presentation of the graphics material so that the subjects could build up an artificial syntax of what the next picture would contain. Knapp, Moses, and Gellman (1982) used double-cue coding and color coding to highlight various parts of displays. Windowing, zooming, and selective call-up of aggregated information was also used to draw attention to various parts of the display (Knapp et al., 1982). Gould and Finzer (1982) describe a program which uses interactive graphics and animation along with a written problem to make algebraic problems more concrete. Graphics will not always be useful, but their usefulness can be enhanced in particular situations if they are optimally designed.

As mentioned previously, spatial manipulations can be used to solve problems. The computer provides an opportunity to present the user with an explicit representation of the task, through graphics, so that the graphics can be internalized and problems can be solved mentally without the use of the computer. As an example, Steamer (Hollan, 1984; Hollan, Stevens, & Williams, 1980; Stevens, Roberts, & Stead, 1983; Williams, Hollan, & Stevens, 1981) is a CAI system that was developed for the Navy to teach students how to control a steam plant such as would be found on board large ships. In an analysis of experts performing the task, Hollan et al. (1980) concluded that the experts had a spatial representation of the plant that was used to solve problems as they occurred. To build up such a representation for novices, they designed the CAI task so that the trainees could easily manipulate graphics on the computer to see how things worked. The idea is that the computer can be used, through its graphics capabilities, to present information to trainees for tasks that would later be done without the use of the computer. In a similar application, Eberts (1984; Eberts & Schneider, 1985) found that presenting subjects with computer graphics was beneficial when the subjects were later asked to solve problems when the graphics were not available.

7.4.3. Problem Solving

Analogies and spatial reasoning were used to solve problems in specific ways. Generally, for the cognitive approach, human computer interaction is seen as presenting problems which have to be solved. Thus, many of the problem-solving theories which have been developed in cognitive psychology can also be applied to the human-computer interaction task. In particular, the influential Newell and Simon (1972) theory of problem solving has been modified in the form of the GOMS model (Card, Moran, & Newell, 1983) to be applicable to human-computer interaction tasks. The GOMS model is based upon a cognitive theory of human information processing, and thus, is appropriate to be discussed in this section. However, the GOMS model also provides us with a predictive modeling technique and, therefore, will be discussed in great detail in the next section. It is an example of a model which crosses both approaches.

7.4.4. Scripts, Plans, and Goals

This particular cognitive approach to human-computer interaction was influenced by Schank and Abelson's Scripts, Plans, Goals, and Understanding in 1977, which attempted to interrelate cognitive psychology and artificial intelligence to determine how humans understand the world. The important concept for this approach is that when people understand events in the world, they incorporate scripts, plans, and goals. A script is a stereotyped knowledge structure that describes appropriate sequences of actions in a particular context. The script is made up of slots and requirements about what can fill those slots. Schank and Abelson illustrate a script through the use of a restaurant situation. They explain that when someone goes to a restaurant, this person has certain expectations about what will occur which allows the person to interpret the actions of others and carry out the appropriate behaviors. A plan, on the other hand, is used in situations where people deal with events they have never encountered before, and is made up of general information about how people achieve goals. When dealing with this unexpected event, a person will try to find a plan that will make sense out of the situation. Finally, goals are used as part of the plans to understand other people's actions by determining their goals and predicting future actions from those goals.

Experimental work has shown that computer users do use the kinds of knowledge structures outlined by Schank and Abelson (1977) when interacting with the computer. In particular, users will utilize the goal structures when solving a problem such as, while using a text editor, changing the word "the" to the word "those". Robertson and Black (1983) analyzed this particular task by breaking it down into proper goals and subgoals. They hypothesized that for a user performing this task, the overall goal of CHANGE (changing "the" to "those") could be broken down into subgoals of DELETE (the) and INSERT (those). DELETE (the) could further be broken down into subgoals of MOVE-POINTER (under t) and USE-KEY (word-delete). INSERT (those) could further be broken down into subgoals of MAKE-ROOM (5 spaces) and TYPE (those). Robertson and Black further hypothesized that the boundaries of the major goals could be characterized by increased human processing because of the goal evaluation and goal initiations which occur at those boundaries. In measuring the times to perform the subtasks, they found that pauses did occur between the major goals and, with practice, plan units were produced. In other words, pauses at various points disappeared with practice so that sequences of actions could be unitized or stereotyped.

This approach to human-computer interaction has the potential to make the understanding of user action by the computer more meaningful. If the computer understands the plans and goals of the user, then the computer can present more meaningful error messages, in terms of plans and goals, than are presently given by the computer. This was one of the stated purposes of the Robertson and Black (1983) experimentation, but they did not specifically present error or help messages based upon their goal structures. Similar plan analyses have been used by Ehrlich and Soloway (1982) and Soloway, Ehrlich, Bonar, and Greenspan (1983) in the composition and understanding of computer programs. Plan analysis has also been used by Soloway et al. (1983) to analyze why some computer programs are more understandable than others.

7.4.5. Mental Models

The term mental model has found increased usage in human-computer interaction. Often, researchers will say that the user has a mental model of the computerized task which is being performed. This mental model is used to control behavior and is based upon the user's previous background and experiences with the particular problem and other problems like it. Researchers say that subjects form a mental model of a task through the application of analogies (Mayer, 1975), through the imposition of a spatial model (Eberts & Schneider, 1985; Hollan et al., 1980), or by their goals and plans (Moran, 1981). The term mental model can mean many different things to different researchers.

Norman (1983) has made some observations on the kinds of mental models that users utilize when interacting with machines. Norman states that mental models are incomplete, unstable, unscientific, and parsimonious. Furthermore, he found that the abilities to "run" them are severely limited and no firm boundaries exist between the separate mental models. In observations that he made of calculator users, he found that the subjects always took extra steps or declined to take advantage of features of the calculator. As an example, he found that subjects wrote down partial results when they could have been stored in memory, they hit the clear key several times to clear the calculator before each problem, and they would not use memory for often-occurring constants. Thus, their mental model of the calculator was rather elementary and they did not use the calculator to its fullest advantage.

Whether the term mental model will be useful as a predictive instead of a descriptive device for human-computer interaction tasks remains to be seen. At this time, researchers use the term more as a catch-all descriptive term; they do not have good techniques for measuring and analyzing users' mental models.

7.4.6. Attentional Resources

When the human processes information, attention is necessary for that information to be processed. Because humans cannot process or apply attention to all the information that is received through the senses, attention is often viewed as having limited capacity (Moray, 1967) or as a processing resource of limited availability (Kahneman, 1973). The human must, therefore, consciously or unconsciously decide which pieces of information will receive attention. Recently, this undifferentiated resource theory of attention has been refined to state that the human has multiple resource "pools" so that more interference will occur in tasks if they pull from the same resource pool (e.g., Navon & Gopher, 1979; Wickens, 1980, see also Wickens this volume). This multiple resource theory of attention has important implications for the design of human-computer tasks, especially the design of tasks in environments such as a nuclear power control room or an aircraft cockpit, where the operator must process information from several different sources and make multiple responses. In these situations the operator is required to time-share or perform more than one task at a time.

Wickens (1980, 1984) has synthesized the results from this area into a model of the structure of processing resources. He argues that this structure may be defined by three relatively dichotomous dimensions: two stage-defined resource pools (encoding/central processing and responses), two modality-defined resource pools (auditory versus visual encoding), and two resource pools defined by processing codes (spatial versus verbal). The implications of this model for human-computer interaction tasks are as follows. For the stage-defined resource pools, information is assumed to be processed by the human in stages, going from encoding through central processing to a response stage. Thus, the human would be able to encode or centrally process information at the same time as previous information is responded to. For the modality-defined resource pools, the human would be able to attend to visual information at the same time that auditory information is attended to. Thus, computer information presentation should be divided between auditory and visual signals, so the user does not become overloaded. Finally, for the processing code resource pools, information can be either presented spatially (e.g., graphics) or verbally (in textual form). Again, the user would be less likely to be overloaded if both of these resource pools are drawn upon instead of presenting all the information in a form that would draw upon just one of the pools.

The predictions from the model for human-computer interaction tasks have been validated by experimental work. In particular, manual and vocal responses can be time-shared with a high degree of efficiency. McLeod (1977), Wickens (1980), and Wickens, Sandry, and Vidulich (1983) have shown that manual tracking and a discrete verbal task are time-shared more efficiently when the verbal task employs vocal as opposed to manual response mechanisms. In addition, Robinson, Base, and Eberts (1985) showed that an auditory verbal display could be encoded more efficiently than visual verbal and visual spatial displays when the subjects were required to perform a manual tracking task at the same time. All of these results can be accounted for by the model of attentional resources.

7.4.7. Advantages and Disadvantages

The advantage of the cognitive approach is that it views the user as a flexible, adaptive information processor actively trying to solve problems when using the computer. This approach has also been used to suggest which design may be appropriate and easy to use instead of merely testing the design after it has been finalized (e.g., Bewley et al., 1983), The success of this approach has been realized in the interface for the Xerox Star, which was in many ways the predecessor of the Apple Macintosh. The Apple Macintosh has, in turn, been very successful and is known for its ease of use. Other possibilities for designs, such as the error and help messages based upon users' plans (Robertson & Black, 1983), look promising but have not yet been implemented.

The only possible disadvantage is that some of the terms used in this approach may not be well-defined and specific enough to offer suggestions for human-computer interface designs. This problem was specifically addressed in the section on mental models. The cognitive approach may be invoked but be unsuccessful because of this problem.

7.5. THE PREDICTIVE MODELING APPROACH

The purpose of the predictive modeling approach is to try to predict performance of humans interacting with computers, similar to the predictions engineering models make for physical systems. The idea is to try to predict which design would be best before the design is prototyped. Two general types of predictive models are currently being used. The GOMS model, mentioned in the cognitive approaches section, has received much attention in recent years. State-transition networking is the other method which has been used.

7.5.1. The GOMS Model

The GOMS model was developed by Card, Moran, and Newell (1983) based upon the earlier work by Newell and Simon (1972) on human problem solving. GOMS is actually a family of models and stands for Goals, Operators, Methods, and Selection rules, which is descriptive of how the model works. GOMS has been applied most extensively to the use of text editors, so this particular application will be considered. When working with a text editor, the user usually has some kind of goal in mind, such as taking out a piece of text and inserting it someplace else. Thus, the first letter of GOMS stands for goals. Next, the user knows the operators or the editing commands, for this example, which can be used to accomplish the goals. Thus, the second letter of GOMS stands for operators, Third, methods or procedures are needed to carry out the operators. As an example, the user may hit the "d" key for removing or deleting a section of text. Thus, the third letter of GOMS stands for method. Finally, the user must select among the available operators to reduce the discrepancy between the state of the text at that particular time and the goal of what the text should look like. In other words, the user must select a strategy for accomplishing the goal. Thus, the fourth letter of GOMS stands for selection rules.

Card, Moran, and Newell also refer to human processor model where they parameterized the cognitive functions of the human processor. The human processor is divided into three interacting subsystems, each with their own memories and processors. The three subsystems are the perceptual system (which perceives the items on the computer screen or auditory signals from the computer), the motor system (which carries out the responses), and the cognitive system (which retrieves information from memory). Each of these subsystems can be described by a few important parameters: the storage capacity in number of items μ, the decay time of an item δ, the main code type κ, and the cycle time τ. To assign numbers to the three numerically-based parameters, Card, Moran, and Newell reviewed the research in cognitive psychology to find estimates of the times, along with a plausible range. Table 7 presents the processors and memories associated with the three subsystems along with the parameters and values for each. In the table, the numbers preceding the brackets are means, and the numbers in brackets represent the range of the parameter.

Predictions for the amount of time and the number of errors associated with a task or interface design can be made by breaking the task into its components and providing values for the parameters associated with each component. The time can be predicted from the numerical parameters. Errors can be predicted from a mismatch between code types, from information decay in memory, or from an analysis of the goals, operators, methods, and selection strategies used by the operator. Card, Moran, and Newell present examples of how the models can be applied. One simple application was provided for choosing the position of the "function" key on a calculator to reduce the amount of time a finger has to travel to its position from the other keys. The GOMS model has also been applied to more complicated tasks.

Card, Moran, and Newell validated the model by comparing the predictions with performance of human subjects on various kinds of tasks. They found that GOMS was quite accurate in predicting time and errors for performing various kinds of computer tasks. One experiment validated the model for a manuscript editing task. They found that

TABLE 7. PARAMETERS OF THE GOMS MODEL

Processor	Memory
Perceptual τ_P = 100 [50 - 200] msec	Visual image store δ_{VIS} = 200 [70 - 100] msec μ_{VIS} = [7 - 17] msec κ_{VIS} = physical Auditory image store δ_{AIS} = 1500 [900 - 3500] msec μ_{AIS} = 5 [4.4 - 6.2] letters κ_{AIS} = physical
Cognitive τ_C = 70 [25 - 170] msec	Working memory μ_{WM} = 3 [2.5 - 4] chunks μ_{WM} = 7 [5 - 9] chunks δ_{WM} = 7 [5 - 226] sec δ_{WM} (1 chunk) = 73 [73 - 226] sec δ_{WM} (3 chunks) = 7 [5 - 34] sec κ_{WM} = acoustic or visual
Motor τ_M = 70 [30 - 100]	Long term memory δ_{LTM} = infinite μ_{LTM} = infinite κ_{LTM} = semantic

GOMS predicted the methods used by the user 80-90% of the time, the methods the operators used were predicted 80-100% of the time, and the predicted time on new tasks was within 36% of the actual times. In another validation of the model, a computer-aided circuit design task, GOMS predicted the time to execute the commands within a 16% error. In another application of GOMS, Douglas and Moran (1983) used GOMS to predict the errors that novice computer users would make when transitioning from a typewriter to a computer text editor. They predicted that the user would incorrectly employ the typewriter operators to the new task. They found that GOMS predicted 62 of the 75 errors made in the misapplication of the typewriter operator to the computer editor.

7.5.2. State Transition Networks

A state transition network shows the states of the system and how to transition from one state to another. Table 8 provides an example from Jacob (1983) for how state transition networks can be applied to a prototype message system. The first column contains labels for all the states of the system. The second column contains the transitions that are provided or taken when transitioning from the state in the first column to the state in the third column. A transition can occur in one of four ways. For an input taken (preceded by an "I" in the table), such as IQUIT, the model will transition depending upon the input from the user. Thus, if the user types in "QUIT" when in the getlog state, the system will transition to the end state. For an output transition (preceded by an "o" in the table, such as oLOGNAME), the system will transition to another state after the particular item is displayed on the screen. As an example, in the promptlog state "LOGNAME" appears on the screen and the system transitions to the getlog state. A nonterminal transition (such as login in the table) operates similar to a subroutine in a computer language; a series of steps is carried out at that particular point. Finally, in a conditional transition (such as cond: testret in the table), the condition must be true for the transition to take place.

A state transition network such as that in Jacob (1983) can be used in the design process. Jacob states that this design process can be broken down into five steps: 1) the input syntax is specified with no actions or outputs; 2) information description of actions and outputs is added at each step: 3) new states are introduced, each action and condition is put into state transitions; 4) each state is represented as a function; and 5) provisions are added for handling errors, help messages, and aborts. Once the design is carried out, a program could go through the network to ensure that no dead-end paths exist and that the same input does not produce two different states. If such mistakes are found, the program would notify the programmers. Subsequently, a bug-free model would be available to help develop the prototype New states and transitions could be added by modifying the model. Roach and Nickson (1983) present a similar method to test computer dialogues before the actual implementation of the prototype.

TABLE 8. STATE TRANSITION NETWORK (from Jacob, 1983)

Start State	Transition	End state
st	oLOGIN	promptlog
promptlog	oLOGNAME	getlog
getlog	login	gotlog
getlog	IQUIT	end
gotlog	cond:testret ("ok")	setup
gotlog	cond:testret ("bad")	badlog
badlog	oBADLOG	st
setup	setup	promptcmd
promptcmd	oCMDNAME	getcmd
getcmd	cmd	ready
getcmd	lLOGOUT	end
ready	oCMD	promptcmd

State transition networks are especially useful for testing the completeness of the code for a design before it is implemented. For this method, then, one design is better than another if the design is complete and bug-free. Unlike the GOMS model, this state transition method does not predict the time and errors that users will make. State transition models are more of a computer science approach than a human factors approach.

7.5.3. Advantages and Disadvantages

The GOMS model presents a significant advance in our ability to predict which design will be better than others, based upon human performance. It provides the advantage that it can make accurate error and time estimates for a task, and it can be used to help predict how the user will carry out a particular task. The fit of the model's predictions with actual data has been quite accurate. Future testing of the model should prove to be interesting.

One possible problem with human processor model is that the values for the parameters may be taken too literally in some situations. The values were obtained from previous cognitive psychology experiments based upon average performance. When applying the model, it must be remembered that individual differences could be quite large. When a task requires several processes, and as a result, several parameters must be added together to obtain a time estimate, the possible range of times would be quite large.

The state transition models are used for a slightly different application from GOMS. They cannot be used to choose between designs, but rather, can be used to make sure that a design is complete, and does not place the computer in a nonrecoverable error state. By automating the test procedures of the networks, a programmer is freed from what could prove to be a long testing period.

7.6. THE ROLE OF HUMAN FACTORS

For each of the four approaches, the human factors engineer assumes a different role in the design and implementation of human-computer interaction tasks. Each of these roles will be delineated.

For the empirical approach, the human factors engineer would be required to design, implement, and analyze the results from the empirical studies. As an experimenter, the human factors engineer must ensure that the experimental variables are not confounded, and that the results are interpretable and generalizable to other situations. In addition, the human factors engineer could also play a role as a keeper of the data base which has been built up from this approach. A data base, such as presented in Tables 2-4, must be familiar to the human factors engineer. This person must be familiar with the experimental designs used so that the generalizability of the results can be predicted. Thus, for the empirical approach, the human factors engineer is an experimenter and a data base keeper.

For the anthropomorphic approach, the human factors engineer would be required to analyze a human-computer interaction task and determine how that task may be different from a human-human interaction task. This could be done by formal methods of task analysis, by speculation, or by running experiments on humans similar to those of Kelly and Chapanis (1977) and Ochsman and Chapanis (1974). The human factors engineer could be required to implement this task analysis in a specification for the design of the human-computer interaction. The human factors engineer would also be required to be familiar with the technology of "naturalness" (such as voice recognition and natural language processing) so that he or she could determine which designs are possible. Thus, for this approach, the human factors engineer is a task analyzer, a designer, and possibly an experimenter.

For the cognitive approach, the human factors engineer is expected to be familiar with theory and advances in cognitive psychology. Such a person would also have to have a talent for applying theoretical perspectives to concrete situations in human-computer interaction tasks. Thus, for this approach, the human factors engineer is a theoretician, a designer, and has a talent for applications.

For the predictive modeling approach, the human factors engineer must be familiar with the models that do exist and how those models can be applied to various tasks. When implementing the GOMS model model, the human factors engineer would perform many of the functions that other kinds of engineers perform: a task is specified, the possible designs are considered, and the optimal design is chosen based upon quantitative predictions from the model. For the state transition models, the human factors engineer would use it to make sure that a design is complete and error-free. Thus, for this approach, the human factors engineer is a modeler.

By taking these four approaches to human-computer interaction, the human factors engineer can contribute at all phases in the design of the human-computer interface. The human factors engineer can be a theoretician, a designer, a data base keeper, an experimenter, and a modeler. The complexities and difficulties of human-computer interaction tasks require that the human factors engineer take these roles in trying to make the interface efficient and easy to use.

REFERENCES

Anderson, J.R., Greeno, J.G., Kline, P.J., & Neves, D.M. (1981). Acquisition of problem-solving skill. In J.R. Anderson (Ed.), *Cognitive skills and their acquisition*. Hillsdale, New Jersey: Lawrence Erlbaum Associates. (pp. 191-230).

Aretz, A.J. (1983). A comparison of manual and vocal response modes for the control of aircraft systems. *Proceedings of the Human Factors Society, 27*, 97-101.

Avner, A.R. (1979). Production of computer-based instructional materials. In H.F. O'Neil (Ed.), *Issues in instructional systems development*. New York: Academic Press. (pp. 133-180).

Badre, A.N. (1984). Designing transitionality into the user-computer interface. In G. Salvendy (Ed.), *Human-computer interaction*. Amsterdam: Elsevier. (pp. 27-34).

Barr, A., Beard, M., & Atkinson, R.C. (1976). The computer as a tutorial laboratory: The Stanford BIP project. *International Journal of Man-Machine Studies, 9*, 373-451.

Barr, A., & Feigenbaum, E.A. (Eds.). (1981). *The handbook of artificial intelligence*. Los Altos, California: Kaufman.

Bevan, N. (1981). Is there an optimum speed for presenting text on a VDU? *International Journal of Man-Machine Studies, 14*, 59-76.

Bewley, W.L., Roberts, T.L., Schroit, D., & Verplank, W.L. (1983). Human factors testing in the design of Xerox's "Star" office workstation. In A. Janda (Ed.), *Human factors in computing systems*. New York: Association of Computing Machinery. (pp. 72-77).

Billingsley, P.A. (1982). Navigation through hierarchical menu structures: Does it help to have a map? *Proceedings of the Human Factors Society, 26*, 103-107.

Boose, J. (1984). A framework for transferring human expertise. In G. Salvendy (Ed.), *Human-computer interaction*. Amsterdam: Elsevier. (pp. 247-254).

Bower, G.H. (1972). Mental imagery and associative learning. In L. Gregg (Ed.), *Cognition in learning and memory*. New York: Wiley. (pp. 51-88).

Brosey, M., & Shneiderman, B. (1978). Two experimental comparisons of relational and hierarchical database models. *International Journal of Man-Machine Studies, 10*, 625-637.

Burton, R.R., & Brown, J.S. (1979). Toward a natural-language capability for computer-assisted instruction. In H.F. O'Neil (Ed.), *Procedures for instructional system development*. New York: Academic Press. (pp. 272-313) (a).

Burton, R.R., & Brown, J.S. (1979). An investigation of computer coaching for informal learning activities. *International Journal of Man-Machine Studies, 11*, 5-24. (b).

Bury, K.F., & Boyle, J.M. (1982). An on-line experimental comparison of two simulated record selection languages. *Proceedings of the Human Factors Society, 26*, 74-78.

Bury, K.F., Boyle, J.M., Evey, R.J., & Neal, A.S. (1982). Windowing versus scrolling on a visual display terminal. *Human Factors, 24*, 383-394.

Calhoun, G.L., Arbak, C.J., & Boff, K.R. (1984). Eye-controlled switching for crew station design. *Proceedings of the Human Factors Society, 28*, 258-262.

Campbell, D.T., & Stanley, J.L. (1966). *Experimental and quasi-experimental designs for research*. Chicago: Rand-McNally.

Card, S.K., Moran, T.P., & Newell, A. (1983). *The psychology of human-computer interaction*. Hillsdale, New Jersey: Lawrence Erlbaum Associates.

Cohill, A.M., & Williges, R.C. (1982). Computer-augmented retrieval of help information for novice users. *Proceedings of the Human Factors Society, 26*, 79-82.

Cook, T.D., & Campbell, J.L. (1976). The design and conduct of quasi-experiments and true experiments in field settings. In M.D. Dunnette (Ed.), *The handbook of industrial and organizational psychology*. Chicago: Rand-McNally. (pp. 223-226).

Cooper, L.A., & Shepard, R.N. (1974). Mental transformation in the identification of left and right hands. *Journal of Experimental Psychology: Human Perception and Performance, 1*, 48-56.

Cooper, L.A. (1975). Mental rotation of random two-dimensional shapes. *Cognitive Psychology, 7*, 20-43.

Cullingford, R. (1981). SAM. In R.C. Schank & C.K. Riesbeck (Eds.), *Inside computer understanding*. Hillsdale, New Jersey: Lawrence Erlbaum Associates. pp. 75-119.

Di Persio, T., Isbister, D., & Shneiderman, B. (1980). An experiment using memorization/reconstruction as a measure of programmer ability. *International Journal of Man-Machine Studies, 13*, 339-354.

Douglas, S.A., & Moran, T.P. (1983). Learning text editor semantics by analogy. In A. Janda (Ed.), *Human factors in computing systems*. New York: Association for Computing Machinery. (pp. 207-211).

Duchnicky, R.L., & Kolers, P.A. (1983). Readability of text scrolled on visual display terminals as a function of window size. *Human Factors, 25*, 683-692.

Dumais, S.T., & Landauer, T.K. (1981). Psychological investigations of natural terminology for command & query languages. In A. Badre & B. Shneiderman (Eds.), *Directions in human/computer interaction*. Norwood, New Jersey: Ablex. (pp. 95-109).

Eberts, R.E. (1984). Augmented displays for problem solving. In G. Salvendy (Ed.), *Human-computer interaction*. Amsterdam: Elsevier. (pp. 261-265).

Eberts, R.E., & Schneider, W. (1985). Internalizing the system dynamics for a second order system. *Human Factors, 27*, 371-393.

Eberts, R.E., & Brock, J.B. (1987). Computer assisted and computer managed instruction. In G. Salvendy (Ed.), *Handbook of human factors/ergonomics*. New York: Wiley.

Ehrlich, K., & Soloway, E. (1982). *An empirical investigation of the tacit plan knowledge in programming (Technical Report #236)*. New Haven, Connecticut: Yale University Department of Computer Science. (pp. 976-1011).

Embley, D.W. (1978). Empirical and formal language design applied to a unified control construct for interactive computing. *International Journal of Man-Machine Studies, 10*, 197-216.

Emmons, W.H., & Hirsch, R.S. (1982). Thirty millimeter keyboards: How good are they? *Proceedings of the Human Factors Society, 26*, 425-429.

Fairweather, P.G., & O'Neil, A.F. (1984). The impact of advanced authoring systems on CAI productivity. *Journal of Computer-Based Instruction, 11*, 90-94.

Fitter, M. (1979). Towards more "natural" interactive systems. *International Journal of Man-Machine Studies, 11*, 339-350.

Gade, P.A., Fields, F.A., Maisano, R.E., & Marshall, C.F. (1981). Data entry performance as a function of method and instructional strategy. *Human Factors, 23*, 199-210.

Gannon, J.D. (1976). An experimental evaluation of data type conventions. *Communications of the ACM, 20*, 585-595.

Gentner, D., & Gentner, D.R. (1983). Flowing water or teeming crowds: Mental models of electricity. In D. Gentner & A.L. Stevens (Eds.), *Mental models*. New York: Academic Press. pp. 253-271.

Gilmore, D.J., & Smith, H.T. (1984). An investigation of the utility of flowcharts during computer program debugging. *International Journal of Man-Machine Studies, 20*, 357-372.

Gomez, L.M., Egan, D.E., Wheeler, E.A., Sharma, D.K., & Gruchacz, A.M. (1983). How interface design determines who has difficulty learning to use a text editor. In A. Janda (Ed.), *Human factors in computing systems*. New York: Association of Computing Machinery. (pp. 176-181).

Gould, J.D., & Drongowski, P. (1974). An exploratory study of computer program debugging. *Human Factors, 16*, 258-277.

Gould, J.D., & Finzer, W. (1982). A study of TRIP: A computer system for animating time-rate-distance problems. *International Journal of Man-Machine Studies, 17*, 109-126.

Gould, J.D., & Alfaro, L. (1983). Revising documents with text editors, handwriting recognition, and speech recognition systems. *Proceedings of the Human Factors Society, 27*, 831-833.

Grant, E.E., & Sackman, H. (1967). An exploratory investigation of programmer performance under on-line and off-line conditions. *IEEE Transactions on Human Factors in Electronics, 8*, 33-48.

Grudin, J., & Barnard, P. (1984). The cognitive demands of learning and representing command names for text editing. *Human Factors, 26*, 407-422.

Hansen, W.J., Doring, R., & Whitlock, L.R. (1978). Why an examination was slower on-line than on paper. *International Journal of Man-Machine Studies, 10*, 507-519.

Hayes, P.J., & Reddy, R. (1983). Steps toward graceful interaction in spoken and written man-machine communication. *International Journal of Man-Machine Studies, 19*, 231-284.

Hendler, J.A., & Michaelis, P.R. (1983). The effects of limited grammar on interactive natural language. In A. Janda (Ed.), *The human factors of computing systems*. New York: Association for Computing Machinery. (pp. 190-192).

Hicks, K.I., & Keller, A. (1982). Editing with contextual and screen-oriented cursors. *Proceedings of the Human Factors Society, 26*, 703-707.

Hollan, J.D., Stevens, A., & Williams, N. (1980). *STEAMER: An advanced computer assisted instruction system for propulsion engineering*. Paper presented at the Summer Simulation Conference, Seattle.

Hollan, J.D. (1984). Intelligent object-based graphical interfaces. In G. Salvendy (Ed.), *Human-computer interaction*. Amsterdam: Elsevier. (pp.293-296).

Huttenlocher, J. (1968). Constructing spatial images: A strategy in reasoning. *Psychological Review, 75*, 550-560.

Innocent, P.R. (1982). Towards self adaptive interface systems. *International Journal of Man-Machine Studies, 16*, 287-299.

Israel, J.B., Wickens, C.D., Chesney, G.L., & Donchin, E. (1980). The event related brain potential as an index of display monitoring workload. *Human Factors, 22*, 211-224.

Jacob, R.J.K. (1983). Executable specifications for a human-computer interface. In A. Janda (Ed.), *Human factors of computing systems*. New York: Association for Computing Machinery. (pp. 28-34).

Jenkins, J.P. (1984). An application of an expert system to problem solving in process control displays. In G. Salvendy (Ed.), *Human computer interaction*. Amsterdam: Elsevier. (pp. 255-260).

Kahneman, D. (1973). *Attention and effort*. Englewood Cliffs, New Jersey: Prentice Hall.

Kelly, M.J., & Chapanis, A. (1953). Limited vocabulary natural language dialog. *International Journal of Man-Machine Studies, 9*, 477-501.

Kelly, G.A. (1955). *The psychology of personal constructs*. New York: Norton.

Kennedy, T.C.S. (1974). The design of interactive procedures for man-machine communication. *International Journal of Man-Machine Studies, 6*, 309-334.

Kernighan, B.X., & Plauger, P.J. (1978). *The elements of programming style*. New York: McGraw-Hill.

Knapp, B.G., Moses, F.L., & Gellman, L.H. (1982). Information highlighting on complex displays. In A. Badre & B. Shneiderman (Eds.), *Directions in human/computer interaction*. Norwood, New Jersey: Ablex. (pp. 195-215).

Kolers, P.A., Duchnicky, R.L., & Ferguson, D.C. (1981). Eye movement measurement of readability of CRT displays. *Human Factors, 23*, 517-527.

Kozar, K.A., & Dickson, G.W. (1978). An experimental study of the effects of data display media on decision effectiveness. *International Journal of Man-Machine Studies, 10*, 495-505.

Langlotz, C.P., & Shortliffe, E.H. (1983). Adapting a consultation system to critique user plans. *International Journal of Man-Machine Studies, 19*, 479-496.

Ledgard, H.F., & Chmura, L.J. (1978). *FORTRAN with style*. Rochelle Park, New Jersey: Hayden.

Lewis, B.N., & Cook, J.A. (1969). Toward a theory of telling. *International Journal of Man-Machine Studies, 1*, 129-176.

Liebelt, L.S., McDonal, J.E., Stone, J.D., & Karat, J. (1982). The effect of organization on learning menu access. *Proceedings of the Human Factors Society, 26*, 546-550.

Love, T. (1977). An experimental investigation of the effect of program structure on program understanding. *Proceedings of the ACM Conference on Language Design for Reliable Software,* , pp. 105-113. New York: Association for Computing Machinery.

Lucas, H.C., & Kaplan, R.B. (1976). A structured programming experiment. *The Computer Journal, 19*, 136-138.

Magers, C.S. (1983). An experiment evaluation of on-line HELP for nonprogrammers. In A. Janda (Ed.), *Human factors in computing systems*. New York: Association for Computing Machinery. (pp. 277-281).

Mayer, R.E. (1975). Different problem solving competencies established in learning computer programming with and without meaningful models. *Journal of Educational Psychology, 67*, 725-734.

McLeod, P. (1977). A dual task response modality effect: Support for multiprocessor models of attention. *Quarterly Journal of Experimental Psychology, 29*, 651-667.

Miller, L. (1977). A study in man-machine interaction. *AFIPS Conference Proceedings, National Computer Conference,* , pp. 408-421.

Miller, L. (1975). Naive programmer problems with specification of flow of control. *In AFIPS Conference Proceedings, National Computer Conference,* , pp. 657-663.

Moher, T., & Schneider, G.M. (1982). Methodology and experimental research in software engineering. *International Journal of Man-Machine Studies, 16*, 65-87.

Molzberger, P. (1983). Aesthetics and programming. In A. Janda (Ed.), *Human factors in computing systems*. New York:

Association for Computing Machinery. (pp. 247-250).

Moran, T.P. (1981). The command language grammar: A representation for the user interface of interactive computer systems. *International Journal of Man-Machine Studies, 15*, 350.

Moray, N. (1967). Where is attention limited? A survey and a model. *Acta Psychologica, 27*, 84-92.

Mountford, S.J., & North, R.A. (1980). Voice entry for reducing pilot workload. *Proceedings of the Human Factors Society, 24*, 185-189.

Moyer, R.S. (1973). Comparing objects in memory: Evidence suggesting an internal psychophysics. *Perception & Psychophysics, 13*, 180-184.

Mozeico, H. (1982). A human/computer interface to accommodate user learning stages. *Communications of the ACM, 25*, 100-104.

Navon, D., & Gopher, D. (1979). On the economy of the human processing system. *Psychological Review, 86*, 214-253.

Neal, A.S., & Darnell, M.J. (1984). Text-editing performance with partial-line, partial-page, and full-page displays. *Human Factors, 26*, 431-441.

Newell, A., & Simon, H.A. (1972). *Human problem solving*. Englewood Cliffs, New Jersey: Prentice-Hall.

Norman, D.A. (1983). Some observations on mental models. In D. Gentner & L. Stevens (Eds.), *Mental models*. Hillsdale, New Jersey: Lawrence Erlbaum Associates. pp. 7-14.

O'Shea, T. (1978). A self-improving quadratic tutor. *International Journal of Man-Machine Studies, 11*, 97-124.

Oberquelle, H., Kupka, I., & Maass, S. (1983). A view of human-machine communication and co-operation. *International Journal of Man-Machine Studies, 19*, 309-333.

Ochsman, R.B., & Chapanis, A. (1974). The effects of 10 communication modes on the behavior of teams during co-cooperative problem-solving. *International Journal of Man-Machine Studies, 6*, 579-619.

Ogden, W.C., & Boyle, J.M. (1982). Evaluating human-computer dialog styles: Command vs. form/fill-in for report modification. *Proceedings of the Human Factors Society, 26*, 542-545.

Ohlsson, K., Nilsson, L.G., & Ronnberg, J. (1981). Speed and accuracy in scanning as a function of combinations of text and background colors. *International Journal of Man-Machine Studies, 14*, 215-222.

Pask, G., Scott, B.C.E., & Kallikourdis, D. (1973). A theory of conversations and individuals (Exemplified by the learning process on CASTE). *International Journal of Man-Machine, 5*, 443-566.

Perlman, G. (1984). Natural artificial languages: Low level processes. *International Journal of Man-Machine Studies, 20*, 373-419.

Pollack, M., Hirschberg, J., & Webber, B. (1982). User participation in the reasoning processes of expert systems. *Proceedings of the 1982 National Conference on Artificial Intelligence,* . Pittsburgh, Pennsylvania: AAAI.

Power, M., Lashley, C., Sanchez, P., & Shneiderman, B. (1984). An experimental comparison of tabular and graphic data presentation. *International Journal of Man-Machine Studies, 20*, 545-566.

Press, L. (1971). Toward balanced man-machine systems. *International Journal of Man-Machine Studies, 3*, 61-73.

Price, L.A., & Cordova, C.A. (1983). Use of mouse buttons. In A. Janda (Ed.), *Human factors in computer systems*. New York: Association of Computing Machinery. (pp. 262-266).

Ramsey, H.R., Atwood, M.E., & Van Doren, J.R. (1983). Flowchart versus program design languages: An experimental comparison. *Communications of the ACM, 26*, 445-449.

Roach, J.W., & Nickson, M. (1983). Formal specifications for modeling and developing human/computer interface. In A. Janda (Ed.), *Human factors of computing systems*. New York: Association for Computing Machinery. (pp. 35-39).

Robertson, S.P., & Black, J.B. (1983). Planning units in text editing behavior. In A. Janda (Ed.), *Human factors in computing systems*. New York: Association for Computing Machinery. (pp. 217-221).

Robinson, C.P., Base, S.J., & Eberts, R.E. (1985). Eyes/ears: spatial/verbal: How should a cockpit talk to a pilot? In R. Eberts & C. Eberts (Eds.), *Trends in Human Factors/Ergonomics II*. Amsterdam: Elsevier. (pp. 51-58).

Robinson, C.P., & Eberts, R.E. (1985). There's a voice in the sky; but can a picture tell you more? *Proceedings of the Human Factors Society, 29*, 61-65.

Roe, C.J., Muto, W.H., & Blake, T. (1984). Feedback and key discrimination on membrane keypads. *Proceedings of the Human Factors Society, 28*, 274-281.

Rosenthal, R. (1963). On the social psychology of psychological experiment: The experimenter's hypothesis as unintended determinant of experimental results. *American Scientist, 51*, 268-283.

Rosinski, R.R., Chiesi, H., & Debons, A. (1980). Effects of visual feedback on typing performance. *Proceedings of the Human Factors Society, 24*, 195-199.

Rosson, M.B. (1983). Patterns of experience in text editing. In A. Janda (Ed.), *Human factors in computing systems*. New York: Association for Computing Machinery. (pp. 171-175).

Rumelhart, D.E., & Norman, D.A. (1981). Analogical processes in learning. In J.R. Anderson (Ed.), *Cognitive skills and their acquisition*. Hillsdale, New Jersey: Lawrence Erlbaum Associates. (pp. 335-359).

Scapin, D.L. (1981). Computer commands in restricted natural language: Some aspects of memory and experience. *Human Factors, 23*, 365-375.

Schank, R.C., & Abelson, C.K. (1977). *Scripts, plans, goals, and understanding*. Hillsdale, New Jersey: Lawrence Erlbaum Associates.

Schank, R.C., & Riesbeck, C.K. (Eds.) . (1981). *Inside computer understanding*. Hillsdale, New Jersey: Lawrence Erlbaum Associates.

Schneider, M.L. (1982). Models for the design of a static software user assistance. In A. Badre & B. Shneiderman (Eds.), *Directions in human/computer interaction*. Norwood, New Jersey: Ablex. (pp. 137-148).

Sheil, B.A. (1981). The psychological study of programming. *Computing surveys, 13*, 101-120.

Sheppard, S.B., Curtis, B., Milliman, P., & Love, T. (1979). Modern coding practices and programmer performance. *Computer, 12*, 41-49.

Shneiderman, B., Mayer, R., McKay, D., & Heller, P. (1977). Experimental investigations of the utility of detailed flowcharts in programming. *Communications of the ACM, 20*, 373-381.

Shneiderman, B. (1982). *Software psychology*. Cambridge, Massachusetts: Winthrop.

Shneiderman, B. (1982). Systems message design: Guidelines and experimental results. In A. Badre & B. Shneiderman (Eds.), *Directions in human/computer interaction*. Norwood, New Jersey: Ablex. (pp. 55-78).

Sime, M.E., Green, T.R.G., & Guest, D.J. (1973). Psychological evaluation of two conditional constructions used in computer languages. *International Journal of Man-Machine Studies, 25*, 105-113.

Small, D.W., & Weldon, L.J. (1983). An experimental comparison of natural and structured query languages. *Human Factors, 25*, 253-263.

Smith, C.H., & Dunsmore, H.E. (1982). On the relative comprehensibility of various control structures by novice Fortran programmers. *International Journal of Man-Machine Studies, 17*, 165-171.

Soloway, E., Ehrlich, K., Bonar, J., & Greenspan, J. (1982). What do novices know about programming? In A. Badre & B. Shneiderman (Eds.), *Directions in human/computer interaction*. Norwood, New Jersey: Ablex. (pp. 27-54).

Soloway, E., Ehrlich, K., & Black, J.B. (1982). Beyond numbers: Don't ask "How many" .... ask "Why". In A. Janda (Ed.), *Human factors of computing systems*. New York: Association for Computing Machinery. (pp. 240-246).

Stern, K.R. (1984). An evaluation of written, graphics, and voice messages in proceduralized instructions. *Proceedings of the Human Factors Society, 28*, 314-318.

Stevens, A., Collins, A., & Goldin, S.E. (1979). Misconceptions in student's understanding. *International Journal of Man-Machine Studies, 11*, 145-156.

Stevens, A., Roberts, B., & Stead, L. (1979). The use of sophisticated graphics interface in computer-assisted instruction. *IEEE Computer Graphics and Applications, 3*, 25-31.

Suther, T.W., & McTyre, J.H. (1982). Effect on operator performance at thin profile keyboard slopes of 5, 10, 15, and 25 degrees. *Proceedings of the Human Factors Society, 26*, 430-434.

Teitelbaum, R.C., & Granda, R.E. (1983). The effects of positional constancy on searching menus for information. In A. Janda (Ed.), *Human factors in computing systems*. New York: Association for Computing Machinery. (pp. 150-153).

Teitleman, W. (1979). A display oriented programmer's assistant. *International Journal of Man-Machine Studies, 11*, 157-187.

Thimbleby, H. (1980). Dialogue determination. *International Journal of Man-Machine Studies, 13*, 295-304.

Thomas, J.C. (1978). A design-interpretation analysis of natural English with applications to man-computer interaction. *International Journal of Man-Machine Studies, 10*, 651-668.

Tullis, T.S. (1981). An evaluation of alphanumeric, graphic, and color information displays. *Human Factors, 23*, 541-550.

Wickens, C.D. (1980). The structure of attentional resources. In R. Nickerson (Ed.), *Attention and performance*. Hillsdale, New Jersey: Lawrence Erlbaum Associates. (pp. 239-257).

Wickens, C.D., Sandry, D., & Vidulich, M. (1983). Compatibility and resource competition between modalities of input, central processing, and output: Testing a model of complex task performance. *Human Factors, 25*, 227-248.

Wickens, C.D. (1984). *Engineering psychology and human performance*. Columbus, Ohio: Merrill.

Williams, M., Hollan, J., & Stevens, A. (1981). An overview of STEAMER: An advanced computer-assisted instruction system for propulsion engineering. *Behavior Research Methods and Instrumentation, 13*, 85-90.

Zwaga, H.J.G., & Duijnhouwer, F. (1984). The influence of a fixed background in a VDU display on the efficiency of colour and shape coding. *Proceedings of the Human Factors Society, 28*, 331-335.

is more apparent or influential than the transformation of the human role from that of an active system manipulator to the job of system supervisor or monitor. These emerging job-demand characteristics are readily seen in the control rooms of many automated and semi-automated processing plants.

A more subtle change is occurring in the content of tasks. In traditional work, the human has been recognized as the determining component of a pre-specified process which emphasized repetitive motoric actions. As a consequence, initial training focused upon sequencing of tasks. Skill and experience reflected the precision and rapidity with which such functions could be replicated and completed. Often, knowledge of the operation itself was superfluous to task completion, although many operators were familiar with the mechanics of the machines they used.

Contemporary systems have begun to emphasize knowledge-based operations where, in addition to consultative interaction with the machine, the human is employed for capabilities such as pattern-recognition and inferential reasoning. When widely defined, these capabilities are intrinsic to problem-solving activities. It is the ability to synchronize the required response aspects of both human and machine into a single coherent unit that is the goal of current interface technology. This cannot be achieved by considering only static characteristics of the interface, but requires instead a dynamic and therefore adaptive interdependence. However, adaptive human-machine interaction implies innovative task allocation protocols.

8.1.4. Task Allocation

Conventional task allocation policies in human-machine systems are based upon the assumption that the adaptive function is intrinsic to the human operator. This individual may be required to cope with a wide variation in input task loading while the machine element performs a deterministic sequence of actions that represent a constant level of load resolution. Traditionally, the human has been required to compensate for the shortcomings and supplement the capabilities of the machine. However, contemporary developments in artificial intelligence (AI), the understanding of complex human-machine systems, and methods to assess mental workload (MWL) encourage the development of dynamic task reallocation which is responsive to momentary changes in operator efficiency and task demand. These developments that are explored in later sections on adaptive human-machine systems.

It is our contention that adaptation is a costly process and is becoming an intolerable burden upon the loaded operator. We suggest that adaptation can be shared, or totally assumed, by an intelligent intermediary interposed between the operator and the machine. However, to achieve this we need to know considerably more about human adaptive response and particularly how humans perform in the face of differential task loadings and stress. We address these issues below. As a first step we examine the recognized superiority in human adaptive capability as one strategy for adaptive interface construction.

8.1.5. The Human as an Adaptive System

Adaptation is a characteristic of living systems. The absence of adaptive capability constitutes a reasonable definition of the end or failure state of organisms. Adaptation is an interactive process implying the response of one entity to the actions of another. Typically the driving agency has been seen as the environment or some specific component thereof. Living systems are commonly recognized as the respondents. It is clear that the human organism is differentiated by the capability to engineer the total environment. On the continuum of adaptive capability, the human dominates. Adaptive capacity might be the pre-eminent characteristic of the human organism and, indeed, our ability to sculpt the environment may lead to change in human functional capabilities in future generations. It is this range or vocabulary of response that should be captured in the structure of an adaptive interface. For both theoretical and pragmatic reasons it is important to explore further the capacity of the human system to respond to fluctuations in the environment. These latter perturbations are often generically identified as stress.

8.2. STRESS AND ADAPTIVE FUNCTIONING

8.2.1. The Origins of Investigations into Stress Effects

Some of the earliest knowledge on human adaptive functioning came from Bernard's classic work upon the regulatory characteristics of diverse physiological systems. This principle was elaborated by Cannon (1932) who first coined the term homeostasis. The concept of stability in the face of dynamic perturbation is also intrinsic in the influential work of Selye (1956) who explored the defense mechanisms of the body to diverse environmental attack. The property of syntoxic agents which lessen the external impact by decreasing organismic sensitivity, and catatoxic agencies which destroy or remove the impinging agent, are of pertinence to later discussion of task strategy response. It is principally through these physiological insights that the notion of arousal was subsequently able to flourish as an approach to understanding more cognitive aspects of stress.

However, like many comparable areas of Human Factors, research into stress effects on performance was accelerated by the recognition of human limitations that came from experience during World War II. Perhaps the most influential studies emerged from the Applied Psychology Unit in Cambridge, England, where a number of eminent researchers engaged in a programmatic attempt to classify stress effects (e.g., Bartlett, Broadbent, Poulton, Welford). One of their legacies is information upon the effect of different forms of stress upon a common performance task, a tactic that still influences research today (Hockey & Hamilton, 1983).

The theoretical backdrop to much of this progress has been the notion of a unitary behavioral arousal continuum whose functional relationship to performance matches the inverted-U description first postulated by Yerkes and Dodson (1908). This law-like property was resurrected largely by Eysenck (1955) and Broadhurst (1957) at a period in time in which a fruitful integration with arousal theory could be realized (see Duffy, 1957; Hebb, 1955; Malmo, 1959). For a number of diverse reasons (see Hancock, 1987), this unitary arousal position has begun to fail as a descriptive account of empirical findings, while its theoretical foundations have always been somewhat suspect. In the vacuum that has followed this conceptual disintegration, a number of novel and synthetic alternatives have been postulated. These developments are relevant not only to stress research in general, but more pertinently to the understanding of operator behavior in controlling complex systems.

8.2.2. Competing Views of the Action of Stress

While the unitary behavioral arousal notion has proved somewhat simplistic and over optimistic in dealing with the variety of behavior recorded under stress-loading conditions, the foundation of physiological arousal remains a solid phenomenon upon which an account of stress effects must be based. Eysenck (1982) indicated the polarity of thinking between Duffy's (1962) implicit acceptance of a direct link between behavior and physiological state as compared to Broadbent's (1971) less sanguine position over such a simple isomorphism.

Hockey and Hamilton (1983) have suggested searching among stress effects for not one, but a number of common patterns. In this effort, while there may be no single unitary effect, a finite set of limited patterns may emerge which suggest perhaps a number of common physiological strategies. Their search is fraught with difficulty. Principal among these is the satisfactory pre-specification of tasks into their primitive components. It is perhaps too optimistic to hope that traditional information-processing elements (e.g., short-term memory, decision-making, response selection, etc.) will be mapped directly into identifiable substrate neurophysiological systems. An additional problem with their approach concerns the identification of these patterns in the face of performance variability which is often found in stress research. This problem is of general concern and does not apply solely to the formulation of Hockey and Hamilton (1983).

Despite some drawbacks, this fractionated arousal position clearly has appeal. The authors integrate usefully a number of discrepant findings within a structure not too far removed from the unitary arousal framework, which many studies were designed to test. Also, as a descriptive approach, their identification of common patterns has considerable practical application for those who design systems which are expected to work under adverse environmental conditions. However, unless a priori reasons for common patterns between stress effects are forthcoming, and more importantly, unless specific predictions are made as to why different forms of stress act disparately, the patterning of stress states is a reactive structuring of experimental findings and suffers some of the same theoretical limitation as its conceptual forebear.

In a somewhat different approach, Sanders (1983) has attempted to integrate the energetic aspects of performance with the classic stages of information processing. Conceptually this necessity has been recognized for some time (Simon, 1967) and has been advocated by those aware of the powerful impact that motivation and anxiety exercise upon behavior (Eysenck, 1982). This effort is one of the more cogent attempts to merge the seemingly qualitatively distinct concepts of linear stage processing and arousal.

To accomplish this integration, Sanders has taken a simple view of informational stages, as suggested by Sternberg, and has integrated into these stages, the processes of arousal, activation, and effort which Pribram and McGuiness (1975) proposed as controls of attention. This composite structure is similar in conception to the view of Wickens (1986) in which he suggests that stress acts to scale a gain value attached to the entry of each stage of processing. Both approaches are reminiscent of Welford's general assertion concerning the variation of internal signal to noise ratio as a function of environmental stress.

Efforts at integration recognize the necessity to incorporate energetics into linear-stage approaches. However, neither component has undergone any qualitative change in order to participate in these hybrid models. Thus, the products seem somewhat forced by the necessity to integrate, and have little basis in the dictionary of experimental observation from which a coherent theory should emerge. Indeed, the view of Koestler (1978) among others suggests that the logical processing of cognitive operations and the notions of energetic variation involved in emotion and stress response are undeniably separable entities and consequently are not amenable to simple integration.

8.2.3. An Emerging Synthesis

We offer a new view of stress, which is consistent with some elements of the foregoing arguments. However, our proposal contains a number of unique components which differentiate it from those previously discussed. This position has been generated (Hancock & Chignell, 1985) and elaborated (Hancock, 1986a; Hancock & Rosenberg, 1987) in a number of recent reports. Our purpose at this juncture is to provide sufficient information to allow the reader to follow our subsequent argument for mental workload as a form of cognitive stress response, and further to follow how such information provides a vital signal for input to an adaptive human-machine interface.

In our approach, a trinity of stress is represented in three descriptive loci. The first of these is an input locus composed of the deterministic physical characteristics, or signature, of the dynamic environmental display. The second locus is that of adaptation undertaken by the responsive individual to compensate for the perturbations introduced by the input stress. The final locus is an output value that represents the efficiency of performance upon an on-going, goal-directed task. As adaptation may be one of these latter goals, and as a task may be regarded as an input stress, it is clear that these three loci may overlap.

The determinancy of first locus is largely dependent upon the sophistication of measurement instrumentation. The second locus provides commonalities based upon the limited number of strategies available to the responsive operator and thus presents nomothetic or lawful trends. The final locus is largely idiographic, or individual specific. As operator actions depend upon individual evaluation, with respect to the diversity of inputs, common response across individuals is unlikely. However, it is possible to constrain response strategies by the nature of the task presented and the input level of stress experienced. We accept the multiplicity and time-varying change in human goal structure (see Eysenck, 1982), although we suggest that goals are in large part contingent upon the input stress encountered. It is upon this three-part description that the nature of adaptive human response to stress is based.

Figure 1 represents the input stress as a unidimensional entity along the x-axis. This ranges from extremes of underload (hypostress) to extremes of overload (hyperstress). Superimposed upon this axis are four zones that indicate adaptive functioning. Centered in these is a normative zone in which active adaptation is maximal, as the demands of the environment do not mandate dynamic compensation. Residence in such a zone is transient as the multi-faceted and multi-dimensional environment generates an input stress that presents an almost constant stream of challenges.

Surrounding this zone is a region of comfort. Behavioral action intended to return the operator to the comfort zone is initiated if the boundary of this zone is violated. Surrounding the comfort zone is an area in which the efficiency of behavioral (goal-directed) action remains high, despite the perturbing influence of the input stress. However, the failure of such psychological adaptability can occur rapidly and is reflective of the dynamic instability that occurs in the failure of physiological systems. The regions of stability and instability for both psychological and physiological functioning may be modeled through the use of negative feedback and positive feedback characteristics respectively.

We have equated psychological adaptability with attentional capacity in this picture. In our previous work (Hancock & Chignell, 1985), we have suggested that catastrophe theory (see Zeeman, 1977) may be useful in describing the transitions across the boundaries of dynamic stability as given in Figure 1, such discontinuities may be dependent upon the spatio-temporal characteristics of the input stress array. The final region represented on the illustration is physiological in nature and describes functioning consistent with homeostatic response (Cannon, 1932). An elaboration of the relationship between psychological adaptive failure and physiological breakdown is given in a recent analysis of sustained attention under stress (Hancock, 1986b). This approach has used the notion of attentional resource capacity (Kahneman, 1973; Wickens, 1984) to represent psychological adaptability, or as we define it, <u>the necessity to reconcile demands placed upon the active central</u>

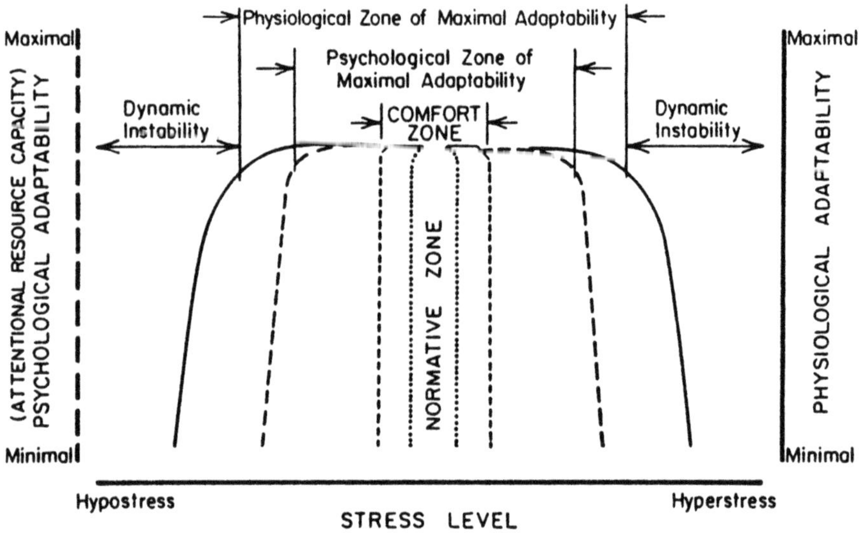

Figure 8-1. Physiological adaptive capability (solid lines) and psychological adaptive capability (outer dashed lines), equated with attentional resource capacity, as a function of stress level. Embedded in these zones is a region of comfort which is sought by the active operator. A central normative zone denotes a location where compensatory action is minimized, as environmental input is insufficient to demand dynamic response. Within zones of maximal adaptability, negative feedback predominates. Outside stable limits, positive feedback induces dynamic instability that proceeds to breakdown of adaptive response and eventual system failure.

nervous system. For systems in general it is possible to place an envelope of physical failure outside the present construction. However, due to the focus on adaptive human-machine functioning this is omitted from the illustration.

A more recent refinement of this description has differentiated the uni-dimensional notion of input stress as represented on the base axis in Figure 1. This two dimensional representation is shown in Figure 2. In this description the base axes are defined as information rate and information structure. Information rate is the perceived temporal flow as assimilated by the active, sentient individual. Information structure connotes the meaning of the environmental display as derived by the active search of the individual. This is a searching process as environmental information may be structured in quantal or matrix form and presents numerous possibilities among which the active perceiver must choose. It is amongst these possibilities that an optimal or acceptable goal-directed strategy must be sought. Clearly, the number of solution paths with respect to any task is constrained by the input presented. At extremes along the two base axes, behavior becomes more stereotypical as solution path density decreases. These axes are not orthogonal in that information on each axis tends to covary. So the content of perception demonstrably influences the perceived duration and vice versa (see Doob, 1971).

The regions defined by the boundaries A, B, C, and D represent normative, comfort, psychological, and physiological zones of adaptability as defined earlier in Figure 1. The vertical axes indicate that individuals seek to retain an optimal information flow via the actions that influence the search for meaningful task solutions. This search strategy is related to the use of attentional resources, denoted as cognitive adaptability in Figure 2. The manner in which multiple forms of input may be equated in this approach using base-plane vector resolution, is addressed elsewhere (Hancock & Chignell, 1985).

8.2.4. The Task as a Stress

We view the above conception as presenting a description that transcends individual levels of analysis. It represents an approach that is as relevant at the physical as at the sociological level and is therefore similar to other cross-level constructs such as Living-Systems Theory (see Miller, 1978). Although an input at any level has a reverberative effect at all other levels, it is the system primarily impacted that exhibits clearest response.

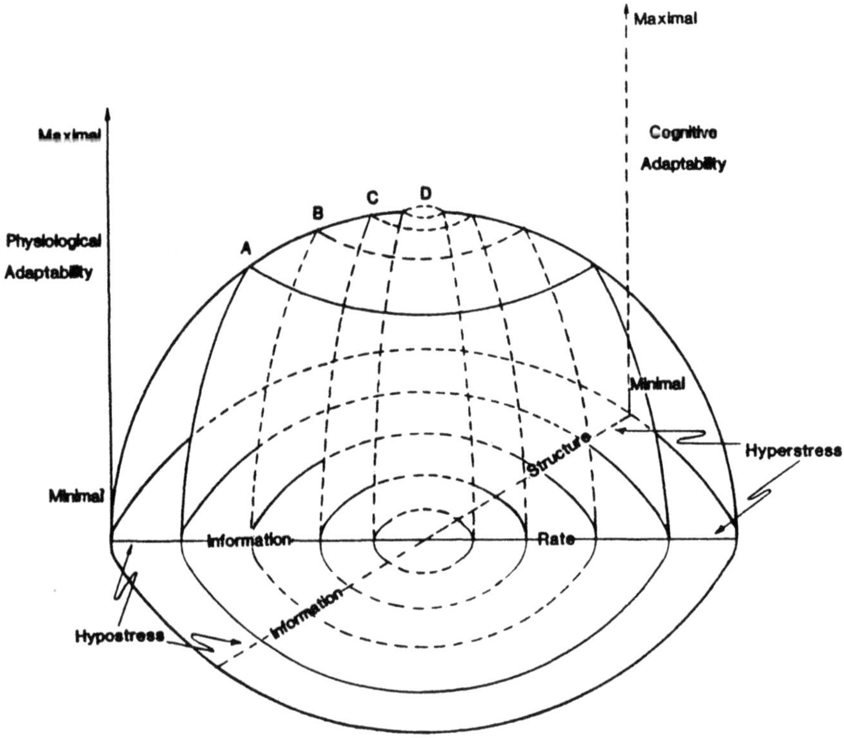

Figure 8-2. Physiological and psychological (cognitive) adaptability as functions of hypostress and hyperstress expressed upon the dual axes of information rate and information structure. Multiple stressors are represented as summated vectors plotted upon the two-dimensional base. Adaptive reaction may be facilitated by the navigation of the overall manifold through the perturbations presented in the environment. The introduction of machine prosthetics magnifies the range of tolerable environments through the augmentation of human adaptation. Surrounding environments cushion the operator from large perturbations.

Typically, stress has been viewed as an environmental property, e.g., heat, noise, which affects primarily physiological systems (see Hockey, 1983). In our description we view the task to be performed as the major source of behavioral stress, a characterization that others have also envisaged (see Hockey, Gaillard, & Coles, 1986). Thus traditional environmental stresses should be viewed as interactive but secondary components of the total input, where task demands assume an overall priority until superseded by simple survival necessities. We examine the stress of task demands in the section which follows concerning mental workload (MWL), where we view MWL as a response to task demands which are expressed as a cognitive input stress.

8.3. USE OF MENTAL WORKLOAD MEASURES AS INPUT TO AN ADAPTIVE SYSTEM

8.3.1. Current Workload Assessment

Human mental workload assessment is of considerable practical importance and has posed a number of theoretical challenges which have largely defeated simplistic information-processing views of human action. Mental workload is much more aligned with energetic aspects of capability such as motivation, appraisal, and attentional resource capacity (Eysenck, 1982; Kahneman, 1973; Lazarus, 1966; Lazarus & Folkman, 1984). This is perhaps one reason why resource theory has proved one of the more effective avenues of attack in the attempt to assess MWL (see Wickens, 1980). It is not the purpose here to examine these contemporary developments. Indeed such an endeavor is superfluous when there is a more clear and cogent assessment of mental workload by Kantowitz (Chapter 3 in this volume), and related discussion of attention by Wickens (Chapter 2 in this volume). It is to these sources that the interested reader is directed. Rather, we present a recent elaboration of our own position on mental workload which adopts a perspective that is clearly influenced by the previous discussion of stress effects and adaptive response in general. Our position is characterized by a focus on time, referred to on Figure 2 as information rate, and it is to the role of time in mental workload that we first turn.

8.3.2. Time As A Factor in Mental Workload

Is the "load" of mental work directly contingent upon the time-span available for effective action? To resolve this question, it is necessary to examine two contrasting views of mental workload (Hancock & Chignell, 1986a). Mental workload can be viewed in terms of task demand and observed response or, alternatively, as the situational appreciation of the operator (Gopher & Donchin, 1986). The first approach requires the pre-specification of the task under consideration and a viable assessment of the environmental background against which performance is set. This necessitates a knowledge of how to express the environment as a physical assemblage and a knowledge of what terms are needed to exhaustively express the requirements of any single task. However, the transduction between input demand and output action implies the existence and characteristics of an entity that can resolve task demands. It is clear also that expression of demands must be made in light of the characteristics of the response entity. As we have insufficient understanding of the cognitive capabilities and response characteristics of the human at the present time, this means-ends analysis of mental workload has yet to be adequately elucidated.

In contrast, it is possible to see MWL as directly dependent upon the situational appraisal of the individual operator. This latter view allows the incorporation of the idiographic adaptive response, and as a result the complexities of individuality are brought into the workload equation rather than being excluded as in the previous approach. The selection between these two alternatives is largely pragmatic in that our concern is currently for the operation of contemporary systems, rather that an exhaustive theory of human cognitive functioning per se.

8.3.3. An Integration: Stress and Mental Workload

From the foregoing arguments it can be seen that we view mental workload as a form of stress which may be described in terms of the tripartite description given earlier. The input form is the demand made by the presented task or group of tasks and has traditionally been approached through task analysis. The process of adaptation is accomplished through the compensatory capabilities of the individual, in our view this is equated in general with attentional resource usage and is related to both behavioral and physiological subjective appreciation. The output is the efficiency on the task as reflected in performance assessment, commonly recognized as the primary task approach. Consequently, MWL may be described in each of the three loci established earlier for stress.

The input-output view of mental workload is that in which the specifics of the task and required performance are identified. Workload is considered the cost of transduction between these two states. As indicated, the flaw in this argument concerns the nature of the transducer. Although appearing neutral, this view actually implies specific human capabilities. An operator centered approach (see also Gopher & Donchin, 1986) takes the perception of the individual as the central focus. This appraisal or adaptive view purposely includes the variation involved in individual capability. It is as an expression of this adaptive function that we see commonalities between mental workload, attention, and the cognitive stress response.

8.3.4. The Cost of Adaptation and Its Place in Cooperative Human-Machine Systems

Consistent with our view of stress we suggest that adaptation in general is a costly process. When widely defined, we can view demand as drawing upon the resources of the operator, or draining adaptive capacity. These resources may be physiological or psychological in nature. MWL is clearly a psychological expression, although measurement may be made through the reflections evident in the change of physiological functioning (Hancock, Meshkati, & Robertson, 1985). The undiluted demands of complex systems rapidly deplete these resources causing degradation in performance efficiency. Chronic exhaustion of resources may be recognized in the fatigue associated with long-term operator overload or underload, where each of these maladaptive loading situations gives rise to an untenable drain on adaptive capability.

8.4. THE ROLE OF ARTIFICIAL INTELLIGENCE IN HUMAN-MACHINE SYSTEMS.

In response to the preceding observations, we propose that the cost of adaptation should be either shared between operator and machine or transfered to the machine element alone. More specifically, the adaptive function may be located in an intelligent interface that stands as intermediary between the machine and its operator. Figure 3 illustrates how the components of an intelligent interface modify the nature of human-machine interaction. The human-machine interface is seen as a cyclic process composed of the following four components:

* system responses which are perceived by the user

* cognitive analysis of the system output by the user who integrates the information into their model of the task and system status.

* selection of operations on the basis of the user's model of the task

* execution of the operations using the physical interfaces

Thus there are four major interactive components of the interface, two representing input to the human and two representing output from the human. Inputs to the human are the perceptual display and cognitive analysis, while outputs are the planning and decision-making elements with response execution, where the latter is subsumed by response selection.

This interactive system of physical and cognitive interfaces can be augmented by adding an intelligent interface which then requires three additional components.

1. The discourse input machine which ideally allows a form of user response input that is compatible with the user's model of the task and the subgoals they are currently attempting to execute.

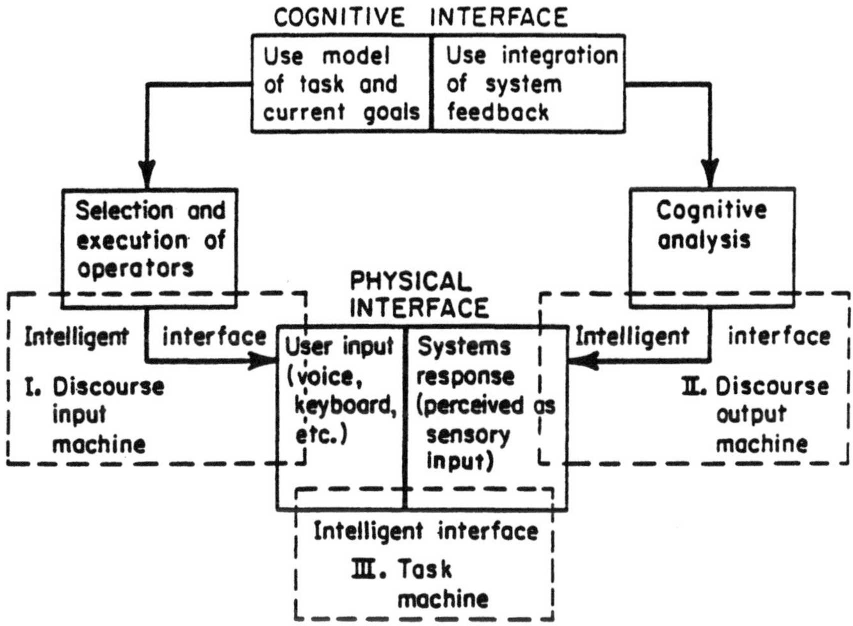

Figure 8-3. A representation of the physical and cognitive interfaces to a human-machine system showing the intelligent interface components that can be included at different stages of the interactive processing cycle.

2. The discourse output machine which is charged with the task of transforming the system response into an output that is compatible with the user's model. The degree of compatibility is inversely related to the amount of cognitive analysis required to incorporate the systems output in terms of the user's model.

3. The task machine which provides the machine reasoning capability necessary at the level of the most appropriate user model for the task.

In the sections which follow we examine the basis of Artificial Intelligence (AI) developments which permit the realization of this functional intermediary.

8.4.1. Artificial Intelligence Inputs to an Adaptive Interface

The human operator has often had to shoulder the responsibility for adaptation when the machine has neither the intelligence nor the flexibility to adapt. Developments in AI can ameliorate these machine deficits. The problem of the human factors designer has become that of allocating and orchestrating task elements rather than that of magnifying and focussing human adaptive capacities.

In spite of improved machine reasoning, the innate differences between humans and machines lead to problems in designing the cooperative and effective human-machine partnership, see the Chapter on human-computer interaction by Eberts in this volume. We distinguish three main parameters of a system which can be improved using automated reasoning and related techniques developed by AI researchers.

1. Communication. In typical human-machine systems there is a gulf between the language of the human and the language of the machine. In the past it has been the responsibility of the human to cope with, and overcome, this disparity. However, it is possible to construct interfaces which act as translational intermediaries between human and machine, thereby improving communication (Chignell, Hancock, & Lowenthal, 1987; Hancock & Chignell, 1986b), see also Figure 3.

2. Control. The problem of control is that of deciding what should be done next and by whom (or what). In human systems, this is the problem of management and the organization of social behavior, see the Chapter by Hendrick in this volume. In automated systems, the problem of control is becoming increasingly important as deterministic processes (e.g., Von Neumann machines) are required to react in near-optimal fashion in uncertain environments. In computer science, there has been evolution from direct control of program execution (the use of the Jump statement in assembly language programming) to high-level goal-oriented control as in declarative programming (Kowalski, 1985). Similarly, in human-machine systems, there has been a transition from direct control of the physical system to supervisory control. Supervision requires the human operator to control the behavior of a complex

system by monitoring it and effecting corrective action only when lower levels of the hierarchy fail to maintain adequate system response.

3. Task Performance. Aside from improving communication and control, AI techniques can enhance the performance capabilities of the system by automating or simplifying tasks, as in the Pilot's Associate Project. Currently, this involves the development of expert systems (Harmon & King, 1985; Hayes-Roth, Waterman, & Lenat, 1983; Waterman, 1986) to perform (or aid in the performance of) tasks such as troubleshooting and navigation.

The development of expert systems for performance enhancement has been the initial focus of AI applications in human-machine systems. Expert systems aiding of supervisory control tasks is discussed by Higgins, Chignell, and Hancock (1987). AI methods also have considerable potential for increasing control (Chignell & Hancock, 1985; Hancock & Chignell, 1986b) and communication (Chignell & Hancock, 1986). The development of appropriate intelligent interfaces for enhanced communication and adaptive mechanisms for improved control requires that relevant knowledge representations and inference tools be available.

8.4.2 Horn Clause Representation of the Human-Machine System

Design of adaptive (and intelligent) interfaces is a difficult task which requires new methods for controlling the interaction between human and machine. Research in artificial intelligence (e.g., Rich, 1983; Winston, 1984), and the success of expert systems (Waterman, 1986), has demonstrated that appropriate representations of knowledge and associated reasoning mechanisms are the keys to intelligent behavior. First order predicate logic (Nilsson, 1980) is one of a number of knowledge representation techniques. A Horn clause is a predicate expressed as a conclusion followed by one or more conditions. A coherent set of Horn clauses acts very much like an AND/OR goal tree (Winston, 1984) where the conclusion becomes the goal and the conditions act as subgoals which may be conjunctively (AND) or disjunctively (OR) satisfied.

The major advantage of Horn clause representations of knowledge is that, as well as being declarations of facts and relationships (predicates), they can be implemented as procedures. Thus the same set of Horn clauses can be used as the basis for both describing and simulating a human-machine system. Recently, logic programming tools have become available which enable Horn clause representations to be used for description and inference. The major logic programming tool currently available is the PROLOG programming language. Knowledge can be represented as facts and logical predicates in PROLOG. One can also reason on the basis of that knowledge, using processes of pattern matching, recursion, and backward chaining inference. In a later section, we illustrate the logic programming approach to knowledge representation and inference in an adaptive interface. We shall begin here with a logic programming representation of a familiar concept. Consider the definition of the arithmetic mean.

Logic programming example: The arithmetic mean.

Declarative definition: The arithmetic mean of a list of values is the sum of that list divided by the number of values in the list.

Pseudocode:

```
MEAN is the_mean_of [List_of_numbers] if
SUM is the_sum_of [List_of_numbers] and
COUNT is the_count_of [List_of_numbers] and
Mean is SUM divided_by COUNT.
```

Prolog Code:

```
mean (List, MEAN) if
sum (List, SUM) and
Count (List, COUNT) and
MEAN = SUM/COUNT.
```

To make the definition complete we should now define "sum" and "count" in the same way.

The above shows how executable PROLOG code that can be used to actually calculate the mean follows naturally from a declaration of the definition of the arithmetic mean. Pseudocode (similar to the SIMPLE language available in Micro-Prolog; Clark & McCabe, 1984) can be used to smooth the transition between declaration and eventual PROLOG code.

8.4.3. Implications for Human-Machine Systems Design

In describing and simulating human-machine systems, traditional approaches oscillate between two contrasting positions. On one hand, verbal descriptions of tasks characterize some semantic features but are incapable of generating explicit procedural representations. Control theory, statistical simulation, and other quantitative methods on the other hand overcome these objections, but have been unable to account for those knowledge-based aspects of the human-machine system which are accessible only to symbolic reasoning methods. Logic programming appears to be an ideal tool for representing and analyzing human-machine systems for two major reasons. First, it provides a common language for implementing both quantitative and symbolic reasoning techniques. Second, its declarative and modular nature is well suited to the incremental development of systems descriptions. One of the major benefits of the logic programming approach is that it encourages a more systematic analysis and provides Human Factors specialists with a means of structuring knowledge about human-machine systems, whether or not complete simulations of those systems are subsequently achieved. Horn clauses, supplemented where necessary with other knowledge representation techniques, can be used to develop explicit models of human-machine systems which include both quantitative and symbolic aspects. These models can then be applied to the problem of allocating tasks and controlling workload within an adaptive interface.

8.5. ADAPTIVE SYSTEMS

8.5.1. The Necessity For And Development Of Adaptive Human-Machine Systems

Conventional views of human-machine systems have the human controlling, or being controlled by, the machine component. For instance, in machine-paced assembly the human is effectively controlled by the machine. In contrast, in driving a car the human appears to be in control, at least under normal operating conditions. We follow an alternative perspective of a human-machine system as a cooperative enterprise. In this view, the human and machine work together to ensure successful system performance and the satisfaction of task demands. The assumption of a synergistic and cooperative relationship between the components of a human-machine system leads to a new type of system design. Firstly, cooperation pre-supposes communication between intelligent entities. Expert system consultants provide low-level examples of this communication. Secondly, an interface (translation process) is required for the communication of needs, requests, and ideas between the entities.

Early human-machine systems forced the human to fit in with the requirements of the machine, (i.e., the human bent while the machine was straight). Human factors engineers design machines and tools which fit human capabilities, in keeping with the maxim "bend the tool, not the person" (McCormick & Sanders, 1982, Chapter 10). The ideal situation would be one where both the person and the machine stood straight (i.e., performed according to their design principles) while a translational interface adapted inputs and outputs in order to render information compatible with the characteristics of each of the collaborative partners.

8.5.2. Task Structuring

The type of adaptive interface necessary for genuine human-machine cooperation would be capable of restructuring the task in accordance with system goals and environmental constraints, and of reallocating task components between human and machine for a given task structure. Figure 4 shows the overall control structure for such an adaptive interface. Task structuring would be carried out by a task definition supervisor. As technology advances and human-machine systems develop, the task will no longer be set as a fixed entity. Instead, the definition of the task will change in accordance with the higher level goals set by some external agency and with changes in the number and type of environmental constraints acting on the human-machine system. Given a particular definition of the task, allocation of tasks to the human and the machine component would be carried out by an adaptive interface.

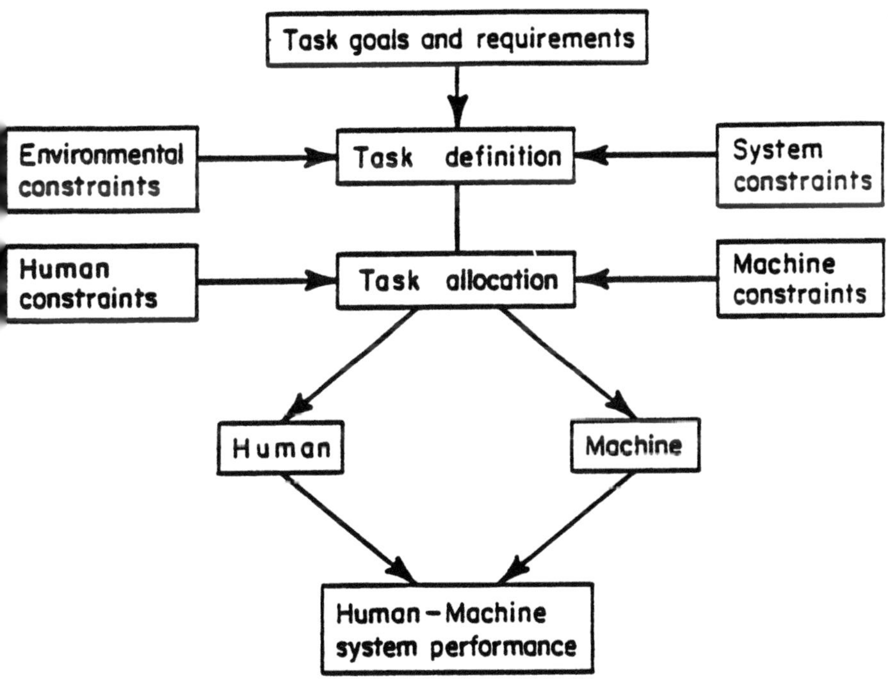

Figure 8-4. An overall control structure for a human-machine system which implements flexible restructuring and reallocation of tasks through an adaptive interface.

8.5.3. Task Allocation

It is clear that optimal allocation of task functions between operator and system requires a knowledge of human versus machine capabilities. At the present, some of the task functions will not be involved in the allocation decision because they are clearly suited only for the machine or only for the human component of the system. Allocation is relevant for tasks which can be switched between human and machine without having a detrimental impact on overall performance.

Price (1985) has developed a method for allocating functions between humans and machines which requires designers to construct a fixed task allocation policy. The optimal allocation will not be static but will be conditional upon the task definition, working environment, and current capabilities of system components. Currently, knowledge about the human abilities that are relevant to the performance of various tasks is incomplete, but there are several major characteristics which should be taken into account. These include human sensitivity to relative rather than absolute change, limitations in attention and memory, compatibility of various input/output modalities for different tasks, performance variability, error correction capabilities, and reactions under stress (Chignell & Hancock, 1985, Hancock & Chignell, 1985; McCormick & Sanders, 1982, see also the information in the companion Chapters of this book). Optimal, or close to optimal, task allocation will require a sophisticated reasoning process based on models of human and machine capability and a detailed analysis of the current task.

8.5.4. Task Description and Task Performance

Recent work in defining and describing tasks has been summarized by Fleishman and Quaintance (1984). The type of description and classification method that is appropriate will depend on the task and the human-machine system. Given a description of the task and an operator with a certain level of skill, adequate performance will require a particular level of utilization for each of a number of task-relevant attentional resources. Wickens (1980) has proposed a multiple resource model of attention where different task aspects draw on different pools of attentional resource. This model can be used to relate person and task characteristics, allowing a prediction of performance efficiency on the basis of an assessment of the current demands in and the status of the human operator. Measurement of resources actually utilized by the operator may be possible using probe secondary tasks and evoked potential techniques, but at the present this type of assessment is problematic (see Kantowitz this volume).

Accurate assessment of the resources used in a task is not yet possible in most cases, since the prior problem of characterizing and describing the task is unresolved. Part of the problem is that every task is unique in certain respects. Nevertheless, techniques are available which should form the basis for general task description methodologies in the future. The overall structure of most tasks can be described in terms of a hierarchy of nested goals and subgoals, see Figure 5 (upper portion). Performance on a task (i.e., the procedures used in performing the task) can often be described as a means- ends analysis where the choice and order of execution of subtasks is based on the pursuit of an overall goal. In routine cases, task performance can be modeled as the elaboration and instantiation of a script (Schank & Abelson, 1977), or procedure, see Figure 5 (lower portion).

The performance of a task may be thought of as movement through a problem space from an initial state to a goal state. In the means-ends analysis approach to problem solving (Newell & Simon, 1972) procedures are selected according to their ability to reduce the estimated difference between the current state and the goal state. When this approach is applied to human-machine task analysis, each state corresponds to a description of the system at a given point in time. The state space representation forms the basis of the main AI techniques for problem solving. From a task-load management perspective, a description of the task is needed which includes an estimate of amount of mental effort required in each step of task performance. Using the problem space formulation, it is possible in principle to develop an explicit accounting of task-induced load, at each step of the task.

<u>Generating moves (node expansion)</u>. The solution of a problem may require transition through a number of states. Specific operators are used to achieve the transitions from state to state. Prior to selecting an

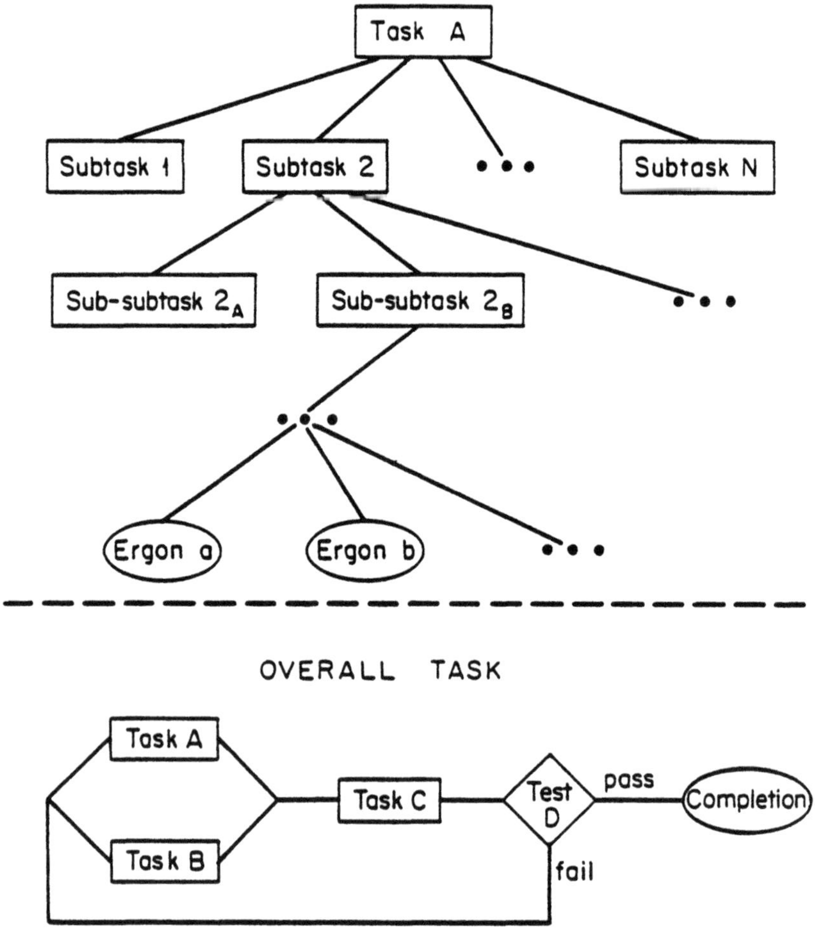

Figure 8-5. Representation of a generic task as an hierarchy as tasks and subtasks. At the lowest level, indivisible task primitives are represented as ergons, which are integrated into the higher level sub-tasks. In the lower portion of the figure, tasks are coordinated as a procedure.

operation, the person carrying out the task must select a subset of operations that are feasible in the current context. Training and experience often determine the ease with which a candidate set of operations can be generated.

Informational Complexity. The number of operations available, possibly weighted according to the prior probability of selection, determine the informational complexity of the choice (Attneave, 1959). A skilled chess player for instance, may only consider a small subset of "promising" moves, thereby reducing the informational complexity associated with each move.

Operation Selection. Once the available operations are known individuals then have to choose the best one, i.e. the move which will bring then closest to the overall goal. The difficulty of this choice will depend on the discriminability of the best operation among the identified candidates. If one choice is clearly superior, the decision is easy. If a number of operations are approximately equivalent in their effect, the choice becomes more problematic and requires greater mental effort.

Action and Subgoal Execution. Once the choice of operation is made, the preparation of the action necessary to implement the choice requires further mental effort. This often includes the eye-hand coordination and motor control necessary to execute the operation. For high level cognitive tasks the operation may in turn be a subtask or subgoal which has its own state space and solution requirements (i.e., we may have one or more hierarchies of nested operations).

8.5.5. Fluctuations in Task Definition and System Response

Although satisfactory description of most tasks is difficult, it is only part of the problem. Allocation policies based on a static representation of a task ignore intrinsic variability in the nature of the task itself and the human- machine system's response. In addition, the perceived difficulty of the task changes with the variation in the environmental context. As a result the mismatch between task demands and available resources will vary over time. On occasion the mismatch may create inadmissible overload or underload with consequent decrement in performance, (see Chignell & Hancock, 1985, Figure 1).

In order to achieve a consistently high level of performance, the human-machine system needs to cope with large task and person fluctuations. In many cases, the human can adapt to mismatches between current task demands and their available capacity. In other cases, where human capability is exceeded, stability (residence in the zone of adaptability) requires dynamic reallocation of the task components, and possibly restructuring of the task. The process of dynamic reallocation is achieved by an adaptive interface which is described below.

8.6. A KNOWLEDGE-BASED ADAPTIVE MECHANISM

Dynamic task reallocation requires an adaptive mechanism which can assess the mismatch between task demands and available capacity and redefine the task so as to reduce this mismatch. This knowledge-based adaptive mechanism (KBAM) will be more than a simple reflexive action (e.g., table lookup) in many cases, since the complexities of human capability and the extent to which it can be redefined requires some degree of knowledge-based reasoning.

The concept of an adaptive interface has been discussed for some time (e.g., Chignell, & Hancock, 1985; Edmonds, 1981; Morris, Rouse, & Ward, 1984). Complex systems require interfaces that adapt to the needs of the task and the operator. Adaptive interfaces will be useful only with certain types of tasks. Firstly, the task must be performed by a human- machine system and, secondly, the task must be of moderate complexity, being neither so difficult as to tax the system as a whole, nor so easy as to allow the system to perform adequately whatever allocation policy is adopted. In general, there will be some combination of task complexity and variability where an adaptive reallocation policy will improve performance significantly. Personnel training and selection will also impact the degree of adaptivity required of the interface.

8.6.1. The Error Signal

The first step in developing an adaptive interface is the identification of the error signal representing the mismatch between the current task demands and the available capacity of the human operator. The purpose of the error signal is to provide dynamic information about the human operator both in terms of instantaneous MWL and the global mismatch between resource requirements and resources utilized. The identification of this error signal in task allocation is then a two-stage process. First, a global assessment of MWL is made, i.e., is MWL too high or too low and by how much? Second, diagnostic information about demands on specific attentional resources is extracted to indicate which particular pool of attention (e.g., spatial encoding) is subjected to unacceptable loading. An aggregate signal derived from the status of all attentional resource pools drives subsequent task reallocation policy. Ideally, task reallocation returns the operator to a more desirable level of MWL, and balances resource usage across different attentional pools.

The basis of successful task allocation is accurate identification of task demands, which can be assessed either by examining task characteristics directly, or indirectly through assessment of mental workload. Mental workload assessment is necessary in many situations because of variability in response to specific task characteristics, both within and between individuals. Currently, the formation of the error signal is problematic. Mental workload measures may be used to derive the error signal directly, but they are likely to be confounded by a number of factors, not the least of which is emotional responses to the task situation and performance feedback. In the ideal situation, the error signal would be based on a theory of attentional resource utilization and the various relationships between task demands, workload, and physiological response (see Hancock, Meshkati, & Robertson, 1985). In the absence of a comprehensive well-validated theory, we favor redundancy in error signal derivation using a combination of measures. In this approach, direct measurement of the error signal via MWL assessment is augmented by calculations of the mismatch between task demands and attentional resources.

Physiological measures appear to be the most useful form of direct MWL assessment for an adaptive interface. Not only do they supply a continuous stream of data which is amenable to statistical analysis, but they also reflect the human's physiological response to the current workload and they do not alter substantively the existing task structure. Performance measures also provide important clues about the operator's ability to perform the task as currently defined. The problem of converting a matrix of physiological and performance measures into a signal which provides the necessary information for task redefinition and reallocation within an adaptive interface can only be solved if a complete specification of the information required is given. The informational requirements of an adaptive interface stem from its purpose which is to optimize system performance.

Fuzzy logic can combine the above perspectives into a comprehensive error signal provided that the associated membership functions of different physiological and performance variables are known. Existing techniques of work measurement and task analysis should be extended to provide estimates of the resources required in different tasks. A variety of methods, like evoked potential measurement (the P300 component in particular), eye movement patterns, may also lead to reasonable estimates of resource utilization. It is possible that the scaling of physiological and performance measures as membership functions on fuzzy subsets such as "excessive mental workload" and "insufficient mental workload" will be based on experimental functions derived from existing empirical data. The final measurement problem would then consist of assigning appropriate weighting coefficients to each of the physiological, primary-task performance, subjective ratings, or dual task measures of MWL. This problem is considered in more detail by Chignell and Hancock (1986).

8.6.2. Task Demands, Available Resources, Task Allocation, and Load Leveling

The key component of an adaptive interface is a reasoning process which selects a task allocation policy that changes the loading on the human in such a way as to improve overall system performance. This process must have access to both overall system goals (a model of the task) and information about what the person and machine components of the system are capable of accomplishing (person and system models). In addition, the reasoning process should decide which loading strategies and task allocation policies are available for selection, and what the implications of the current error signals are.

Models of the task, person, system, and available loading strategies are part of the knowledge base of the adaptive interface, but none of these inputs give information about the current status of the person. Thus, the error signal is a dynamically changing input which has to be combined with the static information provided by the other informational sources available to the loading strategy reasoner. Not only must the error signal provide information about how well the person is performing particular subtasks, but it must also indicate any discrepancies between overall system performance and system goals in order that the available capacity of the person can be integrated with the needs of the entire system. It would be inappropriate to optimize the workload experienced by the operator if the overall system failed or suffered as a consequence.

Once the error signal is derived, it can then be input to an adaptive interface. Models of the task, system, and person then allow prediction of the effect of alternative task redefinitions, while look-up of a database of admissible loading strategies will check whether or not a proposed loading strategy violates available resource capacity. Only tasks which can be switched reasonably between the human and machine components of the system should be considered for reallocation, except under emergency conditions which demand continued performance, with the result of system failure having intolerable consequences.

Adaptive interfaces of varying sophistication can be developed, differing in the amount and complexity of the reasoning used in load leveling. The simplest system would have a look-up table which assigned a loading level to each task definition. The tabulated task loading would be adjusted on the basis of the apparent effort being used by the person (measured either physiologically or as a subjective rating) in performing the task.

Additional complications would be introduced if a multivariate error signal were used, as would occur if the error signal was constructed using a multiple resource model of attentional capacity and multicomponent analysis of the task. Even so, the reasoning process might not be too complex providing that the redefinition of the task was regarded as a classification task. Conventional expert systems are well equipped to handle the classification problem and existing techniques

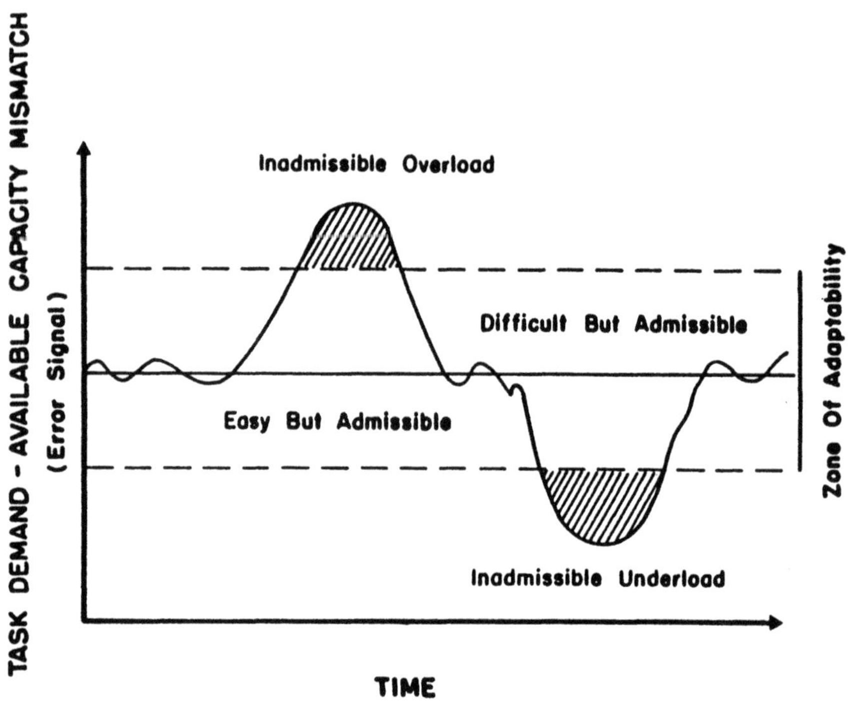

Figure 8-6. Analog signal representing the difference between current task demand and available capacity. The demand level may fluctuate within or beyond a range of operator adaptability. The adaptive interface serves to zero the level of this error signal which it initially receives as dynamic input. This is accomplished by task reallocation procedures as described in the text.

using a well defined knowledge base. A production rule inference engine (e.g., Hayes-Roth, Waterman, & Lenat, 1983) might be sufficient in most cases.

The task redefinition process can be viewed as one of classifying a given error signal in terms of a fixed set of task definition choices, as occurs in table look-up methods. In more complex adaptive interfaces, however, the table look-up procedure of classification would be replaced by rule based inference. The success of this latter strategy will depend to a large extent on the quality of the information stored in the knowledge base, i.e., whether or not the set of allowed task definitions or the specific rules are appropriate. The output of the adaptive interface will be a definition of the task which alters (where necessary) the loading of task components between the human and machine so as to reduce the error signal. Thus, a human-machine interface is developed which acts as a servomechanism minimizing the difference between current demands and available capacity. The overall structure of this interface is shown in Figure 6.

Central to the reasoning process used in an adaptive interface is a knowledge base containing information about all the facts deemed to be relevant to the performance of the task. Figure 6 shows only one of a number of ways in which a KBAM might be designed, although we expect any variation to contain the components identified here, albeit in different configurations. The purpose of an adaptive interface is an allocation of task functions between human and machine which maximizes performance outcome. The process of task reallocation should not be disruptive. A large number of sudden, discrete changes might lead to a worsening, rather than an improvement, in performance. It is likely that smooth and relatively continuous changes in task definition will be preferable.

The manner in which the change in task demands is best communicated to the human operator will depend to some extent on the task being performed. One can distinguish between insidious systems which reallocate tasks without directly warning the human, and conversational systems which signal explicitly each change in the task definition and allocation. Alternatively, the adaptive interface may be consultative, suggesting a better task allocation policy, while allowing the human operator to decide whether or not the suggested task reallocation should take place. In order to minimize a variety of stresses associated with the lack of task autonomy, it is suggested that the latter proposal will generally be more useful for initial use in dynamic human-machine systems.

8.6.3. Knowledge Systems Development

Complete understanding of a complex human-machine system will require a great deal of knowledge. The knowledge required for task reallocation may well be a small subset of this exhaustive information. Initially, it is anticipated that three types of knowledge should be specified, namely, knowledge about how well each subtask can be performed by the machine, knowledge about how well a subtask can be performed by the operator for a given level of resource utilization, and knowledge about the way in which tasks demands and resource availability jointly affect MWL. As discussed earlier, the declarative nature of Horn clauses and their ability to be implemented as procedures makes them highly suitable for representing knowledge about human-machine systems. The top-level clause for an adaptive interface might be:

```
  reallocate([Subtasks], [System_components], [Available_resources])
if
  MWL1 > .8        OR
  MWL2 > .8        OR
  Inad_perf > .8   OR
  Inad_perf * max(MWL1, MWL2) >.5
AND
  allocate([Subtasks], [System_components], [Available_resources]).
```

where MWL1, MWL2, and Inad_perf are facts derived from the current information input. We can now define a set of allocation rules of the following form:

```
  allocate([Subtasks], [System_components], [Available_resources])
if
  allocate_job(Subtask1, [System_components], [Available_resources])
AND
...
AND
  allocate_job(SubtaskN, [System_components],
                          [Available_resources])
```

```
  allocate_job(SubtaskI, human, Human_resources)
if
  performance(SubtaskI, human, Human_resources) is X
AND
  performance(SubtaskI, machine, Machine_resources) is Y
AND
  X > Y.
{else}
  allocate_job(SubtaskI, machine, Machine_resources).
```

The basic information about human and machine capabilities necessary for making allocation decisions can be defined as facts, as in the following example.

```
performance(arithmetic, human, Typical_resources) is mediocre.
performance(arithmetic, machine, Working_condition) is excellent.
```

The previous code fragments show the general outline of a task reallocation system. Further code would prevent allocating additional subtasks to a human who is experiencing high mental workload, or to a machine component which had broken down. The purpose of this section is to show how Horn clauses can implement a task reallocation system, not to describe a fully functioning system (this is currently a research topic in our laboratory).

8.7. CONCLUSION

The development of many human-machine systems can be characterized by the following sequence. First, a machine system is built and evaluated for its capabilities and limitations. Second, the system is reconfigured so that a human fulfills only those tasks that cannot be performed by machine. Third, an interface is built so that the humans may carry out their subtasks and system performance is evaluated. Finally, the human operators are trained and the interface structure is modified until satisfactory performance is achieved.

We advocate a different approach where a hierarchical model of the task is built in terms of prodedural and knowledge-based components.

Conventional engineering approaches can then be used to create the basic operators, procedures and task primitives. Central to our view of human-machine interaction is a view of the task as a knowledge system which requires combined human-machine intelligence along with an interface that permits and controls joint human-machine reasoning (Woods 1985).

In the present chapter we have described some developments in the understanding of human-adaptive response, and how such adaptive capability may be replicated in human-machine systems. We suggest that such adaptation is costly and can, in common with other elements of work, be subsumed by machine elements. Our treatise indicates a number of practical and theoretical stumbling-blocks to such implementation but implies that potential gains from such a development justify the effort in their construction.

As is often the case, this chapter has generated more questions than answers. We have the capabilities required to build adaptive interfaces, but orchestrating the various components, such as logic programming based reasoning, mental workload assessment, task description, and their integration requires a considerable empirical research effort. We now know what an adaptive interface looks like, and how it should be constructed, but many challenges remain in building safe and operational adaptive interfaces.

ACKNOWLEDGEMENT

Preparation of this chapter was supported in part by the National Aeronautics and Space Administration (NASA), Ames Research Center, through Grant NCC 2-379. Dr. Michael Vidulich was the technical monitor for the Grant. The opinions expressed are those of the authors and do not necessarily reflect those of the above agency.

REFERENCES

Attneave, F. (1959). *Application of information theory to psychology*. New York: Holt, Rinehart, and Winston.

Broadbent, D.E. (1971). *Decision and stress*. New York: Academic Press.

Broadhurst, P.L. (1957). Emotionality and the Yerkes-Dodson law. *Journal of Experimental Psychology, 54*, 345-352.

Cannon, W.B. (1932). *The wisdom of the body*. New York: W.W. Norton.

Chignell, M.H., & Hancock, P.A. (1985). Knowledge-based load leveling and task allocation in human-machine systems. *Proceedings of the Annual Conference on Manual Control, 21*, 9.1-9.11.

Chignell, M.H., & Hancock, P.A. (1986). A fuzzy comparison of mental workload and available capacity in complex person-machine systems. In W. Karwowski & A. Mital (Eds.), *Fuzzy methods and techniques in ergonomic research*. New York: Elsevier.

M.H. Chignell, P.A. Hancock, & A. Lowenthal. (Eds.). (1987). *Intelligent interfaces: Theory, research, and design*. Amsterdam: North-Holland.

Clark, K.L., & McCabe, F.G. (1984). *Micro-Prolog: Programming in logic*. Englewood Cliffs, New Jersey: Prentice-Hall.

Doob, L.W. (1971). *The patterning of time*. New Haven: Yale University Press.

Duffy, E. (1957). The psychological significance of the concept of arousal or activation. *Psychological Review, 64*, 265-275.

Duffy, E. (1962). *Activation and behavior*. New York: Wiley.

Edmonds, E.A. (1981). Adaptive man-computer interfaces. In M.J. Coombs & J.L. Alty (Eds.), *Computing skills and the user interface*. New York: Academic Press.

Eysenck, H.J. (1955). A dynamic theory of anxiety and hysteria. *Journal of Mental Science, 101*, 28-51.

Eysenck, M.W. (1982). *Attention and arousal*. New York: Springer-Verlag.

Fleishman, E.A., & Quaintance, M.K. (1984). *Taxonomies of human performance: The description of human tasks*. New York: Academic Press.

Gopher, D., & Donchin, E. (1986). Workload: An examination of the concept. In K. Boff & L. Kaufman (Eds.), *Handbook of perception and human performance*. New York: Wiley.

Hancock, P.A., & Chignell, M.H. (1985). The principle of maximal adaptability in setting stress tolerance standards. In R. Eberts & C. Eberts (Eds.), *Trends in Ergonomics/Human Factors II*. Amsterdam: North Holland.

Hancock, P.A., Meshkati, N., & Robertson, M.M. (1985). Physiological reflections of mental workload. *Aviation, Space and Environmental Medicine, 56*, 1110-1114.

Hancock, P.A. (1986). Sustained attention under thermal stress. *Psychological Bulletin, 99*, 263-281. (a).

Hancock, P.A. (1986). Stress and adaptability. In G.R.J. Hockey, A.W.K. Gaillard, & M.G.H. Coles (Eds.), *Energetic aspects of human information processing*. The Netherlands: Nijhoff. (b).

Hancock, P.A., & Chignell, M.H. (1986). Toward a theory of mental workload: Stress and adaptability in human-machine systems. *Proceedings of the IEEE Conference on Systems, Man, and Cybernetics,* , 378-383. (a).

Hancock, P.A., & Chignell, M.H. (1986). Input information requirements for an adaptive human-machine system. *Proceedings of the Tenth Annual Symposium on Psychology in the Department of Defense, 10*, 493-498. (b).

Hancock, P.A. (1987). Arousal theory, stress and performance: Problems of incorporating energetic aspects of behavior into human-machine systems function. In L.S. Mark, J.S. Warm, & R.L. Huston (Eds.), *Developments in Human Factors/Ergonomics*. New York: Springer-Verlag.

Hancock, P.A., & Rosenberg, S.R. (1987). A model for evaluating stress effects of work with display units. In B. Knave & P.G. Wideback (Eds.), *Work with display units*. North Holland: Amsterdam.

Harmon, P., & King, D. (1985). *Expert systems*. New York: Wiley.

Hayes-Roth, F., Waterman, D.A., & Lenat, D. (1983). *Building expert systems*. Reading, Massachusetts: Addison-Wesley.

Hebb, D.O. (1955). Drives and the C.N.S. (Conceptual Nervous System). *Psychological Review, 62*, 243-254.

Higgins, T.J., Chignell, M.H., & Hancock, P.A. (1987). Augmented supervisory control: An aviation case study. In M.H. Chignell, P.A. Hancock, & A. Loewenthal, A (Eds.), *Intelligent interfaces: Theory, research, and design*. Amsterdam: North-Holland.

G.R.J. Hockey. (Ed.). (1983). *Stress and fatigue in human performance*. New York: Wiley.

Hockey, G.R.J., & Hamilton, P. (1983). The cognitive patterning of stress states. In G.R.J. Hockey (Ed.), *Stress and fatigue in human performance*. New York: Wiley.

G.R.J. Hockey, A.W.K. Gaillard, & M.G.H. Coles. (Eds.). (1986). *Energetic aspects of human information processing*. The Netherlands: Nijhoff.

Kahneman, D. (1973). *Attention and effort*. Englewood Cliffs, New Jersey: Prentice-Hall.

Koestler, A. (1978). *Janus: A summing up*. New York: Vintage.

Kowalski, R. (1985). Logic programming. *Byte, 10*, 161-177.

Lazarus, R.S. (1966). *Psychological stress and the coping process*. New York: McGraw-Hill.

Lazarus, R.S., & Folkman, S. (1984). *Stress, appraisal, and coping*. New York: Springer.

Malmo, R.B. (1959). Activation: A neuropsychological dimension. *Psychological Review, 66*, 367-386.

McCormick, E.J., & Sanders, M.S. (1982). *Human factors in engineering and design*. New York: McGraw-Hill.

Miller, J.G. (1978). *Living systems*. New York: McGraw-Hill.

Morris, N.M., Rouse, W.B., & Ward, S.L. (1984). Human-computer interaction: A conceptual model. *Proceedings of the IEEE International Conference on Systems, Man, and Cybernetics,* , pp. 178-183.

Newell, A., & Simon, H. (1972). *Human problem solving*. Englewood Cliffs, New Jersey: Prentice-Hall.

Nilsson, N.J. (1980). *Principles of artificial intelligence*. Palo Alto, California: Tioga Press.

Price, H.E. (1985). The allocation of functions in systems. *Human Factors, 27*, 33-45.

Rich, E. (1983). *Artificial intelligence*. New York: McGraw-Hill.

Sanders, A.F. (1983). Toward a model of stress and human performance. *Acta Psychologica, 53*, 61-97.

Schank, R.C., & Abelson, R.P. (1977). *Scripts, plans, goals, and understanding*. Hillsdale, New Jersey: Lawrence Erlbaum Associates.

Selye, H. (1956). *The stress of life*. New York: McGraw-Hill.

Simon, H.A. (1967). Motivational and emotional controls of cognition. *Psychological Review, 74*, 29-39.

Waterman, D.A. (1986). *A guide to expert systems*. Reading, Massachusetts: Addison-Wesley.

Wickens, C.D. (1980). The structure of attentional resources. In R. Nickerson (Ed.), *Attention and performance VIII*.

Hillsdale, New Jersey: Lawrence Erlbaum Associates.

Wickens, C.D. (1984). *Engineering psychology and human performance*. Columbus, Ohio: Charles Merrill.

Wickens, C.D. (1986). Gain and energetics in information processing. In G.R.J. Hockey, A.W.K. Gaillard, & M.G.H. Coles (Eds.), *Energetics and human information processing*. The Netherlands: Nijhoff.

Winston, P.H. (1984). *Artificial intelligence*. Reading, Massachusetts: Addison-Wesley.

Woods, D.D. (1985). Cognitive techniques: The design of joint human-machine cognitive systems. *Artificial Intelligence Magazine, 6*, 86-92.

Yerkes, R.M., & Dodson, J.D. (1908). The relation of strength of stimulus to rapidity of habit formation. *Journal of Comparative Neurological Psychology, 18*, 459-482.

Zeeman, E.C. (1977). *Catastrophe theory: Selected papers 1972-1977*. Reading, Massachusetts: Addison-Wesley.

HUMAN FACTORS PSYCHOLOGY, Peter A. Hancock (ed.)
© Elsevier Science Publishers B.V. (North-Holland), 1987

9. HUMAN FACTORS IN ORGANIZATIONAL DESIGN AND MANAGEMENT

Hal W. Hendrick
Department of Human Factors
University of Southern California
Los Angeles, California 90089

INTRODUCTION

Historically, human factors has been concerned with the design of controls, displays, and workspace arrangements. In system design, the operations to be required of the system typically have been analyzed to identify the specific functions that comprise them. The human factors specialist often enters the design process at this point and, based on his or her professional knowledge of human performance capabilities and limitations, assists in allocating these functions to humans or machines. The human factors specialist's knowledge of human-machine interface technology next is applied to designing specific jobs, integrating jobs into work groups, and then designing specific human-machine interfaces. Many of these human factors activities have been covered in previous chapters.

Although applied within a systems context, the above described activities actually are at the individual, team, or at best, subsystem level. They represent what herein shall be referred to as human factors applications at the microergonomic level. The focus of this chapter will be on the application of human factors at the macroergonomic or overall organizational level. Conceptually, it is entirely possible to do an outstanding job of microergonomically designing a system's components, modules, and even subsystems, yet fail to reach relevant goals of system effectiveness because of inattention to the macroergonomic design of the system.

An historic example of this problem was the introduction of the longwall method of mining in a British deep-seam coal mine. The traditional mining method was largely manual in nature. It utilized teams of small, fairly autonomous work groups. Control over work was exercised primarily by the group itself. Each miner performed a variety of tasks, and most jobs were interchangeable. The workers derived considerable satisfaction out of being able to complete the entire "task." Sociotechnically, the psycho-social characteristics, the characteristics of the external culture, the task requirements, and the system's organizational design and method of managing work were congruent. The more automated and technologically advanced longwall method replaced this costly manual method of mining. The workers no longer were restricted to working on a short face of coal. They now could extract coal from a long wall. Unfortunately, this new and more technologically efficient system resulted in an organizational design and management system that was not congruent with the psycho-social and cultural characteristics of the work force. Rather than small autonomous groups, shifts of 10 to 20 men were required. Workers were restricted to narrowly defined tasks. Job rotation was not possible. There now was a high degree of interdependence among the three shifts, and problems from one shift carried over to the next, thus holding up labor stages in the extraction process. This complex and highly rigid design was very sensitive both to productivity and social disruptions. Instead of improved productivity, low production and intergroup rivalry became common (Degreene, 1973). Later, in studies of other coal mines by the Tavistock Institute of Human Relations in London, (Trist, Higgin, Murray, & Pollock, 1963) this conventional longwall method was compared with a composite longwall method in which the system's organizational design utilized a combination of the new technology and features of the old psycho-social work structure of the manual system. In comparison with the conventional longwall method, the composite longwall system reduced the interdependence of the shifts, increased the variety of skills utilized by each worker, and permitted self selection by the workers of their team members. Production was significantly higher than for either the conventional longwall or the old manual system. Absenteeism and other measures of poor morale dropped significantly. The Tavistock Institute studies led Emery and Trist (1960) to conclude that different organizational designs can utilize the same technology. The key is to select that design, or subset of designs and related management systems that are most effective in terms of (a) the people who comprise the human portion of the system and (b) relevant external environments, and then employ the available technology in a manner that achieves congruence.

Prior to being employed in the design of a system, a given technology potentially can accommodate a broad spectrum of organizational designs. Technology, once it is employed, however, does constrain the subset of possible designs. With the introduction of microelectronics and related automation into managerial, administrative, production, logistical, marketing and other facets of our modern complex systems, organizational designs and related managerial approaches to decision-making and control become far more constrained than in more traditional, labor-intensive person-machine systems. Because of this progressively increasing automation, it has become increasingly important to first consider what is the optimal macroergonomic design of the system, including its management structure, prior to proceeding with the microergonomic design of person-machine modules. In summary, "a top-down ergonomic approach is essential to insure that the dog wags the tail, and not visa versa" (Hendrick, 1984a).

9.1. THE CONCEPT OF ORGANIZATIONAL STRUCTURE AND DESIGN

An organization may be defined as "the planned coordination of two or more people who, functioning on a relatively continuous basis and through division of labor and a hierarchy of authority, seek to achieve a common goal or set of goals" (Robbins, 1983). Breaking this definition down, we can note the following:

a. Planned coordination of collective activities implies management. Activities do not just emerge, but are premeditated. Our modern, complex, high technology organizations of today require information and decision support systems to aid people in managing them. The design of these management support systems must be compatible with the intended design of the organization, including its management structure and process.

b. Because organizations are comprised of more than one person, individual activities must be designed and functionally allocated so as to be complementary, balanced, harmonized and integrated to insure effective, efficient and relatively economical functioning as a system. This design activity first requires a macroconception of the overall organizational design, including how the organization optimally should be managed.

c. Organizations accomplish their activities and functions through a division of labor and a hierarchy of authority. Thus, organizations have structure. How this structure is designed is likely to be crucial to an organization's functioning.

d. The collective activities and functions of an organization are oriented toward achieving a common goal, or set of goals. From a system design standpoint, this implies that these goals can be explicitly identified, measured and utilized in evaluating feasible alternative designs and management approaches for the organization.

9.1.1. Organizational Structure

We noted in "c" above that the concept of organization, with its division of labor and hierarchy of authority, implies structure. The structure of an organization may be conceptualized as having three components: Complexity, formalization, and centralization (Robbins, 1983).

a. Complexity refers to the degree to which organizational activities are differentiated, and the extent to which integrating mechanisms are utilized to coordinate and facilitate the functioning of the differentiated components.

b. Formalization is the degree to which an organization relies on rules and procedures to direct the behavior of the people as part of its management system.

c. Centralization refers to the extent to which the locus of decision making authority is either centralized or dispersed downward within the hierarchy.

The nature of each of these components, including operational measures and design guidelines will be discussed later.

9.1.2. Organizational Design

Organizational design refers to the design of an organization's structure, including the structure of its management processes of planning, directing, coordinating, and controlling the activities of the system to accomplish its goals. From a macroergonomic standpoint, as part of the system design process, organizational design involves: (a) identifying the organization's goals and making them operationally explicit for use as design criteria, (b) systematically developing the design of the three major components of organizational structure, (c) systematically considering the system's technology, psycho-social and relevant external environmental variables as moderators of organizational structure and managerial approach, and (d) deciding on the general type of organizational and managerial structure for the system. These four macroergonomic "steps" of the design process will comprise the remainder of this chapter. It should be noted that this process can be used for modifying an existing organizational and managerial system as well as for designing a new one. In fact, far more opportunity exists for improvement on existing systems--particularly when they are in the process of updating their technology--than for designing one from scratch.

9.2. ORGANIZATIONAL GOALS

Goals may be defined as the objectives of an organization. They are the desired states of affairs that organizations are designed to achieve. Goals may be classified by criteria, by focus, or by time frame (Szilagyi & Wallace, 1983).

9.2.1. Classification by Criteria

Raia (1974) has identified the following as among the most frequently used criteria.

a. Productivity. Productivity goals usually are expressed as levels of output per unit or per worker across the organization. Units produced per employee per day, costs per unit of production, or income generated per employee are commonly used operational measures of productivity.

b. Market. Market goals can be defined operationally in a variety of ways. Examples might be to increase the market share for a given product by ten percent (market share goal), or to sell a specified number of units per year (output oriented goal).

c. Resources. Organizations sometimes establish goals in terms of changes to their resource base. Examples might include reducing the company's debt by ten million dollars in five years (financial base goal), increasing production capacity by twenty percent in three years, or decreasing turnover by fifteen percent next year (human resource goal).

d. Profitability. Profit goals usually are expressed as net income, or a ratio, such as earnings per share, or return on investment.

e. Innovation. Innovation goals are particularly useful when, because of rapid technological change, the timely development of new products is essential in maintaining the organization's competitive position. An innovation goal might be to develop a new, more efficient manufacturing process within three years, or to develop a new computer having a specified increased data processing capacity.

f. Social Responsibility. In part, because of culturally based psycho-social changes in the work force as well as potential long term benefits to the organization, social responsibility goals are becoming increasingly important to an organization's effectiveness. These goals might center around such issues as improving the quality of work life, reducing pollution, or providing financial support for research to educational institutions.

9.2.2. Classification by Focus

Szilagyi and Wallace (1983) noted that classifying goals by focus entails describing the nature of the action that will be taken. They identified three frequently used categories of focus goals.

a. Maintenance Goals. These goals state the specific level of activity or action that is to be sustained over time. An example for an airline might be to have eighty percent of its aircraft in service at all times.

b. Improvement Goals. Any goal that includes an action verb is likely to be an improvement goal, because it indicates a specific change that is wanted. Increasing return on investment, decreasing accidents, or increasing market share would be examples of improvement goals.

c. Development Goals. Development goals also include action verbs but, in addition, they refer to some form of growth, expansion, learning or advancement. An example of this would be, increasing the number of new products introduced, increasing plant capacity, or raising the educational level of the organization's management.

A major advantage of expressing goals in terms of focus is that they are readily understood by persons having diverse backgrounds, such as managers, production workers, engineers, and system designers. Focus goals thus facilitate communication among the various groups that must be responsive to them.

9.2.3. Classification by Time Frame

As we shall see later, when one considers environmental influences on organizational complexity, classifying goals by time frame becomes very useful to the organizational design process. In terms of time, goals usually are classified as being short, intermediate, or long-term.

a. Short-term Goals. Short-term goals concern those objectives that cover a period of one year or less. Production goals frequently take this form.

b. Intermediate-term Goals. These goals usually span a period from one to three years. Sales organizations often have goals with an intermediate time orientation.

c. Long-term Goals. Long-term goals normally cover a period from five to ten years. In any event, they will cover a period of three years or longer. Research and development goals often fall into this category.

Any given organizational activity may have a set of goals that fall into all three time-frame categories. Usually though, as suggested by the examples used above, each major organizational activity tends to have a predominant time orientation. Often, it is the nature of the functional environmental demands on a given activity that determines its particular time orientation. For example, in most cases, research and development of a new product takes three to five years. Accordingly, this function has to be oriented at least this far into the future in order to be aware of product requirements to use as design criteria. Failure of the research and development function to have a long-term time orientation might result in the development of products that either are obsolete or unwanted by the time they are produced and marketed.

9.2.4. Hierarchical Nature of Goals

When goals are translated into objectives they become ends. In analyzing goals, organizational designers must evaluate alternatives as to how they will be achieved, or the means. The means selected at one hierarchical level of organization become the goals or ends for the next lower level. The hierarchical flow of means-ends has strong design implications for the structural differentiation of the organization. To a significant extent, the division of labor within the organization will be a direct outcome of the means-ends analysis of the system (Szilagyi & Wallace, 1983). For example, if the overall goal for a new system was to improve transportation in a large metropolitan area, a number of alternatives could be considered, such as a new highway, subway, bus, or light-rail system. The choice of type of system, in turn, serves as the ends at the next hierarchical level, as it will strongly affect selection of the types of organizational units that will be involved in operating and maintaining the system. The types of units (means) chosen by the system designers will, in turn, serve as the goals for the next, or subunit level of organization. The approaches (means) chosen for meeting these ends at the subunit level will, in turn, affect selection of the subunit functions to be designed into the system and, hence, the grouping of activities into subunits. The system design choices for enabling these subunits to meet their goals will, in turn, affect their division of labor.

9.3. DESIGNING THE COMPONENTS OF STRUCTURE

As we noted earlier, the structure of a person-machine system can be conceptualized as having three core dimensions. These are complexity, formalization, and centralization.

9.3.1. Complexity

Complexity refers to the degree of differentiation and integration that exist within an organization. Three major kinds of differentiation are inherent in an organizational structure (Robbins, 1983). These are (a) horizontal differentiation or the degree of horizontal separation between units, (b) vertical differentiation or the depth of organization hierarchy, and (c) spatial dispersion or the degree to which the organization's facilities and personnel are dispersed geographically. Increasing any one of these dimensions increases the organization's complexity. Integration refers to the extent to which structural mechanisms are included in an organization's design for facilitating communication, coordination, and control across the differentiated elements of the system.

In general, there is a direct relationship between the degree of differentiation that exists within an organization and the extent to which integrating mechanisms are required for effective system functioning. As the degree of differentiation increases, the need for integrating devices also increases. These mechanisms are needed because with greater differentiation of an organization's activities, the difficulty of communication, coordination, and control increases. Some of the more common integrating mechanisms that can be designed into an organization's structure are formal rules and procedures, liaison positions, committees, system-integration offices, and computerized information-and-decision support systems. Vertical differentiation, in itself, also serves as a key integrating mechanism for horizontally and spatially differentiated units

9.3.1.1. Horizontal Differentiation

Horizontal differentiation refers to the degree of departmentalization and job specialization that is designed into the organization. In spite of its intrinsic disadvantage of increasing organizational complexity, job specialization is common to high technology systems because of the inherent efficiencies in the division of labor, as was demonstrated by Adam Smith over 100 years ago (Smith, 1970). He noted that ten men, each doing particular tasks (job specialization), could produce about 48,000 pins per day. On the other hand, if each man worked separately, performing all of the production tasks, those ten men would be lucky to make 200. Division of labor creates groups of specialists. The way in which human factors personnel group these specialists in system design is known as departmentalization. Departments can be designed into an organization's structure on the basis of: (a) function, (b) simple numbers, (c) product or service, (d) clientele, (e) geography, and (f) process. Most large organizations use all six (Robbins, 1983).

a. Function. Grouping specialists by function is the most frequently used method of departmentalization. A typical manufacturing company might have departments devoted to production, sales, research and development, accounting, etc.

b. Simple Numbers. This is the simplest way of grouping activities. For example, if we have thirty employees and three supervisors, we could simply divide the employees into groups of ten and assign a supervisor to each. Grouping by simple numbers is most useful in small organizations, at the lowest level of more complex organizations, and with unskilled or semiskilled jobs.

c. Product or Service. Grouping by product or service is an effective approach for large organizations. Each product division, such as Chevrolet, Pontiac, Oldsmobile, Buick, and Cadillac in the case of General Motors, operates with considerable autonomy, supported by its own functionally grouped departments. This approach to departmentalization is particularly useful when there are diverse or rapidly changing product or service lines.

d. Clientele. In some cases, grouping activities by the type of client or customer serviced can be the most effective basis for ergonomically structuring the system. For example, an aerospace manufacturing corporation might have to be differentially responsive to military, commercial, and private aviation customers and markets. Thus, having separate sales forces for each might be the best approach.

e. Geography. Grouping by geography or territory is a particularly useful approach for facilitating product or service distribution, or for any situation where localized knowledge enhances decision-making or responsiveness. For these reasons sales organizations often are organized by territory.

f. Process. Organizing by process is beneficial in cases where a customer or product must go through a series of units because of specialized equipment or personal needs. In this approach, different activities and skills are grouped in terms of the process that must be followed. For example, on U.S. Air Force bases, personnel service activities, such as motor vehicle registration and the issuance of identification cards, are grouped in a customer service office. These offices are comprised of military police, personnel administrators, and other occupational specialties.

The degree of horizontal differentiation required can be quantitatively assessed using two basic measures: The number of occupational specialties that will be required , and the level of training to be designed into the jobs. The greater the number of occupations and the longer the period of training to be required by the system's design, the greater the need for horizontal differentiation in the design of the organization's structure.

9.3.1.2. Vertical Differentiation

The measure of vertical differentiation simply is the number of hierarchical levels separating the chief executive position from the jobs directly involved with the system's output. In general, as the size of an organization increases, the need for vertical differentiation also increases (Mileti, Gilespie, & Haas, 1977). For example, one study of vertical differentiation in organizations found size alone to account for between 50 and 59 percent of the variance (Mileti, Gilespie, & Haas, 1977). The primary factor underlying in this size-vertical differentiation relationship appears to be span of control. Put simply, the number of subordinates that can be controlled effectively by any one supervisor is limited (Robbins, 1983). As the number of subordinate positions required to meet an organization's goals increases, the number of first level supervisory positions also must increase. This increase, in turn, increases the number of second, third, and higher-level positions required. Thus, the number of higher levels needed also ultimately increases. It is important to note here that, although span of control limitations do underlie the size-vertical differentiation relationship, these limitations can be quite varied, depending on a number of factors. As the span of control can be increased, the number of levels of vertical differentiation required for a given number of employees can be reduced. A major factor affecting effective span of control is the degree of professionalization (education and skill requirements) that has been designed into the positions making up the organization. In general, as the degree of professionalization of the employees increases, the employees are better able to function autonomously with only a minimum of direct supervision. As a result, the span of control of a given supervisor or manager can be increased. Related to professionalization is the fact that certain types of jobs require less direction and control than others. Several of the other more important kinds of variables that can affect vertical differentiation requirements are the type of technology, environmental factors, and psycho-social factors. These will be discussed separately later.

9.3.1.3. Spatial Dispersion

Spatial dispersion refers to the degree to which an organization's activities are performed in multiple locations. In a sense, it is an extension of both vertical and horizontal differentiation in that it is possible to separate both power and task centers geographically. There are three major measures of spatial dispersion and they are used in combination. They are: (a) the number of geographical locations within the organization, (b) the average distance to the separated places from the organization's headquarters, and (c) the number of employees in these separated locations in relation to the the number in the headquarters location (Hall, Haas, & Johnson, 1967). In general, as any of these three measures of geographical dispersion increases, the complexity of the organization's design increases.

9.3.1.4. Relation of Horizontal, Vertical, and Spatial Differentiation

If we were to look only at very large and very small firms, we could note a high intercorrelation among the three types of differentiation. Large high technology corporations, such as General Motors and ITT are characterized by a high level of all three kinds of differentiation. On the other hand, small businesses such as the corner grocery store and neighborhood service station have very little differentiation. However, when we look at the broad spectrum of firms in between these extremes, the picture is quite different. Some fairly large organizations, such as an army battalion, are characterized by a high degree of vertical differentiation but very little horizontal differentiation or geographical dispersion. Universities typically have little vertical or geographical dispersion, but considerable horizontal differentiation (Hage & Aiken, 1967). Some small retail chains have little horizontal or vertical differentiation, but are quite dispersed geographically. Put simply, the degree of one type of differentiation required to optimize an organization does not determine the other two. Each type of differentiation has to be independently assessed in terms of the variables that affect it. Some of these variables were noted in discussing the measurement of each type of differentiation. Other key variables will be discussed later in the chapter as we consider technical, environmental, and psycho-social moderators of organizational and management system design.

9.3.1.5. Integration

Once the nature and degree of each type of differentiation has been determined, human factors specialists need to pay particular attention to the resultant integration needs of the organization. In part, the degree of (formalization) and centralization of the organization's management system will determine the nature of the integration devices, as they are themselves integrating mechanisms. The extent to which computer-based information and decision support systems form part of the system's technology also will be a factor, as these also are integrating mechanisms and thus form part of the organization's design and management system. Careful human factors attention to the ergonomic design of these computer-based systems, including software design, is essential to insuring an effective organizational and managerial system structure.

As we shall see later, other aspects of an organization's technology, an organization's relevant external environments, and psycho-social factors also affect the choice of integration mechanisms.

9.3.2. Formalization

From a human factors perspective, formalization may be defined as the degree to which jobs within organizations are standardized. In highly formalized organizations, jobs are designed to allow for little employee discretion over what is to be done, when, or in what sequence tasks will be performed, and how they will be accomplished (Robbins, 1983). The management system includes explicit job descriptions, extensive rules, and clearly defined procedures covering work processes. Often, design of the system hardware and person-machine interfaces in themselves restrict employee discretion.

Organizations characterized by low formalization allow employees more freedom to exercise discretion. Jobs are designed to permit considerable autonomy and self management. Instead of extensive rules and procedures, employee behavior is relatively unprogrammed. The design of the person-machine interfaces usually allow for considerably greater use of one's mental capacities than in more formalized systems. One major impact of low formalization is to increase the education and skill requirements or professionalization of individual jobs.

In general, the simpler and/or more repetitive the jobs to be designed into the system, the greater is the utility of formalization for effective system functioning; the greater the professionalism of the jobs, the less is the need for high formalization. In fact, with a high level of professionalism, formalization is likely to inhibit both employee motivation and effective system integration and functioning. Related to the issue of professionalism and formalization is the kind of work to be performed. For example, production jobs with stable repetitive characteristics lend themselves to high formalization; research and development or sales positions often require flexibility, innovation and quick responsiveness to change, and thus would be stifled by a high degree of formalization. The degree of predictability and stability versus uncertainty and change in a firm's relevant external environments is also a critical determinant of the desired level of formalization in the organization's design and management structure. In general, formalization is compatible with predictable and stable environments, but inhibits effective and responsive functioning under conditions of uncertainty and change. More will be said about this later.

9.3.3. Centralization

Centralization refers to the degree that formal decision-making is concentrated in an individual, unit or level (usually high in the organization) thus permitting employees (usually low in the organization) only minimal input into decisions affecting their jobs (Robbins, 1983). Centralization, then, is concerned with decision discretion. It is important to note that when decisions are delegated downward, but policies or other formalization mechanisms exist to constrain the discretion of lower-level positions, there is, in reality increased centralization. Conversely, the transference of information in systems requires filtering. Decisions often are made at intermediate organizational levels as to what information gets passed upward to higher management. To the extent that the system is designed to permit information to be reduced, summarized, selectively omitted or embellished, the less is the extent of actual decision-making control at the top and, thus, of actual centralization.

From a system design standpoint, then, it is not the location of the actual decision-making that determines the degree of centralization. Rather, it is the degree of control that various hierarchical positions hold over the decision-making process that is the true measure of centralization. Even after the decision is made it is important to recognize that it is only the intended action. As the decision is passed down through the hierarchical levels of the organization to those who will actually carry it out, it too may undergo filtering and interpretation. The design of the system may even formally permit modification, delay, or rejection of the decision. This kind of decentralized filtering is a desirable ergonomic design feature where conditions require safeguards to prevent implementation of a decision in the light of conditions of change, perhaps unknown to the decision-maker, that could be detrimental to the organization. By the same token, it is equally important ergonomically to design the system so as to prevent filtering of a decision where the filtering is likely to disrupt effective organizational functioning (i.e., where it is likely to dilute the desired level of true centralization).

In general, centralization is desirable (a) when a comprehensive perspective is required, such as in strategic decision-making, (b) when operating in a highly stable and predictive environment, (c) for financial, legal, and other decisions where they clearly can be done more efficiently when centralized, and (d) when significant economies clearly can be realized. Decentralization is to be preferred (a) when operating in a highly unstable or unpredictable environment, (b) when the design of a given manager's job will result in taxing or exceeding human information processing and decision-making capacity, (c) when more detailed "grass roots" input to decisions is wanted, (d) for providing greater motivation to employees by allowing them to participate in decisions that affect their jobs, more fully utilize their mental capacities, and increase their sense of personal control and psychological significance to the organization, (e) for gaining greater employee commitment to, and support for decisions by involving them in the process, and (f) for providing greater training opportunities for lower-level managers. Various environmental, technical and psycho-social factors that affect the desired level of centralization will be discussed later.

9.3.4. Relation of Centralization, Complexity, and Formalization

Centralization and Complexity. Organizational and managerial research has generally shown an inverse relationship between centralization and complexity (Child, 1972; Hage & Aiken, 1967; Robbins, 1983). As the number of occupational specialties and related training requirements designed into an organization increase, the expertise required to make sound decisions also increases, thus forcing decentralization for effective system functioning. Conversely, the simpler or more repetitive the jobs, the greater is the efficiency and value of centralized decision-making.

Centralization and Formalization. No clear or simple relationship has been found between centralization and formalization. Other factors appear to moderate their relationship. For example, with a high level of professionalization, both low formalization and decentralization seem to be optimal (Hage & Aiken, 1967). However, the type of decision involved moderates this relationship. Professionals expect decentralization and little formalization of decisions that directly affect their work, but not decisions concerning personal issues (such as salary and performance appraisal procedures) where the predictability that comes with standardization is desired. In addition, professionals expect most strategic decision-making, which requires a comprehensive perspective and also has only little direct impact upon their daily activities, to be centralized.

Formalization and Complexity. This relationship tends to be a function of both the direction of differentiation and the degree of professionalization (Robbins, 1983). High horizontal differentiation, if achieved by increasing the number and kinds of routine repetitive tasks, will result in the need for a high degree of formalization (Pugh, Hickson, Hinings, & Turner, 1968). If horizontal differentiation is achieved by increasing the number and kinds of highly skilled, complex positions (professionalism) than low formalization is needed, along with decentralized decision-making (Hage, 1965). High vertical differentiation can be a desirable outcome of designing in an increased number of managerial and technical specialists (professionalism) and nonroutine tasks. In this event, low formalization should be designed into the organization for these positions.

9.4. MODERATING ORGANIZATIONAL DESIGN

In order to more adequately convey the nature of complex person-machine systems, Emery and Trist (1960) coined the term sociotechnical system. The sociotechnical system concept views organizations as open systems engaged in transforming inputs into desired outcomes. Organizations are viewed as open because they have permeable boundaries exposed to the environments in which they exist. These environments thus are able to enter the organization along with the inputs to be transformed. The three primary ways in which environmental changes enter the organization are through the people who work in it, through its marketing or sales function, and through its materials or other input functions (Davis, 1982). As transformation agencies, organizations are in constant interaction with their external environments, receiving inputs from them, transforming these inputs into desired outputs, and exporting these outputs back to the external world.

Organizations bring two major sociotechnical system components to bear on the transformation process: Technology in the form of a technological subsystem, and people in the form of what, in human factors parlance, is known as a personnel subsystem. The design of the technical subsystem defines the tasks to be performed. The design of the personnel subsystem prescribes the ways in which the tasks are performed. These two subsystems interact with one another at every person-machine interface. They thus are interdependent and operate under joint causations, meaning that both subsystems are effected by causal events in the environment. The technical subsystem, once designed, is relatively stable and fixed. It therefore falls to the personnel subsystem to further adapt to environmental change.

Joint causation gives rise to a related sociotechnical system concept of joint optimization. Joint optimization means that, since the technological and personnel subsystems respond jointly to causal events, optimizing one subsystem and fitting the second to it will result in suboptimization of the joint system. Joint optimization thus requires joint design of the technical and personnel subsystems, given the objectives and requirements of each, and the nature of the relevant external environments.

As inferred above, the design of an organization's structure and related management system involves consideration of three major sociotechnical system components that interact and affect system functioning: (a) the technological subsystem, (b) the occupational roles and relationships or personnel subsystem, and (c) characteristics of the relevant external environments that permeate the organization. Each of these major sociotechnical system components has been studied in relation to its effect on the elements of the fourth major component, organizational structure, and empirical models have emerged.

9.4.1. Technology

Technology, as a determinant of organizational and management system design, has been operationally defined in three ways: (a) by the mode of production, or production technology, (b) by the action individuals perform upon an object to transform it or knowledge-based technology, and (c) by the strategy selected for reducing the uncertainty that is determined by technology, or technological uncertainty. From each of these empirically developed classification schemes a major generalizable model of the technology-organizational structure relationship also has been empirically developed.

Woodward: Production Technology

The first major series of studies of technology as a determinant of organizational structure was conducted by Joan Woodward and her associates (Woodward, 1965). All together, one hundred manufacturing firms were studied. These were located in Essex, England, and had at least one hundred employees each. The organizations varied greatly in terms of size, type of industry, managerial levels (two-to-twelve), span of control (two-to-twelve at the top; twenty to ninety at the first line supervisory level), and ratio of line employees to staff personnel (less than one-to-one to more than ten-to-one). Through interviews, observations and review of company records, the following factors, among others, were noted for each firm: (a) the company's mission and significant historical events, (b) the manufacturing processes and methods utilized, (c) the organizations' success, including changes in market share, relative growth or stagnation within its industrial field, and fluctuation of its stock prices.

In the course of their analysis, Woodward and her colleagues identified three modes of technology: (a) unit, (b) mass, and (c) process production. These modes were seen as representing categories on a scale of increasing technological complexity. Unit and small batch producers were the least complex. These companies produced custom made products. Next came the mass production or large batch manufacturers, such as those that produce automobiles and other more-or-less standardized products using predictable, repetitive production steps. The most technologically complex were the long-run, heavily automated process production firms, such as oil and chemical refiners. Three important organizational design variables were found to increase as technological complexity increased. These were vertical differentiation, the ratio of administrative support staff to industrial line personnel, and the span of control of the top line managers. For each class of technology, the successful firms tended to have the median number of hierarchical levels for that class. This optimum was three levels for unit producers, four for mass producers, and six for process firms. Woodward's other findings for the successful organizations in each technology mode were as follows.

a. Unit production companies had low complexity; first line supervisors had relatively narrow spans of control; there was little line and staff differentiation; jobs were widely rather than narrowly defined; formalization and centralization were both low.

b. Mass production firms had high complexity; first line supervisors had wide spans of control; there was clear line and staff differentiation;

jobs were narrowly defined; formalization and centralization both were high.

c. Process production organizations had high vertical differentiation; wide spans of control existed at all managerial levels; there was little line and staff differentiation; formalization and centralization were both low.

Table 9-1: Summary of Woodward's Findings on the Design Features of Effective Organizations

Organizational Structure	Mode of Production		
	Unit	Mass	Process
Complexity			
Vertical Differentiation	Low	Moderate	High
Horizontal Differentiation	Low	High	Moderate
Formalization	Low	High	Low
Centralization	Low	High	Low

Several follow-up studies have been conducted, and they have lent support to Woodward's findings (Harvey, 1968; Zwerman, 1970). However, there are a number of observations which argue for caution in interpreting these results. First, Woodward implies a causational relationship, when her data and methodology really only establish a correlation. Second, it should be remembered that Woodward's data was collected from within a single culture, and at a single point in time. In other cultures or on other occasions, the psychosocial and environmental conditions might be very different, and thus result in different interactions with technology mode in terms of organizational design effects.

Perrow: Knowledge-Based Technology

A major shortcoming of Woodward's model is that it applies only to manufacturing firms, and manufacturing establishments comprise less than half of all organizations. Perrow has empirically developed a more generalizable model of the technology-organizational design relationship that utilizes a knowledge-based rather than a production concept of technology (Perrow, 1967). In this classification scheme, Perrow begins by defining technology as the action that one performs upon an object in order to change that object. This action requires some form of technological knowledge; hence, technology can be categorized by the required knowledge base. Using this approach, he identified two underlying dimensions of knowledge-based technology. The first of these is task variability, or the number of exceptions encountered in one's work. The second has to do with the type of search procedures one has available for responding to task exceptions, or task analyzability These search procedures can range from "well-defined" to "ill-defined." At the "well-defined" end of the continuum, problems involving task exceptions can be solved using logical and analytical reasoning. At the "ill-defined" end of the scale, there are no readily available formal search procedures for dealing with task exceptions, and problem solving must be based on experience, judgment, and intuition. The combination of these two dimensions, when dichotomized, yields a two-by-two matrix having four cells. Each cell represents a different knowledge-based technology, as is shown in Table 2.

a. Routine technologies have few exceptions and well defined problems. Mass production units most frequently fall into this category. Routine technologies are best accomplished through standardized coordination and control procedures, and are associated with high formalization and centralization.

Table 9-2: Perrow's Knowledge-Based Technology Classes

Problem Analyzability	Task Variability	
	Routine With Few Exceptions	High Variety With Many Exceptions
Well Defined and Analyzable	Routine	Engineering
Ill Defined and Unanalyzable	Craft	Non-Routine

b. Nonroutine technologies have many exceptions and difficult to analyze problems. Aerospace operations often fall into this category. Most critical to these technologies is flexibility. They therefore require decentralization and low formalization.

c. Engineering technologies have many exceptions, but they can be handled using well defined rational-logical processes. Thus, they lend themselves to centralization, but require the flexibility that is achievable through low formalization.

d. Craft technologies typically involve relatively routine tasks, but problems rely heavily on experience, judgment and intuition for decision. Problem solving therefore needs to be done by those with the particular expertise. Thus, decentralization and low formalization are required for effective functioning. Perrow's model has been supported by considerable empirical research, both in the private and public sectors (Hage & Aiken, 1967; Magnusen, 1970; Robbins, 1983; Van de Ven & Delbecq, 1974).

Thompson: Technological Uncertainty

Thompson (1967) has demonstrated that different technologies require different strategies for reducing uncertainty. He has further shown that different structural arrangements are required of different strategies for optimal system functioning. He identifies three types of technologies, based on the tasks that an organization performs: (a) long-linked, (b) mediating, and (c) intensive. Each of these requires a unique strategy.

a. Long-linked. A long-linked technology accomplishes its tasks by a sequential interdependence of its units. An automobile assembly line would be an example. Long- linked technology involves a fixed sequence of repetitive steps. Therefore, the major uncertainties are at the input and output points. Management responds to uncertainty accordingly, by controlling inputs and outputs. This is accomplished primarily through planning and scheduling, and suggests a moderately complex and formalized structure.

b. Mediating. A mediating technology is one that links clients on both the input and output sides of the system, thus performing a mediating or interchange function. Examples would include banks, utility companies, and post offices. These organizations link units that otherwise are independent. A mediating technology therefore is characterized by a pooled or parallel interdependence of the different units. These otherwise independent units are bound together by rules, regulations, and standard operating procedures. Thus, they function best with low complexity and high formalization.

c. Intensive. An intensive technology is characterized by a customized response to a diverse set of contingencies. A variety of techniques are drawn upon to transform an object from one state to another. At least in part, the particular techniques employed in a given situation are based upon feedback from the object itself. The classic example of this is a hospital, where the object being transformed is the patient. The available techniques are numerous and varied. The particular techniques that are employed will be selected based on the patient's condition, and response to previously used techniques. As is illustrated by the hospital example, the major uncertainty is the object itself, and flexibility of response is essential for effective system functioning. Thus, an intensive technology operates best with an organizational and management system design characterized by high complexity and low formalization.

As yet, Thompson's model has not been fully tested empirically, so no definitive conclusion can be drawn regarding the model's validity (Robbins, 1983). The one study of consequence that has been done analyzed 297 subunits in seventeen business and industrial firms (Mahoney & Frost, 1974). This study provided partial support for Thompson's model by demonstrating that long- linked and mediating technologies were associated with formalization and advanced planning, whereas intensive technologies were characterized by mutual adjustments to other units.

9.4.2. Environment

The survival of an organization depends on its ability to adapt to its external environment. Stated in open systems terms, organizations require monitoring and feedback mechanisms to follow and sense changes in their relevant task environments, and the capacity to make responsive adjustments. "Relevant task environment" refers to that part of the organization's external environment that is made up of the firm's critical constituencies--those that can positively or negatively influence the organization's effectiveness. Neghandi (1977), based on field studies of 92 industrial firms in five different countries, has identified five external environments that significantly impact on organizational functioning.

a. Socio-Economic. especially the degree of stability, the nature of the competition, and the availability of materials and workers.

b. Educational. the availability of educational facilities and programs, and the educational aspirations of workers.

c. Political. particularly the degree of stability; but also the governmental attitudes toward (1) business (friendliness versus hostility), (2) control over prices, and (3) "pampering" of industrial workers.

d. Legal. particularly the degree of stability.

e. Cultural. social status and caste system; values and attitudes toward work, management, etc.; and the nature of trade unions and union- management relationships.

The composite of relevant task environments will differ for each organization in terms of composition, qualitative nature, and relative importance of the specific environments. For a given organization or subunit, the particular weighted combination of relevant task environments can be thought of as its specific task environment.

One of the major determinants of a firm's specific task environment is its domain, or the range of products or services offered and markets served (Robbins, 1983). The domain of an organization is important because it determines the points at which it is dependent upon its specific environment (Thompson, 1967).

A second major determinant of an organization's environment is its stakeholders (Davis, 1982). These include stockholders, lenders, users, government agencies, employees, and the local community, among others. Each stakeholder has interests in the organization. Each also has a potential for action that can affect the organization's nature, efficiency, and indeed its very survival. The design of an organization's structure, including its management system, must enable it to be capable of responding to the objectives and actions of its stakeholders. For example, as a transformation agency, an organization has to respond to the technical, economic, and market aspects of its domain; but it also has to meet governmental regulations for such things as pollution control and safety, it has to meet the investment objectives of its stockholders, it has to be a responsible participant in community affairs, and it has to satisfy the motivational needs and career aspirations of its employees if it is to be successful over time.

Of particular importance to us is the fact that all specific task environments vary along two dimensions that influence the effectiveness of an organization's design. These are the degree of: (a) change and (b) complexity (Duncan, 1972). The degree of change is the extent to which a given task environment is dynamic, or remains stable over time. The degree of complexity refers to whether the components of an organization's specific task environment are few or many in number. These two dimensions in combination determine the environmental uncertainty of an organization (See Table 3).

Two major models have been derived empirically for assessing environmental dimensions as determinants of organizational design. The first model, developed by Burns and Stalker (1961) focuses solely on environmental uncertainty. The second model, developed by Lawrence and Lorsch (1969) treats environmental uncertainty as the most important of several key environmental dimensions affecting organizational structure.

Table 9-3: Environmental Uncertainty of Organizations

Degree of Complexity	Degree of Change	
	Stable	Dynamic
Simple	Low Uncertainty	Mod. High Uncertainty
Complex	Mod. Low Uncertainty	High Uncertainty

Burns and Stalker: Environmental Uncertainty

Based on studies of twenty English and Scottish industrial firms, Burns and Stalker found that the type of organizational structure that worked effectively in a relatively stable and simple organizational environment was very different from that required for a more dynamic and complex environment. For stable and simple environments (low uncertainty) mechanistic structures, characterized by high complexity, formalization and centralization were the most efficient. Mechanistic organizations typically had routine tasks, programmed behaviors, and were slow in their capacity to respond to change. For relatively unstable and complex environments (high uncertainty), organic structures, characterized by flexibility and adaptability were the most successful. Organic firms placed emphasis on lateral rather than vertical communication, information exchange rather than directives from above, authority based on expertise and knowledge rather than on authority or position, conflict resolution by interaction rather than by superiors, and loosely rather than tightly defined responsibilities. To facilitate these processes, a structure characterized by low vertical differentiation, low formalization and decentralization was found to be optimal.

As part of their research on sociotechnical systems, Emery and Trist (1965) also have studied the effect of environmental instability on organizational structures. Their findings were similar to those of Burns and Stalker, and thus lend support to the model.

Lawrence and Lorsch: Subunit Environment and Design Complexity

To cope with a complex task environment, organizations typically develop specialized subunits to deal with specific parts of the environment. Lawrence and Lorsch conducted a series of field studies to determine what type of organizational design was best for coping with different economic and market environments. They studied six plastics firms and, later, investigated a highly effective and a relatively ineffective firm in three different industries: plastics, food, and containers. These industries varied considerably in their degree of environmental uncertainty. From their study of these organizations Lawrence and Lorsch identified five major variables that can be assessed regarding subunit environments to determine the optimal degree of horizontal differentiation. These are (a) uncertainty of information (low, moderate, or high), (b) time span of feedback (short, medium, or long), (c) pattern of goal orientation (focus or task), (d) pattern of time orientation (short, medium, or long), and (e) pattern of interpersonal relationships (task or social). In general, the more dissimilar functions are on one or more of these dimensions, the greater is the likelihood that they should be differentiated into separate subunits for effective

Table 9-4: Design Features of Effective Mechanistic and Organic Organizations

Mechanistic	Design Feature	Organic
	Complexity	
High: Small spans of control	Vertical differentiation	Low: Large spans of control
High: Highly specialized tasks	Horizontal differentiation	Low: Low task specialization
High: Many needed because of high differentiation & low autonomy	Integrating mechanisms	Low: Few needed because of low differentiation & high autonomy
	Centralization	
High: Decision-making reserved for management		Low: Decision-making relegated to lowest level where competency & skill exists
	Formalization	
High: Low autonomy & high differentiation requires many rules and procedures		Low: High autonomy & low differentiation requires few rules

system operation (Lawrence & Lorsch, 1969).

A second major finding was that the greater the differentiation, the greater the need for integrating mechanisms. Lawrence and Lorsch noted that differentiation tends to encourage different viewpoints and, thus, greater conflict. Integrating mechanisms are needed in order to resolve these conflicts to the organization's benefit. They further noted that the more that the tasks of the major subunits are interdependent, the more that information processing is required for effective integration (Lawrence & Lorsch, 1969).

A third major finding was that the level of uncertainty of the external environment was of foremost importance in selecting the structure best suited for effective functioning. Effective subunits with more stable environments (e.g., production) tended to have high formalization, whereas those having less predictable environments (e.g., research and development) had low formalization.

In summarizing from Lawrence and Lorsch's findings, we can note that when ever an organization's design does not fit its mission, environment or resources, the organization is likely to suffer. A mismatch between an organization's task and degree of differentiation results in a loss of relevant information. Differentiation, shifts in mission or resources, and environmental changes each can create integration problems unless adequate integration mechanisms are designed into the organization's structure, including its management system. High levels of interdependence between subunits require particularly careful human factors attention to information processing mechanisms to insure effective integration.

9.4.3. Personnel Subsystem

At least three major aspects of the personnel subsystem are critical to an organization's design. These are: (a) the degree of professionalism, or the skills and training requirements of the organization and its constituent subunits, (b) demographic characteristics of the work force employed, or to be employed in the organization, and (c) psychosocial characteristics of the work force.

9.4.3.1. Degree of Professionalism

It should be noted that formalization can take place either on the job or off (Robbins, 1983). When done on the job, formalization is external to the employee. Rules, procedures, and person-machine interfaces are designed to limit employee discretion. This tends to characterize unskilled and semiskilled positions, and is defined by the term "formalization". In contrast, professionalism creates internalized formalization of behavior through a socialization process which is an integral part of formal training and education. In this case, values, norms, and expected behavior patterns are learned before the employee enters the organization. From a human factors standpoint, in system design there is a tradeoff between formalizing the organizational design and professionalizing the jobs and related person-machine interfaces. As positions in the organization or selected subunits are designed to require persons with considerable training or education (i.e., professionalism), they can, and should be designed and integrated to allow for considerable employee discretion and a management system based on low formalization. To the extent that positions and related person-machine interfaces are designed to be performed with little skills training or education, formalization also should be built into the system's design.

9.4.3.2. Demographic Characteristics

Although there are a number of demographic factors that characterize work forces, our concern here is with those factors that are undergoing systematic change in industrialized countries. Even here, our interest is limited to those characteristics undergoing change that impact upon organizational design. For example, one noticeable systematic change in many countries is the increase in the number of women who are entering what traditionally have been male occupations and positions. As yet, however, there is no clear indication as to how having more women in these positions will impact on organizational design, if indeed women will in any systematic fashion.

There are at least two demographic dimensions undergoing systematic change in industrialized countries that clearly do impact on organizational design. These are (a) the "greying" of the work force, and (b) psycho-social shifts.

a. Greying of the work force. In the United States and other industrialized societies, there was a dramatic upturn in the birthrate immediately following World War II, followed by a return to a lower birthrate in the 60's. The result has been the so called "baby boom" bulge in the population. This population bulge now is moving through the work force, coupled with other relevant factors, such as improvements in medicine and diet, that are enabling all workers to live longer healthy and productive lives. The baby boom bulge is causing the average age of the force to increase at the rate of about six months per year. This "greying" began in the 1970's and will continue well into the 1990's, resulting in a progressively more experienced, more mature and better trained work force. In other words, as a result of this demographic age shift the work force is becoming more professionalized. As we have noted earlier, as the level of professionalism increases it becomes important for organizational structures to become less formalized. It is likely that organizations will have to accommodate by decentralizing tactical decision-making if these more professionalized employees are to feel fully utilized and remain motivated toward their work. Management support system designs similarly will have to accommodate. System designers will need to carefully consider these human factors in modifying bureaucratic organizations where high formalization and centralization traditionally have been the norm.

b. Psycho-social shifts. Based on extensive longitudinal studies of work force attitudes and values in the United States, Yankelovich (1979) notes that those workers born after World War II have very different views and feelings about work than their predecessors, and that these conceptions and values will profoundly affect work systems in America in the next two decades. because their perspectives are so fundamentally different, Yankelovich refers to these younger workers as the "new breed". He reports that this "new breed" of worker has three principal values that distinguish them from the mainstream of older workers: (1) the increasing importance of leisure, (2) the symbolic significance of the paid job, and (3) the insistence that jobs become less depersonalized and more meaningful. When asked what aspects of work are more important, the "new breed" person stresses "being recognized as an individual", and "the opportunity to be with pleasant people with whom I like to work". From an organizational design and management standpoint, these values translate into a need for more decentralized and less formalized organizational structures than those which presently typify most bureaucratic organizations. In terms of ergonomics, person-machine interfaces will need to reflect the designing-in of a relatively greater degree of professionalism. These design characteristics allow for greater individual recognition and respect of an employee's worth, and enhance meaningful social relationships on the job.

9.4.3.3. Psycho-Social Characteristics

Harvey, Hunt, and Schroder (1961) have identified a higher-order structural personality dimension, concreteness- abstractness of thinking or cognitive complexity as underlying different conceptual systems for perceiving reality. We all start out in life as relatively concrete in our functioning. As we gain experience, we become more complex in our conceptualizing, and this changes our perceptions and interpretations of our world. What is important to us here is the fact that different cultures or subcultures tend to foster reaching relatively different levels of cognitive complexity and related conceptualizing. In general, the degree to which a given culture or subculture (a) provides, through affluence, education, communications and transportation systems, an opportunity for exposure to new experiences or diversity, and (b) encourages, by its child-rearing and educational practices, an active exposure to this diversity, the more cognitively complex the people of that cultural group will become. Active exposure to diversity increases one's opportunity to develop new conceptual categories, and shades of grey within categories in which to store experiential data, or differentiation. With an active interaction with one's exposure to diversity, one also learns new rules and combinations of rules for integrating information and deriving more insightful conceptions of problems and solutions. A lack of exposure to diversity precludes this opportunity. Similarly, a closed-minded approach to new experience also blocks conceptual development, even though the opportunity for diverse experience is present. Abstract or cognitively complex functioning then, is characterized by a relatively high degree of differentiation and integration in one's conceptualizing.

Relatively concrete adult functioning consistently has been found to be characterized by a high need for structure and order and for stability and consistency, closedness of beliefs, authoritarianism, absolutism, paternalism, and ethnocentrism. Concrete persons tend to see their views, values, norms and institutional structures as relatively static and unchanging. In contrast, abstract adults tend to have a low need for structure and order, openness of beliefs, relativistic thinking, a high degree of empathy and a strong people orientation. They tend to be less authoritarian and to have a dynamic conception of their world--they expect their views, values, norms, and institutional structures to change (Harvey, 1963; Harvey, Hunt, & Schroder, 1961).

Hendrick (1979, 1981, 1984b) has found evidence to suggest that relatively concrete work groups and managers function best under high centralization, vertical differentiation, and formalization, or mechanistic organizational designs. In contrast, abstract or cognitively complex work groups and managers function best with low centralization, low vertical differentiation and little formalization, or organic organizational structures.

Since World War II, in the United States and, to a greater or lesser degree in other industrialized countries, there has been: (a) a shift to more permissive child-rearing practices, which tends to encourage relativistic rather than absolutist thinking and not to inhibit the exploring of one's environment, (b) a higher average education level, which provides a structured exposure to diversity, (c) greater affluence, which provides greater opportunity for exposure to diversity, and (d) the development of far superior communications and transportation systems that provide exposure to diversity. As the apparent result of these and other factors, the majority of the work force born after World War II is functioning at a more cognitively complex level than the mainstream of their older colleagues. Whereas approximately eighty percent of those who grew up prior to World War II are functioning at a relatively concrete level, this is true of less than half of their post war "baby boom" colleagues. As the pre-World War II work force moves out of our organizations, and more of the "new breed" move in, the trend toward a progressively more cognitively complex work force will continue (Hendrick, 1981). The result is likely to be a progressively increasing demand for more organic organizational structures and supporting management systems.

9.4.4. Convergence of Personnel Subsystem Data Design Implications

Much of the data on personnel subsystem variables as determinants of organizational and management system design are in the form of attitude survey results or are projections of psycho-social and demographic studies. However, in spite of their somewhat tenuous nature, there is a convergence of these data dealing with different personnel subsystem dimensions that lends credence to the conclusion that collectively, these data indicate that organizations and related management systems of the future will need to be more vertically undifferentiated, decentralized, and less formalized than their bureaucratic counterparts of today.

9.5. SELECTING THE TYPE OF ORGANIZATIONAL STRUCTURE

Thus far, we have looked at the first three "steps" of the organizational design process. We have considered (a) the kinds of organizational goals that can be used as design criteria, (b) the three major components of organizational structure and the relevant variables to assess in the process of developing or modifying these three major components, and (c) the system's technology, environment, and psycho-social characteristics as moderators of organizational and management system structure. What remains is the final step of integrating the outcome of these first three steps into an overall structural form. In accomplishing this integration step, the system designers have a variety of types of overall structural form from which to choose. Just as the design of the individual dimensions of organizational and management system structure can enhance or inhibit organizational functioning, and just as the design of these individual dimensions can have particular ergonomic design implications, the same also is true for the particular type of overall design chosen for the organization. The following section will: (a) review the four general types of organization most commonly found, (b) discuss the advantages and disadvantages of each, and (c) provide guidelines for determining when each type is and is not likely to be an appropriate choice.

The four general types of organizational form discussed herein are: the classical or machine bureaucracy, the professional bureaucracy, the matrix organization, and the free-form design. Large, complex organizations often have different, relatively autonomous units that will have differing organizational forms. The smaller the organization, however, the greater is the likelihood that it will possess a single overall type of form.

9.5.1. Classical or Machine Bureaucracy

The classical form of bureaucracy had its roots in two streams of thought: (a) scientific management and (b) Weber's ideal bureaucracy.

a. Scientific Management. The last two decades of the Nineteenth Century were characterized by the accumulation of resources and a rapidly developing technology in American and European industry. During this period labor became highly specialized, and the field of industrial engineering emerged to help design industrial systems and to optimize efficiency. These engineers were involved in the designing of the equipment, installation layout, and work procedures; they also made suggestions for managing the work force (Lawrence, Kolodny, & Davis, 1977). One of these engineers, Frederick W. Taylor, has had a major impact on the shaping of classical organizational theory through his concepts of "scientific management" (Taylor, 1911). Taylor's concepts of organization are implicit in his four basic principles of management (Szilagyi & Wallace, 1983).

First. Develop a science for each element of man's work that replaces the old rule-of-thumb method.

Second. Scientifically select and then train, teach, and develop the workman. In the past he chose his own work and trained himself as best he could.

Third. Heartily cooperate with the men in order to insure all of the work is being done in accordance with the principles of the science that has been developed.

Fourth. There is almost an equal division of work and responsibility between the management and the workmen; in the past, almost all the work and the greater part of the responsibility were thrown upon the men.

As may be seen, Taylor advocated scientific analysis, rather than pure common sense and intuition, as the means for designing both jobs and organizations. He also emphasized the importance of cooperation and developed scientific principles to achieve this in the design of organizations and their components. Finally, he advocated clear job definition through specialization (Szilagyi & Wallace, 1983).

b. Weber's "Ideal Bureaucracy" Much of the classical bureaucratic design was conceptualized by Max Weber at the beginning of the Twentieth Century. Weber recommended that organizations adhere to the following design principles (Weber, 1946).

1. All tasks necessary to accomplish organizational goals must be divided into highly specialized jobs. A worker needs to master his trade, and this expertise can be more readily achieved by concentrating on a limited number of tasks.

2. Each task must be performed according to a "consistent" system of "abstract rules." This practice allows the manager to eliminate uncertainty due to individual differences in task performance.

3. Offices or roles must be organized into hierarchical structure in which the scope of authority of superordinates over subordinates is defined. This system offers the subordinates the possibility of appealing a decision to a higher level of authority.

4. Superiors must assume an impersonal attitude in dealing with each other and subordinates. This psychological and social distance will enable the superior to make decisions without being influenced by

prejudices and preferences.

5. Employment in a bureaucracy must be based on qualifications, and promotion is to be decided on the basis of merit. Because of this careful and firm system of employment and promotion, it is assumed that employ ment will involve a lifelong career and loyalty from employees.

Weber assumed that strict adherence to these organizational and management system design principles was the "one best way" to achieve organizational goals. By implementing a structure that emphasized efficiency, stability and control, Weber believed that organizations could achieve maximum effectiveness (Szilagyi & Wallace, 1983).

These theoretical principles of Taylor's "Scientific Management" and Weber's "Ideal Bureaucracy" have culminated in what today is referred to as the machine bureaucracy type of organizational and management system design. Its basic structural characteristics are as follows.

a. Division of labor. Each person's job is narrowly defined, and is comprised of relatively simple, routine, and well-defined tasks.

b. A well-defined hierarchy. A relatively tall, clearly defined, unambiguous formal structure of positions and offices in which each lower office is under the supervision and control of a higher one. Tasks and positions tend to be grouped by function. Line and staff functions are clearly distinguished and are kept separate.

c. High formalization. A dependence on formal rules and procedures for ensuring uniformity and regulating employee behavior.

d. High centralization. Decision making is reserved for management, with relatively limited opportunity for employee discretion.

e. Career tracks for employees. Employees are expected to pursue a career within the organization, and career tracks form an integral part of the organization's design for all but the most unskilled positions.

Advantages:

The major advantages of the machine bureaucracy are efficiency, stability, and control over the organization's functioning. Narrowly defined jobs with a clear set of routine tasks help minimize the likelihood of error, better enable individuals to know their own function and its relation to the roles of others, requires relatively few prerequisite skills, and minimizes both training time and costs. Formalization insures operational stability and a smooth, integrated pattern of functioning. Centralization enables tight control and thus further enhances stability.

Disadvantages:

There are at least two major disadvantages to the machine bureaucracy. First, this type of design tends to result in jobs that are lacking in intrinsic motivation, and that fail to fully utilize the mental and psychological capacities of the workers. Second, machine bureaucracies tend to be inefficient in responding to environmental change and nonroutine situations.

Appropriate Use:

The machine bureaucracy form of organization tends to be ideal when each of the following conditions exist:

a. When the education and skill levels of the available labor pool are relatively low.

b. When system operations can largely be routinized.

c. When the relevant external environments are stable and predictable.

To the extent that the above stated conditions do not exist, one or another of the other three forms of organization is likely to be more effective.

9.5.2. Professional Bureaucracy

"Professionalism" was defined earlier as the degree of training and education, and related internalized formalization of behavior, required by the design of specific jobs. The professional bureaucracy form of organization relies heavily on a relatively high degree of professionalism in the jobs that comprise the system's design. It differs from the machine bureaucracy in three major aspects:

a. Job design. In professional bureaucracies, jobs are more broadly defined, less routinized, and allow for greater employee decision-making discretion.

b. Centralization. Professional bureaucracies are characterized by decentralized tactical decision-making.

c. Formalization. Since, as a function of greater professionalization, formalization is largely internalized, there is less need for formal rules and procedures. In fact, a high degree of formalization can become quite dysfunctional both to effective employee motivation and organizational functioning.

Like machine bureaucracies, in professional bureaucracies positions are grouped functionally, they are hierarchical, and strategic decision making usually remains centralized.

Advantages:

In comparison with machine bureaucracies, the professional bureaucracy type of organizational form has at least three major advantages:

a. Professional bureaucracies can more effectively cope with complex environments and nonroutine tasks.

b. Jobs tend to be more intrinsically motivating, and to make fuller use of the mental and psychological characteristics of employees.

c. Less managerial control and tactical decision making is required. Management thus is freed to give greater attention to long range planing and strategic decision making.

Disadvantages:

In comparison with machine bureaucracies, professional bureaucracies have the following shortcomings:

a. They are not as efficient for coping with simple environments.

b. They require a more highly skilled labor force, and attendant higher wages and greater training time and expense.

c. Control is not as tight, and both the line and staff delineation and procedures are likely to be less clear.

d. The managerial skills required are likely to be more sophisticated (e.g., the greater is the reliance on a tolerance for ambiguity, and on persuasive and facilitation skills rather than on a simple and direct authoritarian style).

Appropriate Use:

A professional bureaucratic design is to be preferred under the following conditions:

a. The relevant external environments are generally complex.

b. Many of the functions and tasks are of a nonroutine nature.

c. There is an available applicant pool of professionalized workers to staff the organization, and the costs of competing for these human resources are not exorbitant.

d. A strong psychological commitment of the work force to their jobs and the organization is considered critical to its success, and\or the employee's cultural values and attitudes demand intrinsically motivating and psychologically satisfying jobs.

This form of organization is less desirable then the machine bureaucracy if the available management pool is highly concrete and authoritarian, rather than cognitively complex in its functioning. Although the professional bureaucracy is somewhat better able to cope with uncertainty than is the machine bureaucracy, neither form is likely to be optimal if the relative external environments are highly dynamic and unpredictable.

9.5.3. Adhocracy Designs

As noted above, a major disadvantage of bureaucratic designs is that they tend to be inefficient in responding to highly dynamic or unpredictable relevant external environments. It is primarily for meeting this need that the two more recent forms of organization, the matrix and the free form designs have emerged. Collectively, these two new forms have been referred to as adhocracy designs. An adhocracy can be defined as a "rapidly changing, adaptive, temporary system organized around problems to be solved by groups of relative strangers with diverse professional skills." Structurally, adhocracies are characterized by (a) moderate to low complexity, (b) low formalization, and (c) decentralization (Robbins, 1983).

a. Low complexity. Adhocracies are staffed predominantly by professionals. Horizontal differentiation tends to be high. Because of professional staffing, however, the need for supervision is minimized, and vertical differentiation thus tends to be low. This low vertical differentiation reflects the low formalization and decentralized decision making, and the related need for flexibility in tactical responsiveness to a dynamic environment. Many layers of administration would greatly inhibit this responsiveness. With low vertical differentiation the need for integrating mechanisms also tends to be reduced.

b. Low formalization. As we previously noted, with the greater internal formalization provided by professionalism, less external formalization is required, and may even be detrimental. Thus, rules and procedures one finds in adhocracies tend to be informal and often are unwritten. Flexibility of response is more essential than adherence to formal procedure. It is this lack of formalization that most distinguishes adhocracies from bureaucracies. Even in professional bureaucracies problems usually can be classified into some category and treated in a largely routinized manner. In adhocracies, the nature of many of the problems confronted tends to be unique and not subject to a routinized decision process.

c. Low centralization. Adhocracies tend to depend on decentralized "teams" of professionals for decision making (Robbins, 1983). Consequently, the ergonomic requirements for information management and decision support systems tend to be quite different than for bureaucracies.

In summary, adhocracies are characterized by flexible, adaptive structures in which multidisciplinary teams of professionals are formed around specific problems or objectives. They tend to have constantly changing units (as opposed to the relatively stable functional departments of bureaucracies). New units are formed to deal with new problems or objectives. Old units either are dissolved as problems are solved, or change their makeup as different stages of the project are reached.

Advantages:

Compared with bureaucracies, adhocracies have at least two related advantages.

a. Ability to be adaptive and innovative; to respond rapidly to changing situations and objectives.

b. Ability to quickly bring the talents of professionals possessing different specialties together to respond collaboratively to a problem or objective, or to nonprogrammed tasks.

Disadvantages:

All adhocracies have at least three major disadvantages (Robbins,

1983).

a. Conflict. There are no clear boss-subordinate relationships; the lines of authority and responsibility are ambiguous, thus frequently causing conflict.

b. Psychological and sociological stress. Because the structure of teams or units is temporary, work role interfaces are also not stable. The establishment of human relationships is a slower psychological process, and is stressed any time there is a significant change in role relationships. Relatively concrete functioning employees, in particular, are likely to be strained by these stresses.

c. Inherent inefficiency. Adhocracies lack both in precision and in the expediency that comes with routinization of function and structural stability.

Appropriate Use:

An adhocratic design should be used only when the inherent inefficiency of this structural form is more than offset by the gains in efficiency in terms of responsiveness or innovation. Even then, it may not be desirable if the work force and/or management team are relatively concrete in their functioning.

9.5.4. The Matrix Design

Of the two major forms of adhocracy, the matrix design has been the most widely used. This form retains the departmentalization by function, characteristic of bureaucracies, but combines it with departmentalization by product or project. The functional departments tend to be lasting, whereas product or project teams are more dynamic and temporary. Members of the functional departments are farmed out to product or project teams as new product lines or projects are developed, and the technical expertise of the individual departments are required. As the need for a given department's technical expertise is no longer required, or the level of effort needed reduces, individuals return to their "home" department or transfer to another team. The product or project manager supervises the team's interdisciplinary effort, but each team member also has a functional department supervisor. The matrix design thus breaks the fundamental design concept of bureaucracy, unity of command.

Advantages:

The principle advantage of matrix designs is that they combine the stability and professional support depth of functional departmentalization with the interdisciplinary response capability of ad hoc teams, such as characterize free-form designs.

Disadvantages:

The major disadvantage of matrix designs is that of having to serve two bosses. One is their functional department supervisor who tends to be relatively long term and somewhat remote from the team member's immediate tasks; the other is the product line or project team director who is short term, but immediate to the employee's tasks. Serving two masters with overlapping supervisory responsibility and different goals, responsibilities and needs frequently creates conflict, and can be very disruptive to effective organizational functioning. Another problem with this form of organization is that if one is assigned too long outside of one's functional department, it is difficult to keep technically current, and one may lose contact with one's colleagues. Both of these consequences can adversely affect one's career.

Appropriate Use:

The matrix form is particularly well suited for those situations requiring a rapid response to new technology or needs when both (a) an interdisciplinary effort, and (b) functional depth in the individual disciplines are required.

9.5.5. Free-Form Designs

The free-form design is the newest of our general types of organizational structure (Pascucci, 1968). This type of design resembles an amoeba in that it continually is changing in shape in order to survive (Szilagyi & Wallace, 1983). The major focus of free-form organizations is responsiveness to change in highly dynamic, complex, and competitive environments.

In free-form designs, the functional departmentalization of bureaucracies is replaced by a <u>profit center</u> structure. Profit centers are managed as teams, and are highly results oriented. Free-form designs thus are characterized by very low hierarchical differentiation and formalization, and decision making is highly decentralized. A very heavy emphasis is placed on professionalism. As with matrix organizations, project teams are created, changed, and disbanded as required to meet organizational needs. Unlike matrix organizations, there is no functional "home" structure. Managers and employees alike require a great deal of flexibility, tolerance for ambiguity, and the ability to handle change as a way of life. They thus collectively need to be somewhat cognitively complex, as opposed to concrete in their conceptual functioning.

Advantages:

The major advantage of free-form designs is that they respond very rapidly and adaptively to highly competitive, complex and dynamic environments.

Disadvantages:

Free-form organizations have essentially the same disadvantages as matrix adhocracies, only to a greater extent. Conflict, socio-psychological stress, and inherent administrative inefficiency are an integral part of a continuously changing and amorphous organizational and management system design. They thus require a highly professionalized and flexible work force to succeed. To some degree, these disadvantages can be reduced through the ergonomic design of information and decision support systems, work stations, and facilities. Here, the key to effective ergonomic design is responsive flexibility.

Appropriate Use:

Free form designs should be considered when ever the organization's survival depends on speed of response and innovation. These conditions are most likely to exist in highly competitive, complex, and dynamic environments. To successfully implement this form of organization requires a highly professionalized work force.

9.5.6. Human Factors Challenges from the Different Forms of Organization

The choice of a particular type of organizational design brings with it a unique set of human factors challenges that transcend those of specific structural dimensions and sociotechnical system variables, as noted earlier. In general, these have to do with how to design person-machine interfaces to overcome or minimize the disadvantages of a given organizational form, and to enhance its advantages. Some of the more obvious challenges are the following.

Machine Bureaucracies

How can human factors specialists help in enriching the structure of jobs and in increasing employee decision discretion so as to (a) make jobs more intrinsically interesting and to more fully utilize the employees' psycho-social and mental capabilities, yet (b) maintain the stability, control and relatively low personnel and training costs that have made this form of organization and management so successful? As the work force continues to age, to become better educated and experienced, and less cognitively concrete in its functioning, these issues will have to be addressed.

Professional Bureaucracies

How can human factors specialists help in designing person-machine interfaces on a systems level so as to take better advantage of the professional skills and internalized formalization of the employees? In particular, how can human factors help professional bureaucracies become more responsive to dynamic environments and nonroutine problems?

Matrix Organizations

How can human factors specialists help reduce the conflict and attendant stress stemming from the differing objectives of interdisciplinary project teams and functional departments? In particular, how can human factors contribute to minimizing the problems inherent in dual membership and attendant disruptions of organizational functioning?

Free-Form Designs

How can human factors specialists contribute to person-machine interface design at the systems level to provide the needed flexibility and fluidity of task assignment? Can human factors design input also simultaneously reduce role ambiguity and attendant conflict and stress? How can human factors contribute to making these organizations more inherently efficient in the light of their design fluidity?

The answers to these and related challenges will vary with the technology, external environment, and personnel subsystem characteristics of the organization. By approaching these issues at a macroergonomic level, the science and practice of human factors should be able to make significant contributions to both productivity and the quality of work life within organizations. As the field of human factors matures at the macroergonomic level, it should begin contributing directly to organizational theory and advancing the art of organizational and management system design.

REFERENCES

Burns, T., & Stalker, G.M. (1961). *The management of innovation*. London: Tavistock.

Child, J. (1972). Organization structure and strategies for control: A replication of the Aston study. *Administrative Science Quarterly, 17*, 163-17.

Davis, L.E. (1982). Organizational design. In G. Salvendy (Ed.), *Handbook of industrial engineering*. New York: Wiley.

Degreene, K. (1973). *Sociotechnical systems*. Englewood Cliffs, New Jersey: Prentice-Hall.

Duncan, R.B. (1972). Characteristics of organizational environments and perceived environmental uncertainty. *Administrative Science Quarterly, 17*, 313-327.

Emery, F.E., & Trist, E.L. (1960). Sociotechnical Systems. In C.W. Churchman & M. Verhulst (Eds.), *Management science: Models and techniques*. Oxford: Pergamon. Vol. 2.

Emery, F.E., & Trist, E.L. (1965). The causal texture of organizational environments. *Human Relations, 18*, 21-32.

Hage, J. (1965). An axiomatic theory of organizations. *Administrative Science Quarterly, 10*, 289-320.

Hage, J., & Aiken, M. (1969). Routine technology, social structure and organizational goals. *Administrative Science Quarterly, 14*, 366-377.

Hage, J., & Aiken, M. (1967). Relationship of centralization to other structural properties. *Administrative Science Quarterly, 12*, 72-91.

Hall, R.H., Haas, J.E., & Johnson, N.J. (1967). Organizational size, complexity and formalization. *American Sociological Review, 32*, 905-912.

Harvey, O.J., Hunt, D.E., & Schroder, H.M. (1961). *Conceptual systems and personality organization*. New York: Wiley.

Harvey, O.J. (1963). System structure, flexibility and creativity. In O.J. Harvey (Ed.), *Experience, structure and adaptability*. New York: Springer.

Harvey, E. (1968). Technology and the structure of organizations. *American Sociological Review, 33*, 247-259.

Hendrick, H.W. (1961). Differences in group problem solving as a function of cognitive complexity. *Journal of Applied Psychology, 64*, 518-525.

Hendrick, H.W. (1981). Abstractness, conceptual systems and the functioning of complex organizations. In G.W. England, A.R. Negandhi, & B. Wilbert (Eds.), *The functioning of complex organizations*. Cambridge, Massachusetts: Oelgeschiager, Gunn, and Hain. (pp. 25-50).

Hendrick, H.W. (1984). Cognitive complexity, conceptual systems and organizational design and management. In H.W. Hendrick & O. Brown, Jr. (Eds.), *Human factors in organizational design and management*. Amsterdam: North Holland. (a).

Hendrick, H.W. (1984). Wagging the tail with the dog: Organizational design considerations in ergonomics. *Proceedings of the Human Factors Society, 28*, 899-903. (b).

Kerr, C., & Rosow, J.M. (1979). *Work in America: The next decade*. New York: Van Nostrand Reinhold.

Lawrence, P.R., & Lorsch, J.W. (1969). *Organization and environment*. Homewood, Illinois: Irwin.

Lawrence, P.R., Kolodny, H., & Davis, S. (1977). The human side of the matrix. *Organizational Dynamics, 6*, 43-61.

Magnusen, K. (1970). *Technology and organizational differentiation: A field study of manufacturing corporations*. Doctoral dissertation, University of Wisconsin, Madison, Unpublished as of this date.

Mahoney, T.A., & Frost, P.J. (1974). The role of technology in models of organizational effectiveness. *Organizational Behavior and Human Performance, 11*, 122-138.

Mileti, D.S., Gillespie, D.S., & Haas, J.E. (1977). Size and structure in complex organizations. *Social Forces, 56*, 208-217.

Montanari, J.R. (1976). *An expanded theory of structural determination: An empirical investigation of the impact of managerial discretion on organizational structure*. Doctoral dissertation, University of Colorado, Boulder, Unpublished as of this date.

Negandhi, A.R. (1977). A model for analyzing organization in cross-cultural settings: A conceptual scheme and some research findings. In A.R. Negandhi, G.W. England & B. Wilbert (Eds.), *Modern organization theory*. Kent, Ohio: Kent State University Press.

Pascucci, J.J. (1968). The emergence of free-form management. *Personnel Administration, 31*, 33-41.

Perrow, C. (1967). A framework for the comparative analysis of organizations. *American Sociological Review, 32*, 194-208.

Pugh, D.S., Hickson, D.J., Hinings, C.R., & Turner, C. (1972). Dimensions of organization structure. *Administrative Science Quarterly, 13*, 75.

Raia, A. (1974). *Managing by objectives*. Glenview, Illinois: Scott-Foresman.

Robbins, S.R. (1983). *Organizational theory: The structure and design organizations*. Englewood Cliffs, New Jersey: Prentice-Hall.

Smith, A. (1970). *The wealth of nations*. London: Penguin. Originally published in 1776.

Szilagyi, A.D., Jr., & Wallace, M.J., Jr. (1983). *Organizational behavior and performance (3rd ed.)*. Glenview, Illinois: Scott-Foresman.

Taylor, F.W. (1911). *Principles of scientific management*.

New York: Harper and Brothers.

Thompson, J.D. (1967). *Organization in action*. New York: McGraw-Hill.

Trist, E.L., Higgin, G.W., Murray, H., & Pollock, A.B. (1963). *Organizational choice*. London: Tavistock Publications.

Van de Ven, A.H., & Delbecq, A.L. (1974). A task contingent model of work-unit structure. *Administrative Science Quarterly, 19*, 183-197.

Weber, M. (1946). *Essays in sociology*. New York: Oxford. Trans. H.H. Gerth & C.W. Mills.

Woodward, J. (1965). *Industrial organization: Theory and practice*. London: Oxford University Press.

Yankelovich, D. (1979). (1979). *Work, values and the new breed*. New York: Van Nostrand Reinhold.

Zwerman, W.L. (1970). *New perspectives on organizational theory*. Westport, Connecticut: Greenwood.

INDEX

Index

www.ingramcontent.com/pod-product-compliance
Ingram Content Group UK Ltd.
Pitfield, Milton Keynes, MK11 3LW, UK
UKHW020228290926
15470UKWH00001B/418

9 780444 703194